Praise for *The Lagoon*

"For several years in his middle age, Aristotle lived on the island of Lesbos, where he studied the creatures in an inland sea known as Kolpos Kalloni. It was here, Leroi argues in this vivid travelogue and scientific history, that the philosopher pioneered a method of thinking about the natural world that amounted to the invention of science. Breaking with the speculative theories of his naturalist predecessors, Aristotle insisted on rooting claims about the purposes and causes of living beings in observation. The vast catalogues that resulted from his work are messy and filled with unassimilated data, but, as Leroi, a biologist, demonstrates, their basic methodology has filtered down through the ages." —*The New Yorker*

"Armand Marie Leroi's reappraisal of [Aristotle], *The Lagoon*, is one of the most inspired and inspiring I have read. . . . Leroi's ambitious aim is to return Aristotle to the pantheon of biology's greats, alongside Charles Darwin and Carl Linnaeus. He has achieved it." —*Nature*

"*The Lagoon* is an intellectual homage—an admiring, deeply researched, and considered reconstruction of Aristotle's thinking about living things. . . . Marvelous . . . A work as important to a historian and philosopher of science as it is informative to a biologist and entertaining to the general reader. As compelling as Stephen Jay Gould's best work, it will long outlast most nature writing of recent years." —*New Scientist*

"A fascinating new book . . . Leroi argues that Aristotle developed many of the empirical and analytical methods that still define scientific inquiry. . . . Leroi is a brilliant guide to the history of science. He traces the history of ideas with skill and care, and he avoids the smug certainty of many contemporary science writers." —*The Daily Beast*

"Leroi clearly adores Greece and he uses his detailed local knowledge to splendid effect, evocatively re-creating the experiences of the peripatetic philosopher. Leroi is absolutely right to say that even those sections of Aristotle's work we no longer believe to be correct have affected the knowledge that we have today." —*Literary Review* (London)

"In this lush, epic, and hugely enjoyable book, biologist Armand Marie Leroi explores the idea that it was another ancient Greek giant whose shoulders we may all stand upon. Leroi is a beautiful writer and it's been too long, a decade, since his last outstanding book." —*The Observer* (London)

pirical claims. Leroi's own uncompromising investigation gives us a flavor of his subject's indefatigable explorations. Leroi does not upstage Aristotle's descriptions with modern anatomical illustrations, though his attractively illustrated discussions draw on much scholarship that has been expended on editing and interpreting Aristotle's ideas about nature. Leroi's scholarship is impeccable and consistently generous. Only an expert biologist with broad cultural sympathies and a deep feeling for history could have created such a compelling reappraisal of Aristotle's place in the history of science. What's in a name, indeed; in marshalling the facts and ideas that support Aristotle's scientific credentials in exuberant detail, Leroi must be accounted the king."

—*The Times Literary Supplement* (London)

"Armand Marie Leroi opens Aristotle's classical cabinet of curiosities to discover the genesis of science inside. In elegant, stylish, and often witty prose, he probes the near-legendary, almost primeval lagoon, which inspired the ancient Greek's *Historia Animalium* and animates it anew with his own incisive observations. From snoring dolphins to divine bees, Leroi shows us how Aristotle invented taxonomy two and a half millennia before Linnaeus. That, in fact, out of poetry and metaphysics, blending the mythic with the mundane, Aristotle foresaw our contemporary dilemmas of definition and description. *The Lagoon* is a heroic, beautiful work in its own right, an inquiring odyssey into unknown nature and the known world, which science has created out of it." —Philip Hoare, author of *Leviathan, or The Whale*

"A remarkable recovery of an ancient thinker's daringly original enterprises—and mind-set." —*Booklist* (starred review)

"Leroi calls on his expertise and his experience as a BBC science presenter to explain why Aristotle's writings on science are still relevant today. . . . A wide-ranging, delightful tour de force." —*Kirkus Reviews* (starred review)

"Leroi lovingly rescues the reputation of Aristotle's alternately meticulous and bizarre studies of animal behavior from the ruins left in the wake of derision during the Scientific Revolution. Leroi brings modern sensibility to, yet evokes an air of timelessness with, his gorgeous descriptions." —*Publishers Weekly* (starred review)

ABOUT THE AUTHOR

Armand Marie Leroi is a professor of evolutionary developmental biology at Imperial College London. He is also a broadcaster and the author of *Mutants: On the Form, Variety and Errors of the Human Body* (2003), which has been translated into nine languages and won *The Guardian* First Book Award. He lives in London.

www.armandmarieleroi.com

THE
LAGOON

How Aristotle Invented Science

Armand Marie Leroi

with translations from the Greek by Simon MacPherson
and original illustrations by David Koutsogiannopoulos

PENGUIN BOOKS

To my parents

Antoine Marie Leroi
(1925–2013)
&
Johanna Christina Joubert-Leroi

PENGUIN BOOKS

An imprint of Penguin Random House LLC
375 Hudson Street
New York, New York 10014
penguin.com

First published in Great Britain by Bloomsbury Publishing Plc. 2014
First published in the United States of America by Viking Penguin,
a member of Penguin Group (USA) LLC, 2014
Published in Penguin Books 2015

Illustration credits appear on page 487.

THE LIBRARY OF CONGRESS HAS CATALOGED
THE HARDCOVER EDITION AS FOLLOWS:
Leroi, Armand Marie.
The lagoon : how Aristotle invented science / Armand Marie Leroi ; with translations from the
Greek by Simon MacPherson and original illustrations by David Koutsogiannopoulos.
pages cm
Includes bibliographical references and index.
ISBN 978-0-670-02674-6 (hc.)
ISBN 978-0-14-312798-7 (pbk.)
1. Biology—History. 2. Biology—Philosophy—History. 3. Aristotle. I. Title. II.
Title: How Aristotle invented science.
QH331.L5285 2014
570.1—dc23
2014021366

Printed in the United States of America
1 3 5 7 9 10 8 6 4 2

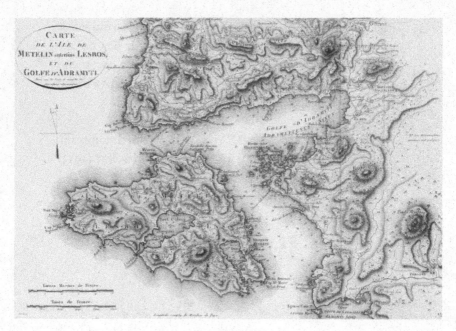

LESBOS (MYTILENE)

Among the isles of Greece there is a certain island, *insula nobilis et amoena*, which Aristotle knew well. It lies on the Asian side, between the Troad and the Mysian coast, and far into its bosom, by the little town of Pyrrha, runs a broad and sheltered lagoon.

D'Arcy Wentworth Thompson,
On Aristotle as a biologist (1913)

CONTENTS

AT
ERATO

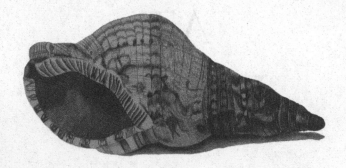

kēryx — ATLANTIC TRUMPET — *Charonia variegata*

I

THERE IS A bookshop in old Athens. It is the loveliest I know. It lies in an alley near the Agora, next to a shop that sells canaries and quails from cages strung on the façade. Wide louvres admit shafts of light that fall upon Japanese woodblock prints propped on a painter's easel. Beyond, in the gloom, there are crates of lithographs and piles of topographical maps. Terracotta tiles and plaster busts of ancient philosophers and playwrights do duty as bookends. The scent is of warm, old paper and Turkish tobacco. The stillness is disturbed only by the muted trills of the songbirds next door.

I have returned so often, and the scene is so constant, that it is hard to remember when, exactly, I first walked into George Papadatos' bookshop. But I do recall that it was the drachma's last spring, when Greece was still poor and cheap and you landed at Ellinikon where the clacking flight boards listed Istanbul, Damascus, Beirut and Belgrade and you still felt as if you'd travelled east. George – lank grey hair, a bookman's paunch – sat at his desk reading an old French political tract. Years ago, he told me, he had taught at Toronto – 'But in Greece, they still had poets.' He returned and named his store for the lyric muse.

Scanning his shelves, I saw Andrew Lang's *Odyssey* and three volumes of Jowett's *Plato*. They were the sort of books that might have belonged to an Englishman, a schoolmaster perhaps, who had retired to Athens, lived on his pension, and died there with some epigram of Callimachus on his lips. Whoever he was, he also left, in a row of Clarendon blue, the complete *Works of Aristotle Translated into English*, edited by J. S. Smith and W. D. Ross and published between 1910 and 1952. Ancient philosophy had never held much interest for me; I am a scientist. But I was idling and in no hurry to leave the calm of the shop. Besides, the title of the fourth volume in the series had caught my eye: *Historia animalium*.* I opened it and read about shells.

> Again, in regard to the shells themselves, the testaceans present differences when compared to one another. Some are smooth-shelled, like the solen, the mussel and some clams, viz. those that are nicknamed

---

* The traditional Latin title. In Greek: *Historiai peri tōn zōiōn*; in English: *Enquiries into Animals*.

'milk-shells', while others are rough-shelled, such as the pool-oyster or edible oyster, the pinna and certain species of cockles, and the trumpet shells; and of these some are ribbed, such as the scallop and a certain kind of clam or cockle, and some are devoid of ribs, as the pinna and another species of clam.

The shell, for me it is always *the* shell, had sat in the sunlight of a bathroom windowsill, buried in sedimentary layers of my father's shaving talc, seemingly for ever. My parents must have picked it up somewhere along the Italian littoral, though whether in Venice, Naples, Sorrento or Capri neither could recall. A summer souvenir, then, of when they were still young and newly married; but indifferent to such associations I coveted the thing for itself: the chocolate flames of its helical whorls, the deep orange of its mouth, the milk of its unreachable interior.

I can describe it so exactly for, although this was so many years ago, I have it before me now. It is a perfect specimen of *Charonia variegata* (Lamarck), the shell of Minoan frescos and Sandro Botticelli's *Venus and Mars*. The trumpet of Aegean fishermen, weathered shells with a hole punched through the apex can still be found in Monastiraki stalls. Aristotle knew it as the *kēryx*, which means 'herald'.

It was the first of many: shells, apparently infinitely various, yet possessed of a deep formal order of shapes and colours and textures that could be endlessly rearranged in shoeboxes until finally, seeing that the mania would not cease, my father had a cabinet built to house them all. A drawer for the luminous cowries, another for the thrillingly venomous cones, one for the filigree-sculptured murexes, others for the olives, marginellas, whelks, conchs, tuns, littorines, nerites, turbans and limpets, several for the bivalves and two, my pride, for the African land snails, gigantic creatures that no more resemble a common garden snail than an elephant does a rabbit. What pure delight. My mother's heroic contribution was to type the catalogue and so become Conseil to my Aronnax, an expert in the Latinate hierarchy of Molluscan taxonomy, though her knowledge was entirely theoretical for she could scarcely tell one species from another.

At eighteen, convinced that my contribution to science would be vast malacological monographs that would be the last word for a hundred years (at least) on the Achatinidae of the African forests or, perhaps – for my attention tended to wander – the Buccindae of the Boreal Pacific, I went to learn marine biology at a research station perched on the edge of a small Canadian inlet. There, a marine ecologist, an awesome Blackbeard-like figure whose violent impatience was checked only by kindness to match, showed me how to peel away the layers of a gastropod's tissues, more fragile than rice-paper, with forceps honed to a needle point and so reveal the severe functional logic that lies within. Another, a professorial cowboy-aesthete – the combination seems incongruous yet he was utterly of a piece – taught me how to think about evolution, which is to say about almost everything. I heard a legend speak, a scientist who had Laozi's gaunt cheeks and wispy beard and who, blind from childhood, had discovered the one part of the empirical world that need not be seen and still can be known – shell form, of course – and had told its tales by touch alone. There was also a girl. She had wind-reddened skin and black hair and could pilot a RIB powered by twin Johnson 60s through two-metre surf and not flinch.

All this is, as I said, long ago. I did not, as it happens, ever write those taxonomic monographs. Science always sets you on entirely unpredictable paths and, by the time I walked into George Papadatos' bookshop, I had long put my shells away. Still, it all came back to me when I read Aristotle on shells and when, reading further, I came across his description of the internal anatomy of the creatures that make them:

> The stomach follows close upon the mouth and, by the way, this organ in the snail resembles a bird's crop. Underneath come two white formations, mastoid or papillary in form; and similar formations are found in the cuttlefish also, only that they are of a firmer consistency [in snails] than in the cuttlefish. After the stomach comes the oesophagus, simple and long, extending to the poppy or quasi-liver, which is in the innermost recesses of the shell. All these statements may be verified in the case of the purple murex and the *kēryx* by observation within the whorl of the shell. What comes next . . .

You may wonder that such blunt words can carry beauty, but for me they did. It was not mere nostalgia, though certainly that played its part. No, it was that I understood, understood against all expectation and probability, what he meant. He had evidently walked down to the shore, picked up a snail, asked 'what's inside?'; had looked, and had found what I found when, twenty-three centuries later, I repeated the exercise. We scientists are no more given to rootling in history's byways than we are to metaphysical speculation. We are, by nature, a forward-looking lot. But this was too wonderful to be ignored.

II

THE DISTRICT KNOWN as the Lyceum lay just beyond Athens' stone walls. A sanctuary dedicated to Apollo Lykeios – Apollo of the Wolves – it contained, among other things, a military training ground, a racetrack, a collection of shrines and a park. The topography is uncertain. Strabo is vague, Pausanias is worse, and, besides, one wrote twenty years, the other two centuries after Sulla, a Roman general, had razed the place to the ground. Sulla also chopped down the ancient plane trees that lined its winding paths and built siege engines from their wood. Cicero, visiting in 97 BC, found only a waste. His visit was an homage to Aristotle who, more than two hundred years before, had rented a few buildings and set up his school there. It was said that Aristotle used to walk the Lyceum's shady paths and that, as he did so, he talked.

He talked about the proper constitution of the city: the dangers of tyranny – and of democracy too. And of how Tragedy purifies through pity and fear. He analysed the meaning of the Good, *to agathon*, and spoke of how humans should spend their lives. He set his students logical puzzles and then demanded that they reconsider the nature of fundamental reality. He spoke in terse syllogisms and then illustrated his meaning with endless lists of things. He began his lectures with the most abstract principles and followed their consequences for hours till yet another part of the world lay before them dissected and explained. He examined his predecessors' thought – the names of Empedocles, Democritus, Socrates and Plato were forever on his lips – sometimes with grudging recognition, often with scorn. He reduced the chaos of the world to order, for Aristotle was, if nothing else, a systems man.

His students would have regarded him with awe and, perhaps, a little fear. Some of his sayings suggest an acid tongue: 'The roots of education are bitter, but the fruit is sweet.' 'Educated men are as superior to uneducated as the living are to the dead.' Of a rival philosopher he said: 'It would be a shame for me to keep quiet if Xenocrates is still talking.' There is a description too, and it isn't an attractive one. It's of a dandy who wore lots of rings, dressed rather too well and fussed about with his hair. Asked why people seek beauty in others he replied: 'That's a question only a blind man would ask.' It is said that he had thin legs and small eyes.

This may be mere gossip: the Athenian schools were forever feuding and the biographers are unreliable. But we know what Aristotle talked about, for we have his lecture notes. Among them are the works – *Categories, On Interpretation, Prior Analytics, Posterior Analytics, Topics, Sophistical Refutations,* the *Metaphysics,* the *Eudemian* and *Nicomachean Ethics, Poetics, Politics* – that loom over the history of Western thought like a mountain range. Sometimes clear and didactic, often opaque and enigmatic, riddled with gaps and rife with redundancies, they are the books that have made Aristotle's name immortal. That we have them at all is mostly due to Sulla, who looted the library of a Piraeus bibliophile and took them back to Rome. But these philosophical texts are only a part – and not even the most important part – of what Aristotle wrote. Among the books that Sulla stole were at least nine that were all about animals.

Aristotle was an intellectual omnivore, a glutton for information and ideas. But the subject he loved most was biology. In his works the 'study of nature' springs to life for he turns to describing and explaining the plants and animals that, in all their variety, fill our world.* To be sure, some philosophers and physicians had dabbled in biology before him, but Aristotle gave much of his life to it. He was the first to do so. He mapped the territory. He invented the science. You could argue that invented science itself.

At the Lyceum he taught a great course in natural science. In the introduction to one of his books there is a sketch of the curriculum: first, an abstract account of nature, then the motion of the stars, then chemistry, meteorology and geology in quick order, and then, the bulk of it, an account of living things – the creatures that he knew, among them, us. His zoological works are the notes for this part of the course. There was one book on what we call comparative zoology, another on functional anatomy, two on how animals move, one on how they breathe, two on why they die, one on the systems that keep them alive. There was a series of lectures on how creatures develop in the womb and grow into adults, reproduce and begin the process again – for there's a book on that too. There were also some books about plants, but we don't know what they contained. They are lost along with about two-thirds of his works.

The books that we have are a naturalist's joy. Many of the creatures that he writes about live in or near the sea. He describes the anatomies of sea urchins, ascidians and snails. He looks at marsh birds and considers their bills, legs and

* Aristotle's phrase is *historia tēs physeōs,* of which biology is a part.

feet. Dolphins fascinate him for they breathe air and suckle their young yet look like fish. He mentions more than a hundred different kinds of fishes – and tells of what they look like, what they eat, how they breed, the sounds they make and the patterns of their migrations. His favourite animal was that weirdly intelligent invertebrate, the cuttlefish. The dandy must have plundered fish markets and hung around wharves talking to fishermen.

But most of Aristotle's science isn't descriptive at all: it's answers to questions, hundreds of them. Why do fishes have gills and not lungs? Fins but not legs? Why do pigeons have a crop and elephants a trunk? Why do eagles lay so few eggs, fish so many, why are sparrows so salacious? What is it with bees, anyway? And the camel? Why do humans, uniquely, walk upright? How do we see – smell – hear – touch? What is the influence of the environment on growth? Why do children sometimes look like their parents, and sometimes not? What is the purpose of testicles, menstruation, vaginal fluids, orgasms? What is the cause of monstrous births? What is the *real* difference between male and female? How do living things stay alive? Why do they reproduce? Why do they die? This is not a tentative foray into a new field: it's a complete science.

Perhaps *too* complete, for sometimes it seems that Aristotle has an explanation for everything. Diogenes Laertius, the gossiping biographer who recorded Aristotle's looks (five centuries after his death), said, 'In the sphere of natural science he surpassed all other philosophers in the investigation of causes, so that even the most insignificant phenomena were explained by him.' His explanations penetrate his philosophy. There is a sense in which his philosophy *is* biology – in which he devised his ontology and epistemology just to explain how animals work. Ask Aristotle: what, fundamentally, *exists*? He would not say – as a modern biologist might – 'go ask a physicist'; he'd point to a cuttlefish and say – *that*.

The science that Aristotle began has grown great, but his descendants have all but forgotten him. Throw a stone in some boroughs of London, Paris, New York or San Francisco and you'll be sure to hit a molecular biologist on the head. But, having felled your biologist, ask her – what did Aristotle do? You will be met with – at best – a puzzled frown. Yet Gesner, Aldrovandi, Vesalius, Fabricius, Redi, Leeuwenhoek, Harvey, Ray, Linnaeus, Geoffroy Saint-Hilaire *père et fils* and Cuvier – to name just a few of many – read him. They absorbed the very structure of his thought. And so his thought became our thought, even when we do not know it. His ideas flow like a subterranean river through the history of our science, surfacing now and then as a spring, with ideas that are apparently new but are, in fact, very old.*

This book is an exploration of the source: the beautiful scientific works that Aristotle wrote, and taught, at the Lyceum. Beautiful, but enigmatic too, for the very terms of his thought are so remote from us that they are hard to understand. He requires translation: not merely into English, but into the language of modern science. That, of course, is a perilous enterprise: the risk of mistranslating him, of attributing to him ideas that he could not possibly have had, is always there.

The perils are particularly great when the translator is a scientist. As a breed we make poor historians. We frankly lack the historical temper, the Rankean imperative to understand the past in its own right. Preoccupied with our own theories, we are inclined to see them in whatever we read. The French historian of science Georges Canguilhem put it like this: 'Agreeing to look for, to find, and celebrate precursors is the clearest symptom of a lack of talent for epistemological criticism.' The *ad hominem* tone of the epigram may cause us to doubt its veracity. It also ignores the fact, obvious to any scientist, if not to all historians, that science *is* cumulative, that we *do* have predecessors and that we *should* like to know who they were and what they knew. Still, there's a discomfiting shard of truth there.

All this should be borne in mind as you read this book. But let me also venture a defence, a scientist's *apologia* if you will. Aristotle's great subject was the living world in all its beauty. It seems possible, then, that something might be gained from reading him as a fellow biologist. After all, our theories are linked to his not only by descent but also by the fact that they seek to explain the same phenomena. It may then truly be that they aren't so different from ours.

In the twentieth century, a generation of great scholars began to examine Aristotle's biological works not as natural history but as natural philosophy. David Balme (London), Allan Gotthelf (New Jersey), Wolfgang Kullmann (Freiburg), James Lennox (Pittsburgh), Geoffrey Lloyd (Cambridge) and Pierre Pellegrin (Paris) gave us a new, thrilling Aristotle. Their discoveries appear on every page of this book (though each of them will disagree, or would have disagreed, with much of it, not least

---

* 'You need not know of the doctrines and writings of the great masters of antiquity, you need not have heard of their names, none the less you are under the spell of their authority' – Theodor Gomperz (1911) *Griechische Denker*, vol. 1, p. 419; quoted and translated by Erwin Schrödinger (1954/1996) *Nature and the Greeks*, p. 3.

because they have so often disagreed with each other). And so I make no
great claims to originality here. However, I like to think that a scientist
may, just occasionally, see in Aristotle's writings something that the philol-
ogists and philosophers have missed.

For sometimes he speaks directly to any biologist's heart, as when he
tells us *why* we should study living things. We must imagine him in the
marble colonnades of the Lyceum, addressing a group of truculent students.
He gestures towards a mound of ink-stained cuttlefish decomposing in the
Attic sun. Pick one, he says, cut, open, *look*.

'. . . ?'

Exasperated, he tries to make them understand:

> So we should not, like children, react with disgust to the investigation of
> less elevated animals. There is something awesome in all natural things.
> Some strangers, so the story goes, wanted to meet Heraclitus. They
> approached him but saw he was warming himself by the stove. 'Don't
> worry!' he said. 'Come on in! There are gods here too.' One should, simi-
> larly, approach research on animals of whatever type without hesitation.
> For inherent in each of them there is something natural and beautiful.
> Nothing is accidental in the works of nature: everything is, absolutely, for
> the sake of something else. The purpose for which each has come together,
> or come into being, deserves its place among what is beautiful.

Scholars call this 'The Invitation to Biology'.

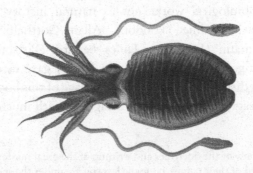

SĒPIA — CUTTLEFISH — *SEPIA OFFICINALIS*

THE
ISLAND

KISTHOS — ROCK ROSE — *CISTUS* SP.

III

THERE IS A mystery here. How did Aristotle think to do biology? How, after all, do you invent a science?

The story was first told by D'Arcy Wentworth Thompson. Or at least he gave it its chronological and geographical bones. Late in life, Thompson became famous for *On Growth and Form*, the eccentric, beautiful book that he wrote about why creatures have the shapes they do. But in 1910 Thompson was a dilettantish failure. Brilliant at Cambridge, he was only twenty-four when he was called to the Chair of Zoology at University College Dundee. Ceaselessly active, he taught, gave working-men's lectures, wrote letters to the *Dundee Courier*, stocked a zoology museum (a platypus was a particular triumph), travelled to the Bering Sea to investigate the seal fishery and submitted philological notes to *The Classical Review* – but published little scientific research. When he was twenty-eight his old Cambridge tutor warned him to do some science before it was too late. When he was thirty-eight another Cambridge friend wrote: 'Let me now suggest to you that you should now shew up some more scientific work.' Thompson agonized and in 1895 published *A Glossary of Greek Birds*, a work in which he collated and identified all the birds mentioned in the ancient Greek and Egyptian texts. His colleagues weren't impressed. So in 1910 Thompson published a translation of Aristotle's *Historia animalium*.

In Thompson's hands Aristotle's worried prose acquires a subdued grandeur: 'All viviparous quadrupeds, then, are furnished with an oesophagus and a windpipe, situated as in man; the same statement is applicable to oviparous quadrupeds and birds, only that the latter present diversities in the shapes of these organs.' Or: 'In the case of oviparous fishes the process of coition is less open to observation.' Or: 'In many places the climate will account for peculiarities; thus in Illyria, Thrace and Epirus the ass is small . . .'

Thompson applied his zoology to identifying the creatures that Aristotle described. In Arabia, Aristotle says, there is a mouse that is much larger than our field mouse 'with its hind-legs a span long and its front legs the length of the first finger-joint'. 'This', Thompson footnotes, 'is the jerboa, *Dipus aegyptiacus* or allied species' – which instantly illuminates. At times his annotations threaten to overwhelm the text: 'ῥινόβατος is probably the modern genus *Rhinobatos*, the *Squatinoraia* of Willughby and other older writers, including *R. columnae*, and other species common in the Greek markets. ῥίνη is probably the angelfish *Rhina squatina* (*Squatina laevis*, Cuv.) which is itself somewhat intermediate between a shark and a skate.' (Years later Thompson would publish a companion to *A Glossary of Greek Birds* called *A Glossary of Greek Fishes*.) As Thompson says, and one detects a note of despair, 'To annotate, illustrate, and criticize Aristotle's knowledge of natural history is a task without an end . . .'

The most important lines in Thompson's *Historia animalium* are in the Prefatory Note. They arrive with so little fanfare that they are easy to miss:

> I think it can be shown that Aristotle's natural history studies were carried on, or mainly carried on, in his middle age, between his two periods of residence in Athens; that the calm land-locked lagoon at Pyrrha was one of his favourite hunting grounds . . .

Pyrrha, Thompson said, was on the Aegean island of Lesbos.*

IV

To THE WEST, Lesbos has the stark clarity of the Cyclades. The landscape is a composition in red, ochre and black. The colours come from volcanic tuffs, eroded pyroclasts and basalts produced by volcanic eruptions 20 million years ago. The plant cover, little though there is of it, is the thorny xerophytic flora of the Aegean phrygana amid which a few skeletal sheep try to graze between stone walls that march across the mountains slopes in geometrical grids. To the east, however, the island is lush and green. The slopes of Mount Olymbos, a massif made of schists, quartzites and marbles, are covered in oak (*Quercus*

---

* With apologies to my Lesvian friends, I call the island 'Lesbos' rather than 'Lesvos', its official modern name, since that it how is was known to Aristotle and is known to most English readers.

ithaburiensis macrolepis and Q. *pubescens*) and, at the highest altitudes, dense stands of sweet chestnut and resinous Turkish pine. Terrapins and eels swim in rivers and storks nest in the chimneys of abandoned ouzo factories. In spring, the valleys are washed yellow by the rare Asian *Rhododendron luteum* and the olive groves of the plains are carpeted in poppies. Poised between the European and Asian continental landmasses, the island draws its flora from both and is exceptionally rich. In 1899 the Greek botanist Palaiologos C. Cantartzis described sixty new endemic species in his *La végétation de l'île de Lesbos* (Mytilène) (Université de Paris, Sorbonne). Nearly all are invalid, but even his more conservative successors count 1,400 plant species, among them seventy-five orchids.

Kolpos Kalloni divides these two worlds. Sheltered from open sea by a narrow, winding strait, it is twenty-two kilometres long and ten wide and cuts the island nearly in two. It is often called a lagoon, but it is really an inland sea of the type that oceanographers call a *bahira*. It is one of the richest bodies of water in the Eastern Aegean. Nutrients flow down the rivers that run from its surrounding hills and feed the phytoplankton that, in the early spring, turn its waters green. The eelgrass beds of its shallows are a nursery for bream and bass and paddle-legged crabs. The gentle slopes of its muddy bed are interrupted only by ancient oyster reefs – but mention Kalloni to a Greek and he will speak of its pilchards that are best eaten salted and washed down with Plomari ouzo.

The salt comes from the works at the northern end of the Lagoon. There a maze of channels carries brine of ever-increasing concentration from pan to pan. The saturated solutions deposit large crystals on branches and stones that glisten beneath swards of marsh samphire and sea lavender. At the innermost pans the salt becomes a harsh, deserted skin that is then broken and heaped into immense white pyramids. Rusting machinery is scattered about but is hardly ever seen at work, salt collection being a restful industry. The ecology of the saltpans is very simple. Halophilic algae are eaten by brine shrimp and brine-fly larvae that, in turn, are sieved and probed by flocks of greater flamingos, black-winged stilts and a miscellany of sandpipers and plovers. Only one fish, the toothcarp, *Aphanius fasciatus*, can live in the bitter, hot brine and it is eaten by the black storks and glossy ibis that wade through the channels and several species of tern that wheel down from the sky. In the spring and autumn, the saltpans, and the marshes that surround them, are a resting place for thousands of migrant birds *en route* between Africa and the north.

V

ARISTOTLE IS NO geographer or travel-writer, but a curiously large number of passages in his work refer to Kalloni, which he knew as Pyrrha, after a town on its eastern shore. It is precisely the frequency of these passages that caused D'Arcy Thompson to suggest that this is where Aristotle did so much of his biological work. Many of them can be found in his great treatise on comparative zoology, *Historia animalium*. They tell of the animals that inhabit the Lagoon. A collation of these passages into a biological Baedeker would read something like this:

> The fishes of Lesbos breed in the Lagoon at Pyrrha. Some of the fishes – mostly the egg-laying ones – are best eaten in early summer; others – the grey mullet and the cartilaginous fishes – are best in autumn. In winter the Lagoon is colder than the open sea so most of its fish, but not the giant goby, swim out of the lagoon only to return in the summer. The white goby is not a marine fish but is also found there. The absence of fish in winter means that edible sea urchins of the strait have more food – which is why they are then particularly rich in eggs and good to eat, although small. There are oysters in the Lagoon. (Some people from Chios came over to Lesbos and tried to transplant them to the waters surrounding their own island.) Once there were also many scallops, but dredging and drought have exterminated them. Fishermen also say that starfish are a particular nuisance near the entrance to the Lagoon. Although the Lagoon contains much life, a number of species are not found: parrotfish, shad, spiny dogfish. None of the other brightly coloured fish are found there either; nor are the spiny lobster, common octopus or musky octopus. The murex snails of Lectum, a mainland cape facing Lesbos, are particularly big.

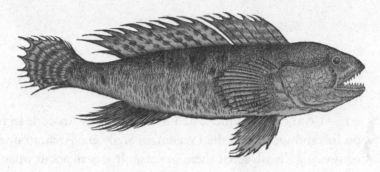

KŌBIOS − GOBY − G*OBIUS COBITIS*

Written this way, Aristotle's remarks about the Lagoon and its creatures make a portrait of the Lagoon as it was twenty-three centuries ago, perhaps the oldest of any natural place that we have.* Little remains now of the ancient town of Pyrrha − Strabo says it was destroyed (by an earthquake in the third century BC) − but the biology still rings true. The Lagoon remains rich in oysters, though today they are exported to Northern Europe by the ton. Until recently there were scallops too. Indeed, a fisherman complained to us that there used to be scallops in the entrance to the Lagoon but that, twenty years ago, dredging had rendered them all but extinct. It seems, then, that the scallop population of Kalloni has been waxing and waning for at least twenty-three centuries, and that the locals have been complaining about it all the while. The fishermen also confirm that fish migrate annually in and out of the Lagoon to breed, and that it contains no parrot-fish, shad or spiny lobsters or spiny dogfish. There have been some changes to the Lagoon's fauna since Aristotle's day. If there were no octopi then, there certainly are now − I have eaten several myself. And, for all their flamboyance, Aristotle does not mention the flamingos − but that is because they arrived at the Lagoon only a few decades ago.

* When Aristotle mentions the sea at Pyrrha, he usually refers to the *euripos* or 'strait', the entrance of Kolpos Kalloni. The Lagoon itself is better described as a *limnothalassa* or 'lake sea'.

VI

B UT ALL GREEKS were interested in fish. Even as Aristotle lectured
on fish and suchlike in the Lyceum, in Sicily one Archestratus was
composing a book about them in verse. It was all about when and
where to catch them, and then how to cook them. If you go to the land of
Ambracia (Western Greece), Archestratus urges, buy the 'boarfish'
(catfish) even if it costs its weight in gold! But get your scallops from
Lesbos, your moray eels from the straits of Italy and your tuna from Byzantium (slice, sprinkle with salt, brush with oil, bake simply and eat while
hot). He titled his book *The Life of Luxury*. For the Greeks fish were about
conspicuous consumption: less objects of philosophy than objects of desire.

So what makes a man stop eating his fish and start dissecting it instead?

VII

I T'S NOT THAT there wasn't any science – or at least natural philosophy – before Aristotle, for there was an abundance of it. By the time
he was born, schools of philosophers deeply concerned with understanding the nature of the physical world had waxed and waned along the
Anatolian and Italian littoral. The Greeks called them *physiologoi*, literally
'those who give an account of nature'. Many were bold theoreticians.
They loved systems that explained, in sweeping terms, the origin of the
world, its mathematical order, the stuff of which it is made and the reasons
why it contains so many different things. Others were empiricists who
tried to measure the heavens or else the intervals of musical scales. Their
writings have some of the ingredients of modern science – though we
rarely get any sense that they challenged their theories with the observations that they made. Their explanations tended to appeal to natural rather
than divine forces.

A comparison of two near contemporaries illustrates the shift in
thought. For the mythographer Hesiod (*fl.* 650 BC) earthquakes are the

consequence of Zeus' wrath; for the first of the natural philosophers, Thales of Miletus (*fl.* 575 BC), they are the result of the earth's precarious location, adrift on an expanse of water occasionally roiled by waves. The difference could not be more clear-cut: on the one hand an explanation that invokes supernatural beings of fathomless antiquity; on the other an explanation that depends on purely physical forces – and never mind if it's wrong.

Yet the comparison is not quite what it seems. For one, we can't be sure if that was really Thales' theory.* No texts by him have survived; for all we know he didn't write any. Seneca the Younger reports the earthquake theory in his *Questions on Nature*. Since he wrote about 500 years after Thales' death and is perfectly vague about his sources we may wonder whether Seneca, or we, have any idea what Thales actually thought about earth-quakes or anything else – though he *is* widely credited with having predicted an eclipse in 585 BC. The same is true for much of the rest of early – 'Pre-Socratic' – Greek thought. The entire corpus has come down to us in fragments buried with the texts of later thinkers who, as often as not, must be suspected of having done with their quotations what they pleased, or even of making them up. Scholars call these texts 'doxographical' and they are their delight and despair.

To be sure, enough fragments can be culled and reconstituted to fill thick books. And those fragments do speak of a new philosophic spirit abroad in fifth-century Greece. But distinctions apparently clear to us now, between science and non-science, philosophy and myth, were less so two millennia and more ago. In the *Metaphysics* Aristotle, himself a rich source of fragments, reviews what earlier thinkers have said about the 'original causes' of the world. He attributes to Thales the theory that everything comes from water. This is a perfectly reasonable, if vague, idea that deserves to be discussed in its own right; Aristotle does so – and doesn't like it. And, he goes on, some think that Thales' view is quite a lot like that held by the 'men in the distant past (well before the current generation) who first gave an account of the gods'.

We are suddenly brought up short. Yes, the myths may be ancient history but evidently not so ancient that they do not deserve an airing in a highly technical discussion about the foundational material of the world.

* Describing Thales' views, Aristotle uses technical terms that were invented after Thales' death (e.g. *arkhē* – origin or principle). This, in itself, makes us unsure that we know what Thales actually meant.

And then, just a few paragraphs later, having left Thales to stew with the ancients, Aristotle decides to analyse a bit of Hesiod – 'Of all things that came to be, the first were Chaos, and broad-bosomed Earth and Love most eminent of the immortals' – to see if any scientific sense can be made of it. Hesiod may be a mythographer, but for Aristotle he's still worth a passing glance.

And that is the problem with making naturalism the hallmark of Pre-Socratic thought. The *physiologoi* do not always 'leave the gods out'; the Divine can usually be found lurking somewhere in their cosmologies. When they asked, What is the origin of the world?, some gave answers that were as creationist as a Christian's; others appealed to more remote forces such as Love itself; yet others again were ardent materialists and thought the world just self-assembled. From Hesiod to Democritus, the Creator advances, retreats or sometimes just curls up and contemplates himself.

Perhaps, then, what marks the *physiologoi* as early scientists is not so much the use of naturalistic explanations for the mysteries that the world presents, as rational ones. They believed that wisdom did not merely have to be received, but that ideas were worth debating and, if need be, discarding. They argued with each other and those who came before them; they were ambitious for their ideas. Here is Heraclitus (*fl.* 500 BC) evaluating some of his predecessors: 'great learning does not teach sense: for otherwise it would have taught Hesiod and Pythagoras and again Xenophanes and Hecataeus'. Nasty – and unmistakably the sound of an intellectual at work.

Most of the *physiologoi* weren't much interested in biology. Empedocles (c. 492–432 BC) was an exception. A Sicilian of noble birth, he was an orator, poet, politician, healer and charismatic seer. In the opening lines of his religious poem *Purifications*, he presents himself as an immortal god and describes how, when he enters a city, thousands flock to him requesting cures and oracles – requests which he satisfied, on at least on one occasion, by raising the dead. Jesus with an ego, then, or Zarathustra with attitude, but he was also an immensely influential natural philosopher who wrote *On Nature*, several thousand lines of verse containing, among other things, a cosmogony, a zoogony, a mechanistic, if implausible, theory of respiration and a four-element chemistry that Aristotle would adopt as his own.

Empedocles' biology reflected the medical lore and practice of his day. So, too, did his appetite for magic and mysticism. Yet even as he pranced around Sicily performing miraculous cures to adoring crowds, on the other side of the Mediterranean Hippocrates (*fl.* 450 BC?) was going to school. In the plateia of Kos town there is a plane, ancient and gnarled, that is – so the label claims – the very tree under which the adult Hippocrates once sat dispensing cures and wisdom. It can't be the same tree, but then the medical writings that are attributed to Hippocrates probably aren't his either. Parts of the *Corpus Hippocraticum*, a medley of some sixty works, are old enough to have been written by him or his pupils, but others date from around the first century A D.

Most of them are sober, professional texts that give naturalistic explanations for disease. Some are simple case studies, but others are more intellectually ambitious. The author of *Fleshes* says he wants to 'explain how man and the other animals are formed, that is, come about, what the soul is, what health and sickness are, what is bad and good in man, and what causes death'. Profound or banal, they're very different from Empedocles' effusions. Here's 'Hippocrates' on curing acute diseases:

> Often, in such cases, you will find 'oxymel', as a drink, extremely useful. It helps bring up sputum and promotes breathing. It is best employed under the following conditions. When strongly acidic, it is particularly effective in cases where there is difficulty in bringing up sputum. By lubricating the sputum, it facilitates expectoration thus clearing the windpipe as if with a feather. This is soothing to the lung and brings relief. And if, in combination, it achieves these effects, it must do much good.

And here's Empedocles' approach:

> Any remedies there may be, defences against harm and age / you alone will know them; I will make sure you know them all. / You'll put a stop to the winds, their tireless might pouncing / upon the land, their whirling breath, a withering force upon the crops. / Equally, if you so choose, you'll bring on whirlings equally strong . . . / Even from Hades you will bring up men, men whose might has withered with time.

Aristotle would call Empedocles' style 'lisping'.

It may seem that all Aristotle needed to do to become a scientist was to broker a marriage between the questing, querulous *physiologoi* and the dourly empirical medics. Which is what he did. That he managed it, however, is a tribute to the power of his mind.

VIII

LITTLE ABOUT ARISTOTLE'S life is certain. The ancient sources, a dozen or so of them, were written centuries after his death and often contradict each other. Muddled in transmission, riddled with gossip and warped by the politics of rival philosophical schools, they have been churned over by centuries of scholars seeking the man behind the works. The results are meagre; the agreed facts could be written on a page.

He was born in 384 BC in Stagira, a coastal town not far from modern Thessalonika. His father, Nicomachus, was an Asclepiad – part priest, part physician. No common quack, he was physician royal to Amyntas III of Macedon. This is less impressive than it sounds. Macedon was a semi-barbaric backwoods state with a court to match. At the age of seventeen Aristotle was sent to Plato's Academy in Athens. He remained there, as student and teacher, for nearly twenty years.

By the time the teenaged Aristotle arrived at Athens to sit at Plato's feet, the tradition of natural philosophy, no more than two centuries old, was dead. Literally so: Democritus of Abdera, the last and greatest of the *physiologoi*, had died just a few years earlier. Years later, Aristotle would see in Democritus a formidable adversary, a foil against which to test the mettle of his own system. Democritus, Aristotle says, made advances. 'But [even] at this time men gave up inquiring into nature, and philosophers diverted their attention to political science and to practical goodness.' He was talking about Socrates.

Socrates (469–399 BC) was a stonemason with a taste for speculative thought. As a young man, he loved natural philosophy. At least that is what Plato makes him say in *The Phaedo*. He puzzled over the origin of life, the physical basis of thought and the motions of the heavens. His efforts were for naught. He followed, or tried to follow, the arguments of the *physiologoi*, now this one, now that, but he only wound up confused. Did 1 + 1 = 2? By the time he was done, he could no longer say for sure. He was, he concluded, 'uniquely unfitted for this sort of inquiry'. Besides, it seemed to him, the *physiologoi* never gave the right kind of answers – or even asked the right kind of questions. When they explained why the earth is flat or round or whatever shape they supposed it to be, they should have explained why it is best that it be so. But they never did. Instead, they appealed to 'natures' — and those are not true causal explanations at all. ('Fancy being unable to distinguish between the cause of a thing and the condition without which it could not be a cause!')

Disillusioned by the *physiologoi*'s singular lack of interest in discussing why the universe was good, Socrates turned away from the study of the natural world. Xenophon picks up the tale:

> Unlike most others Socrates did not discuss the nature of the universe, and investigate the state of what intellectuals call the cosmos or the features necessary to bring celestial phenomena into existence. Instead he argued that those who speculated about such things were wasting their time. The first question he would ask was whether this sort of speculation was based on a conviction that they already had a thorough understanding of human affairs. Did they really think it was appropriate to focus their investigations on the divine at the expense of the human?

The *physiologoi*, with their welter of mutually inconsistent theories, were like 'madmen'. They were social parasites too:

> He raised a further point about these people. Those who study human affairs, he said, think that their subjects will be productive for themselves and other potential beneficiaries. Do those researching celestial phenomena really believe that discovering the features necessary for things to come into existence will allow them to produce, on demand, winds, waters, seasons and any others they might add to the list? Or do they, in fact, have no such expectation but are quite satisfied with discovering how things of this sort come about?

Scientists disagree, therefore they are foolish; who are they to play God?; what good is their work to me? — all this is the authentic voice of anti-science through the ages; it is its first breath. Ethics are so much more useful. 'Socrates called philosophy down from the heavens, placed it in cities, introduced it into families, and obliged it to examine into life and morals, good and evil.' That was Cicero's judgement — and he meant it as praise.

IX

SURROUNDED BY A wall, the Academy had a gymnasium, a sacred olive grove and a garden. Its foundation stones can be seen in a Piraeus park, but the wire, wilting trees and litter make it hard to reconstruct the place. Plato, who had bought the property, founded his school there around 387 BC. Diogenes Laertius lists some of Plato's pupils: Speusippus of Athens, Xenocrates of Chalcedon, Dion of Syracuse and a dozen more from around the Hellenic world including two women. It was less a modern college than a philosophical club. Students didn't pay fees. That in itself made it a very different kind of enterprise from the schools run by the sophists and rhetoricians who were in the business of teaching Athens' youths how to speak nicely, get ahead and win in court.

When Aristotle arrived Plato himself was packing his bags for a two-year trip to Sicily. He probably left his nephew, Speusippus, in charge. Fortyish and famously bad-tempered, he is said to have thrown, in a fit of pique, his favourite dog down a well. Yet he may have taken the youngster under his wing; there are traces of his thought in Aristotle's. Even so, if Plato's dialogues, the doxography of the Academicians and Aristotle's recollection are reliable guides to the scope of the talk in the Academy's garden, then natural philosophy was off the curriculum. Or, if it was there at all, it was so in a peculiar form.

Socrates' interest in moral theology had become Plato's. Of course, it is hard to separate the two since Socrates wrote nothing, Plato wrote much, and much that Plato wrote is voiced by 'Socrates'. Yet while Plato's Socrates is not as crudely anti-scientific as Xenophon's, Plato's mature philosophy is no less inimical to science than Socrates' jibes; far more so because he wrote so beautifully and because his works have survived complete.

The Republic, Plato's most famous dialogue, gives his views on the aims and methods of natural philosophy. Glaucon and Socrates are discussing the education of Philosopher Kings. Should the young study astronomy? Yes, says Glaucon, it's useful for all sorts of things: agriculture, navigation and war. Socrates gently disabuses him of this 'vulgar' utilitarianism. Well, then, replies Glaucon, perhaps they should study astronomy because it 'compels the soul to look upwards'. This, he hopes, is the sort of answer that Socrates is looking for, but, once again, he is disabused. Glaucon is being much too literal-minded: the only study that turns the soul's gaze upwards, says Socrates, is that which deals with 'being and the invisible' – by which he means the true reality that lies behind the superficial appearance of things. Studying the stars, he continues, help us to do this, but not very much. The actual movements of the stars are only an imperfect representation of the invisible realities; you might as well search for geometrical figures in a picture. And these realities can 'be apprehended only by reason and thought, but not by sight, or [Glaucon] do you think otherwise?'

Glaucon doesn't think otherwise. He capitulates entirely to Socrates' – Plato's – anti-empiricism. And, a few pages later, when the talk turns to the study of harmony, the two men join in jeering at those *physiologoi* 'who vex and torture the strings' of their instruments, 'laying their ears alongside, as if trying to catch a voice from next door' in an effort to understand the rules of harmony and the limits of musical perception. These 'worthies' (the musical *physiologoi*) 'do not ascend to generalized problems and do not consider which numbers are inherently concordant and which not and why in each case'.* They fiddle about with harps when they should be working out a general, formal theory for the musical order that they dimly perceive; a theory that would account for the beautiful and the good that we hear in music; a theory that would unify the harmonies of music with the movement of the stars. 'A superhuman task', comments Glaucon – which may seem to us an understatement.

Plato should have left it there. Had he done so then we could at least credit him with becoming modesty. He didn't. Late in life he wrote a work that purports to describe and explain the natural world – all of it. For all its ambition, it is a quarter as long as *The Republic*. The brevity is telling.

X

PLATO'S *TIMAEUS* RECOUNTS the creation of the cosmos and all that it contains: time, the elements, the planets and stars, humans and animals. Although short, it aspires to be encyclopaedic, covering

---

* Note how these jeers are directed at an attempt to solve a serious scientific problem, the cognitive basis of harmonic perception.

ontology, astronomy, chemistry, sensory physiology, psychiatry, pleasure, pain, human anatomy and physiology – with an aside on why the liver is the source of prophecies – and the origin of disease and sexual desire. All this makes it look like a work of natural philosophy.

If so, then it is a very strange one. Devoid of scholarly citation, empirical evidence or even much reasoned argument, The Timaeus is a drawing-room monologue that delivers, with bland assurance, one implausible assertion after another. Deeply religious, it aims to reveal why a divine workman, the Dēmiourgos, constructed the world. It is also a work of political propaganda that shows what the ideal city of The Republic would actually look like. Indeed, it's not even clear that Plato intended The Timaeus as a contribution to natural philosophy. He claims to want to give an account of the visible world; however, he begins by cautioning us that he will deliver only an eikōs mythos – a plausible tale. In part this is because he's really after an account of the world that lies beyond the senses; and any account of this flawed, but visible, world will bear an uncertain correspondence to that perfect, but invisible, one. But it is also because he's not terribly interested in giving a rational account of even this world.

Plato gives the game away with his account of the origin of the animals. Once, he says, there existed men who were to varying degrees depraved or just foolish. They were transformed into the various animals – creepy-crawlies, shellfish and the like – according to their diverse vices. Birds 'sprang by a change of form from the harmless but light-witted men who paid attention to the things in heaven but in their simplicity supposed that the surest evidence in these matters is that of the eye'. He's talking about astronomers.

Did Plato really believe that birds were reincarnated natural philosophers? Or did he simply seize the chance to crack a poor joke? Let us be charitable and assume the latter, for the former is too bizarre even by the elastic standards of fourth-century zoology. But that joke betrays the true nature of The Timaeus: it is not a work of natural philosophy at all, but a poem, a myth, a ponderous jeu d'esprit that revels in its own ambiguity.

The assessment may seem harsh. Plato shared the Pythagoreans' fascination with geometry, and The Timaeus contains one of the first attempts to use mathematics to describe the natural world. 'Let no one ignorant of geometry enter here' is said to have been inscribed on the lintel of the Academy's entrance; the same phrase is written above the swipecard-sealed doors of any Department of Physics, even if you can't see it. Then, too, if Plato's science is barely distinguishable from theology so, to judge by the pronouncements of some physicists, is modern science: 'If we discover a complete theory, it would be the ultimate triumph of human reason – for then we should know the mind of God.' Plato? No, Hawking.

The comparison doesn't save Plato. Here is an example of his style of mathematical modelling: 'The second species of solid is formed out of the same triangles, which unite as eight equilateral triangles and form one solid angle out of four plane angles, and out of six such angles the second body is completed. And the third body is made up of one hundred and twenty triangular elements, forming . . . [etc.]' That's a passage about the elements, one written by a man evidently deeply in thrall to the mystery of Number.

Nor may we simply excuse Plato as being the product of his age. To be sure, the *physiologoi* also had a taste for grand theorizing free of the constraints of empirical evidence. But they, at least, meant what they said. They do not snigger or dodge behind the shelter of myth. Moreover, just a few years after Plato had composed *The Timaeus*, one of his own students would commence a relentless, reasoned assault on the citadel of reality, *this* reality, that in modern print runs to more than a thousand pages: an exhaustive, not to say exhausting, analysis of what his predecessors thought about the causes and structure of the natural world, why those predecessors (more often than not) are wrong, what he thinks they are and the empirical evidence for thinking so. Aristotle would turn his back on his teacher's idealism and see the world, our world, for what it is: a thing that is beautiful and so worth studying in its own right. He would approach it with the humility and seriousness that it deserves. He would observe it with care and be unafraid to dirty his hands doing so. He would become the first true scientist. That he made of himself this after having been taught by one of the most persuasive intellects of all time – *that* is the mystery of Aristotle. All he ever said by way of explanation is: 'piety requires us to honour truth above our friends'.

XI

IN 348 OR 347 BC Aristotle suddenly left Athens. There are at least two accounts that attempt to explain why.

In the first he leaves out of pique. For twenty years he's worked in Plato's Academy. His colleagues call him 'The Reader', but he's original too.

Perhaps *too* original. Plato, with a hint of asperity, called him 'The Foal' – he meant that Aristotle kicked his teachers as a foal kicks its dam. Aelian, writing centuries later, tells a story that isn't particularly to Aristotle's credit and hints of power-struggles at the Academy. One day the elderly Plato, doddery and no longer that sharp, is wandering in the Academy's gardens when he comes across Aristotle and his gang who give him a philosophical mugging. Plato retreats indoors and Aristotle's posse occupies the garden for months. Speusippus is useless against the usurpers, but Xenocrates, another loyalist, finally gets them to move on. Who knows if this is true; but it is certain that when Plato died the top job didn't go to Aristotle but rather to Speusippus and that, coincidentally or not, this is when Aristotle heads east.

In another version, politics rather than pique causes Aristotle to flee. Aristotle has close connections to the Macedonian court. Amyntas' son, Philip II, is flexing his military muscles in the Greek hinterland. He's just razed Olynthus, an ally of Athens, to the ground and sold its citizenry – along with a garrison of Athens' soldiers – into slavery. In Athens, Demosthenes is rousing the citizenry to new heights of xenophobia; Aristotle gets out while he can.

The ancient sources do agree that when Aristotle left Athens, he went east: across the Aegean to the Asian Minor littoral, the edge of the Hellenic world, where micro-states swam precariously in the currents of Athenian, Macedonian and Persian power. Among these was Assos, a city-state on the southern coast of the Troad peninsula. Assos and its sister *polis* Atarneus were ruled by Hermias, a local strongman. Little is known about him except that he was born in obscurity, held power briefly and died horribly. He is said to have started life as the slave of a banker, the incumbent Tyrant of Assos, who, recognizing his talents, freed him and finally made him his heir. He is said to have been educated at Plato's Academy. He is said to have been a eunuch. Much of this may be gossip designed to boost or blacken his reputation – the ancient sources are rarely impartial. Whatever his origins, it seems that he was something of an intellectual for when he became Tyrant in 351 he invited several Academicians to his court, Aristotle among them.

In *The Republic* Plato speaks of how, in the ideal state, political power would be tempered by the wisdom of philosophy. In pursuit of this ideal, Plato had travelled to Sicily to play the sage to the dissolute Dionysos II of Syracuse, a project that had nearly cost him his life. Perhaps, then, Hermias was another try by the Academicians at the manufacture of a Philosopher King; a late biographical fragment suggests that the three years Aristotle spent in Assos did much to soften the rigour of the Tyrant's rule. If so, then this project ended badly too. Hermias was sympathetic to Macedon. In 341, threatened by Macedonian expansionism, Athens told Philip to pull his troops out of the Troad. He did. Hermias was left dangling and the Persians, Athens' temporary allies, trapped, tortured and killed him. Aristotle felt the loss keenly. Years later he erected a statue to him at Delphi which bore the inscription:

> His treatment was outrageous, flouting all respect for divine justice. His killer? The king of the bow-bearing Persians. It was no public contest, no fight to the death by a spear that brought him low. Just the dishonesty of a man he chose to trust.

It also said that each day he would chant a paean for his murdered friend – perhaps the hymn that Diogenes Laertius records in his *Lives of the Philosophers*. The sentiment may seem extravagant, but it is also known that Aristotle married a girl called Pythia who was Hermias' niece or perhaps even his daughter. He was thirty-eight or thirty-nine years old; his bride was probably very young. (In the *Politics* Aristotle says that the best age for a man to marry is thirty-seven; the best age for a woman, eighteen.) 'A spray of myrtle and beauty of rose / were happiness in her hands, and her hair / fell as darkness on her back and shoulders . . .': so Archilochus on another girl, from another place and another time, but I fancy she was like that.

A GREEK GIRL

XII

THE RUINS OF ancient Assos are set upon an extinct volcano that rises steeply from the plain and shore below. A temple to Athena with five standing Doric columns crowns the Acropolis; the foundations of the stoa, bouleuterion, gymnasium, agora and a theatre lie below on the sea-facing slope. In his *Voyage pittoresque de la Grèce* (1809) Choiseul-Gouffier wrote, 'Few cities are blessed with a situation as happy and spectacular as that of Assos . . .' and gave a delightful, if wildly inaccurate, reconstruction of what it looked like in its prime. William Martin Leake said it was the most perfect idea of a Greek city.

Walk up the slopes of the citadel at dusk, through the Turkish village, jump the fence that surrounds the ancient ruins, and you can still see how beautiful Assos must have been. You cannot, however, see what Choiseul-Gouffier and Leake saw. In 1864 the Turkish government demolished much of the still-intact ancient city and used the stone to build the docks of Istanbul's Arsenal. By then the French had taken, as a gift of the Sublime Porte, the temple reliefs and put them in the Louvre. This was just as well. In 1881 an American team, excavating what was left, had to cope with villagers carting off newly dug up walls and stoning a marble centaur that the French had missed.

The temple at Assos was about 180 years old in Aristotle's day, but the theatre is Hellenistic. The view from the citadel cannot have changed much. The massively immovable eastern wall still stands. The surrounding hills are covered with native scrub and the valleys with oaks – the tourist resorts are further down the coast and there aren't even many olive groves. Nothing disturbs you bar a Turkish F-16 arrowing above, testing the fragile airspace frontiers, and the occasional bleat of a goat. But it is the island that compels your attention. Lesbos lies directly before you, astonishingly close, in mounting layers of grey and blue. You feel you could swim there and the urge to do so is almost irresistible, though the Strait of Mytilene is, at its narrowest, nine kilometres wide. You cannot see Lesbos and not want to go. It *promises* discovery.

Above: Assos, restored. *Below:* Lesbos from the citadel of Assos, August 2012

XIII

I N 345, WHILE HERMIAS still ruled, Aristotle took his bride to live on
Lesbos. Thompson, a romantic, called the two years that Aristotle
spent on the island 'the honeymoon of his life'. Perhaps it was; but, in
truth, nothing is known about what, exactly, he did there for he left us no
diaries or notebooks and the ancient biographers are silent. Yet, if D'Arcy
Thompson is right, it was on Lesbos that Aristotle began the great work of
charting, and understanding, the world of living things.

It may have been a conversation; a chance comment that prompted an
excited reply. And then more talk, and yet more, until a vision of the whole
enormous, daunting, thrilling thing emerged. It's an appealing thought –
that biology began so. And it's not an implausible one. For when Aristotle
went to Lesbos it seems that he had at least one other philosopher to talk
to: a man who would become one of his closest friends and who would
inherit his intellectual wealth.

Tyrtamos was born in Eresos, a town on the south-west coast of Lesbos.
The valleys around Eresos (modern Erresos) were green with vineyards;
the town was famous for its wine. Today those same valleys are dry and
uncultivated, yet the remains of ancient terraces can still be seen. We do
not know when and how Tyrtamos and Aristotle met. It's possible that the
younger man – he was thirteen years Aristotle's junior – was one of
Aristotle's pupils at the Academy who had followed him to Assos. If so
then Tyrtamos was now introducing his master to his native land. Or
perhaps Tyrtamos was never in Athens at all and only met Aristotle in
Lesbos – a fluent young local, out to impress and catch an eminent visitor's
ear. We're not even sure what his name was: Strabo has it 'Tyrtamos',
Diogenes Laertius, 'Tyrtanios'. Actually, the spelling doesn't matter since
Tyrtamos/Tyrtanios is quite forgotten. Aristotle renamed the youth
Theophrastus which means 'Divine Speech'. He would become Aristotle's
closest collaborator. Socrates–Plato–Aristotle–Theophrastus: we have
met the next link in a golden chain.

'Divine Speech' is an odd name for a man whose writings, for all their
importance, are as dry as the summer soil. One of his surviving books is
Characters, an encyclopaedia of people you'd want to avoid – the Boor, the

Penurious Man, the Chatty Man and so on interminably. It's as dull as it sounds. Theophrastus also wrote books on logic, metaphysics, politics, ethics and rhetoric – the whole Aristotelian gamut in fact – but they haven't survived. His botany, however, has. It is superb.

Theophrastus wrote two botanical works. One, *Enquiries into Plants*, is descriptive. In it Theophrastus identified the parts of plants and used them to classify plants into groups – trees, shrubs, sub-shrubs, herbs – groups that persisted to the Renaissance. The other, *Explanations of Plants*, is about how plants grow. It examines the effects of environment on their growth, discusses the cultivation of trees and crops and investigates the diseases of plants and why they die. Together, these works are to the study of plants what Aristotle's works are to the study of animals – the founding documents of their science.*

It is a charming conceit to think of the two philosophers strolling in an olive grove, not too far from the Lagoon, dividing up the natural world between them; agreeing, as any two scientists might, to collaborate rather than compete: 'You do the plants, I'll do the animals – and together we'll lay the foundations of biology.' Charming, but too simple. Theophrastus wrote books about animals and Aristotle wrote at least one about plants; but in both cases they have been lost. That botanists look to one as the founder of their science and zoologists the other is, it seems, largely due to the vicissitudes of history – which texts the monks chose to save. Yet it cannot be a coincidence that Aristotle took to studying animals in the native land of the other great biologist of antiquity. Their research programmes and lives are deeply intertwined. Theophrastus succeeded Aristotle as the head of the Lyceum and inherited his most valued possession: his library.

Yet they are very different thinkers. Where Aristotle rarely shies from a bold explanation, Theophrastus is cautiously empirical; where Aristotle is synoptic, Theophrastus prefers to worry at *difficulties*. Given this, it's often supposed that Aristotle dominated the collaboration, and certainly he must have been hard to resist. Even so, placing them both on Lesbos does make one wonder which of them first had the idea to study living things. Who persuaded whom?

---

* The book I call the *Enquiries into Plants* is traditionally known as *Historia plantarum*; the book I call *Explanations of Plants* is traditionally called *Causis plantarum*. By extension, I *should* call *Historia animalium* '*Enquiries into Animals*', but the book's traditional, Latin title is the name that I first learnt to love it by, and so I have kept it.

XIV

To GET TO LESBOS take the evening ferry from Piraeus. If you are young or poor or hardy, travel deck class – thirty euros will take you across the Aegean. You will have to find a place among the gypsy families encamped in the stairwells, the soldiers returning to their island garrisons who occupy the bar, or else the farmers returning to their olive groves who have taken over the lounge. Or you may want to take a cabin – it's a twelve-hour trip.

Athens falls away and you're in the blue. At three in the morning the ship docks at Chios. She's as large as the harbour is small and so, turbines thrashing, she rotates on her own 135-metre axis to get in. Under flood-lights white-uniformed Port Police shrill their whistles and wave their arms to choreograph the container trucks and the frankly uncontrollable foot passengers. Yet it's all implausibly efficient. Thirty minutes later she sounds her horn over the sleeping town, rotates again and faces the Aegean once more.

Dawn silhouettes the Turkish coast black against red. Lesbos appears in the growing light, first pine-clad Mount Olymbos and then the rocky Southern Shore. Cape Malea is rounded: Lesbos lies to port, Assos off the starboard bow and, soon, Mytilene is before you, the cathedral's marble dome stark white in the morning sun.

I have a Mytilenean ritual. As the ship docks, I call Giorgos K. to meet me at a harbour café. A mathematical ecologist at the local university, he is my oldest and dearest friend on the island. The arc of our conversation is always the same: first science, then women – progress and difficulties with both. He has a wayward sensuality, all too generous charm and does not, his friends agree, deserve his beautiful wife. We could mark the years by those talks.

I mention him now because it was he who first took me to Kalloni. We drove north out of Mytilene, skirted the Gulf of Gera, Kalloni's grey little sister, and then cut south-west through the pine-covered lower slopes of Olymbos, emerging at Achladeri where the Lagoon unfolds before you surprisingly vast. There is an excellent fish taverna there, olive groves and, it is said, a few remains of the ancient town of Pyrrha that once stretched down the coast to some neighbouring villages, but I've never found them.

Archaeology, however, doesn't make the argument: the book and the island do. Of all the places in the Eastern Aegean where Aristotle lived, Lesbos is the loveliest. Here, as nowhere else, on this bleak, baked coast the natural world is richly present and seductive; and in Lesbos nowhere more so than by Kalloni. To go down to the quay of one of the villages that dot the shores of Kalloni on a spring morning is to see *Historia animalium* spring to life. You can see Aristotelian fishes – *perkē, skorpaina, sparos* and *kephalos* – gasping in the back of the buyers' pick-up trucks.* Those are the names that Aristotle used and, for these fish at least, they'll still work if you want to buy some to grill. You can also buy a bucket of cuttlefish and, following his text, dissect them. You can lean over the side of a quay, reach down and bring up sea squirts, sea anemones, sea cucumbers, limpets and crabs – all of which he describes. The decks of the fishing boats are littered with the shells and egg cases of the murex snails that infest the bottom of the Lagoon and whose reproductive habits puzzled him so. You can walk along the marshes by the saltpans and see the grebes, ducks, ibises, herons and stilts whose anatomies and habits fascinated him so. You can see European bee-eaters, loveliest of the spring migrants, with their turquoise, gold, ochre and green plumage, nesting in the sand banks, just as he says they do. This is how Thompson put it: 'He will be a lucky naturalist who shall go some day and spend a quiet summer by that calm lagoon, find there all the natural wealth, ὅσσον Λέσβος... ἐντὸς ἐέργει† and have around his feet the creatures that Aristotle knew and loved.' I have done so. He is right.

---

* *Ancient* – modern Greek names: *perkē* –perka; *skorpaina* – skorpiomana; *sparos* – sparos; *kephalos* – kephalos. See Glossaries (Animal Kinds Mentioned) for English vernacular and Latin binomial names.
† 'all that Lesbos has on it': *Iliad* XXIV.

THE
KNOWN
WORLD

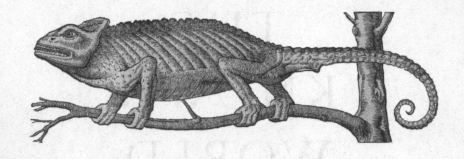

CHAMAILEŌN – CHAMELEON – C<small>HAMAELEO CHAMAELEON CHAMAELEON</small>

XV

To ASSERT THAT ARISTOTLE was a scientist is to suppose that we can recognize one. Sociologists and philosophers have long tried to get the creature in their sights, with indifferent results, for so diverse are their activities and preoccupations that it is hard to find a definition that will embrace them all yet exclude astrologers. Scientists, who are much less exercised about definitions, simply recognize their kin but, if pressed, might offer something like 'A scientist is someone who seeks, by systematic investigation, to understand experienced reality.' This definition, a generous one, allows room for theoretical physicists and coleopterists and some sociologists too; and, though we may quibble about the edges, it narrows the field of human activity considerably, excluding gardeners and physicians (no systematic investigation), literary critics and philosophers (no experienced reality), as well as homeopaths and creation-'scientists' who fail on both counts. It includes Aristotle, whose investigations were nothing if not systematic and who was deeply committed to understanding experienced reality. To be sure, Aristotle never called himself a 'scientist', but he did have a term for 'natural science' – *physikē epistēmē*, literally the 'study of nature'. And he called himself not merely a *physiologos* – 'one who gives an account of nature' – but a *physikos* – 'one who *understands* nature'.

XVI

IN THE COLLECTION OF treatises now called the *Metaphysics*, Aristotle investigates fundamental reality. His ideas are not easy to understand: exegesis of its fourteen books has kept scholars busy for hundreds of years and will certainly do so for hundreds more. Happily we do not have to follow them to appreciate the luminous quality of its opening words:

> All men, by nature, desire to know. An indication of this is the delight that we take in our senses; for even apart from their usefulness they are

loved for themselves; and, above all others, the sense of sight . . . The
reason is that this, most of all the senses, acquaints us with, and brings
to light, many differences between things.

Aristotle does not mean 'know' *just* in the sense of 'understand'; he also
means 'perceive'. Thus in the first instance we should read his words as the
claim that men take pleasure in the exercise of their senses, and the reason
why they do so is because it allows them to perceive all the different things
of which the world is composed. This is merely an opening gambit. For
Aristotle goes on to argue that 'knowing' in the sense of 'perceiving' is the
foundation of 'knowing' in the sense of 'understanding' – indeed, is a
requirement for wisdom. The reason, then, that this statement comes at
the very start of the *Metaphysics* is plain. Aristotle is raising his battle standard and declaring war on the Academy's idealism. His project is not Plato's,
for it concerns *this* world – and he wants us to know it.

To get from perception to wisdom, Aristotle gives us a hierarchy of
understanding. When we perceive something, he says, we acquire a memory
of it. And many memories of a given kind of thing allow us to generalize
about it. Memories of Socrates and Plato, say, allow us to generalize about
'men'. This is Aristotle opening another front *contra* Plato who held that we
are born with all the knowledge that we have – indeed all the knowledge
that we *could* have, that is, all the knowledge in the world. It's just that,
unfortunately, we have forgotten it; our task, then, is to retrieve that knowledge. Such an epistemology is, of course, a call to empirical quietism. If we
already know everything, then we need not actually investigate the world;
perhaps if we talk about it enough it will all come back to us. It is no accident that Plato wrote dialogues.

But talk, for Aristotle, is cheap. Even experience, although necessary for
art and science, is not enough. Aristotle explains why it isn't by imagining
a not very bright, but practically minded, physician, the sort of physician
who supposes that since a remedy worked on one man it will probably work
on another as well, but who doesn't understand or care why it works at all.
Brute empiricism of that sort is useful, says Aristotle, but really not that
admirable. In fact, he's very severe on mere empiricism and compares
labourers undertaking tasks learnt by rote to 'lifeless things': they do what
they do merely because that is what they do.* Master-workmen who under-

---

* It is not working with one's hands *per se* that Aristotle finds objectionable, but rather
lack of understanding. This is evident from his frequent use of craftsmanship as a

stand the *whys* of their craft are 'more honourable and know in a truer sense and are wiser' than such machine-men. (*Politics* 1253b31: 'A slave is a living tool...')

The man who can teach is superior to a man who cannot because he *understands*. This is a very natural view for someone who spent his life doing just that. He can also invent things and, Aristotle continues, inventors are admirable. But – and you can see where this is going – some inventors are more admirable than others. Inventors that produce useful things are inferior to those who produce inventions 'directed to entertainment'. This sounds perverse, but he simply means that the production of pure knowledge is better than the production of useful knowledge. Here, as throughout this argument, he extends invidious distinctions in the forms of understanding to the men who have them. And so he falls into the frank snobbery – extinct now, but extant within our lifetimes – of the pure scientist towards the engineer and the engineer towards the gardener. It is an attitude that sits ill with our own egalitarian instincts, but I would ask the irritated reader to recall that Aristotle is launching a new kind of philosophy: one that is neither concerned with the search for absolute values nor predicated on a perfect world beyond the senses. His philosophy will embrace dirt, blood, flesh, growth, copulation, reproduction, death and decay – the daily experience of the farmer and the fishmonger. He has to persuade *his* listeners, the elite of a highly stratified society, that the knowledge that comes from contemplating such things is of a high order and that those who pursue it are too.

XVII

A RISTOTLE'S SCIENTIFIC METHOD is all of a piece with his epistemology. We have to begin, he says, with the *phainomena* –whence comes our 'phenomena', but perhaps the best translation is 'appearances', for he means by this not only what he sees with his own eyes, but also what other people have seen, and their opinions about it. He favours reports from 'wise' and 'reputable' people. He's conscious that one

metaphor in his biology as well as the reference to the 'master-workman' who presumably also uses his hands, but understands what he is about.

man can't see everything; sometimes you just have to trust what other people tell you (the Greeks inherited huge astronomical catalogues from Babylon and Egypt).

Whatever its source, such data generally consist of many observations of a broad class of objects, say, animals – *zõia*. Once assembled, it has to be ordered into smaller classes: birds, fish, animals with horns, animals without blood and so on. Aristotle's appetite for data is insatiable and his zeal for ordering it tireless. He hoovers up observations about animals, plants, rocks, winds, geographies, cities, constitutions, personalities, plays, poems – the list is partial – processes them, and returns them ordered now one way, now another way, in book after book. For all that, he thinks that this initial inductive phase of research isn't really science, but just the empirical rock upon which scientific reasoning stands.

Aristotle assembles his animal data in *Historia animalium*. A random passage gives a sense of the style:

> Some animals are live bearing, some egg laying, some larva bearing. Live-bearing animals include humans, horses, seals and any other animals with fur; and, among the water-animals, cetaceans – such as dolphins – and the so-called 'selachians'. Some [blooded water animals], e.g. dolphins and whales, do not have gills but do have blowholes. Dolphins' blowholes are located on the back, whales' on the brow. Animals with visible gills include selachians such as smooth dogfish and rays.

The world that Aristotle knew was bound by the Straits of Gibraltar to the west, the Oxus to the east, the Libyan desert to the south and the Eurasian plains to the north. Within it lived more than 500 different kinds of animal, or at least that is about how many he names. Everything about them interests him. He speaks of the reproduction of lice, the mating habits of herons, the sexual incontinence of girls, the stomachs of snails, the sensitivity of sponges, the flippers of seals, the sounds of cicadas, the destructiveness of starfish, the dumbness of the deaf, the flatulence of elephants and the structure of the human heart; his book contains 130,000 words and around 9,000 empirical claims.

The animal world is a vast subject and Aristotle started from scratch. Some medical writings aside, there is no evidence that anyone wrote a zoological treatise before he did. So where did he get all these facts from? The answer appears to be: from just about anywhere that he could.

Some of them came from books. Aristotle is coy about his sources, but it's possible to identify a few of them from glancing allusions. Given the stated scientific nature of his enterprise, a few of the works that he does name are rather odd. Homer crops up occasionally; and he quotes a verse by Aeschylus on the plumage of hoopoes – but that's the Reader at work. The surprise is what is missing. Not that much anatomy seems to come from the Hippocratic treatises, and yet Aristotle's father was a doctor. Here one suspects him of failing to give his predecessors credit. Plato is never cited as a source of factual information – no loss there – though his speculations pervade Aristotle's theory. The *physiologoi* contribute few facts; they, too, are mostly sparring partners in theory. We learn, Aristotle once said, 'by pressing on those in front, and not waiting for those behind'.

There is a suspicion that some of Aristotle's data on mammalian anatomy came from hieroscopic texts – books about prophecy by entrails. He pays an unreasonable amount of attention to the gall bladder, an otherwise insignificant organ that loomed large in the undergrowth of prophetic belief. He's an expert on the astragalus, a minor foot bone used as a die by gamblers and prophets. If Aristotle did indeed get some of his data from sources like this, then he kept the anatomy but ditched the prophecies. Plato did the reverse.

A prophetic manual also probably supplied quite a bit of ethology. 'This is where diviners get their terminology of "alignment" and "non-alignment": animals at war are "non-aligned" while those at peace count as "aligned".' He goes on to describe how eagles fight vultures (and snakes, and nuthatches and herons); how hunting wasps and geckos fight spiders; how snakes fight weasels; how wrens fight owls and so on for pages in a war of nature that is almost Darwinian in its violence. There's a lot of low-quality data here. That wrens, larks, woodpeckers and nuthatches feed on the eggs of other birds would come as a surprise to ornithologists. And if, in Aristotle's day, the ass was at war with the lizard because 'the lizard sleeps in his manger and gets up his nostrils and so stops him eating', then modern asses can rest easy for modern lizards appear to have given up this nasty habit.

Should he have included such material? Perhaps not. Aristotle's sense of empirical reality is as firm as any modern scientist's, and soothsayers' manuals seem unlikely sources of facts. But before we censure him we should pause and consider the difficulties that he faced. Popular culture was steeped in myth; the medical schools knew little human anatomy; country folk were a rich mine of misinformation about the animals that they daily saw. As he constructed the empirical foundation of his science he must have gleaned, and silently suppressed, vast amounts of dubious data.

There is, in his books, only a hint of the thickets of fable and myth that he hacked through. He rejects, or at the very least doubts, tales – the word he uses is *mythoi* – about cranes that carry stones for ballast and that, when vomited, can transmute ordinary matter into gold; lionesses that eject their wombs when giving birth; Ligyans (from Western Greece) who have only seven pairs of ribs; and heads that continue to talk after having been severed from their bodies. In the third century AD, Aelian would fill books with this sort of stuff.

The way in which Aristotle deals with the last of these questions – the talking heads – is instructive. Many people, he says, believe that a struck-off head can talk, and they cite Homer in support. Also, he says, there is an apparently credible description of just such a case. In Caria (Anatolia) a priest belonging to the cult of Zeus Hoplosmios was decapitated. The grounded head named its murderer as one Cerides. A Cerides was accordingly found and put on trial. Aristotle does not comment on the fate of the man, nor even on the possible miscarriage of justice, but he dismisses the story on the grounds that: (i) when barbarians chop people's heads off the heads don't speak; (ii) when animals get their heads chopped off, *their* heads don't make any sounds, and given that, why should human heads be able to do so?; (iii) speech requires breath from the lungs via the windpipe, which it can hardly supply to a severed head. All of this is admirably sane. We should never take such sanity for granted.

XVIII

SEVERED HEADS MAY not vocalize, but fishes certainly do. In a section devoted to animal sounds Aristotle says that the *kokkis* and the *lyra* (both gurnards) make a kind of grunting sound, while the *khalkeus* (John Dory) makes a kind of piping sound. He then goes on to explain that since fish don't have lungs, these sounds aren't a 'voice' of the sort that birds or mammals have; rather the sound is caused by the movement of some internal parts that 'have air or wind inside them'.*

---

* These fishes produce sounds by drumming a specialized 'sonic' muscle against their swim bladders. The sound of the John Dory, *Zeus faber*, has been described by marine biologists as something between a 'bark' and a 'growl'.

KHALKEUS – JOHN DORY – *ZEUS FABER*

Historia animalium is filled with fishy facts, some of them rather recondite. Athenaeus of Naucratis, who wrote a guide to civilized dinner-table conversation circa AD 300, a surprising amount of which apparently revolved around fish, waxed sarcastic:

> But frankly, I'm amazed at Aristotle. Just when did he learn it all? And from whom? Some Proteus or Nereus who'd come up from the depths? What fish do, how they sleep, how they spend their time – that's the sort of stuff he's written about. All so he can amaze the idiots, as the comic poet said!

There was nothing to marvel at: Aristotle's Nereus was simply some fisherman. Aristotle himself doesn't scorn popular wisdom. He often says that we should *begin* investigations by considering what most people think, for they are often right. The problem is that people are prone to telling tall tales. Some fishermen say that fish fertilize their eggs by eating sperm. That can't be right, says Aristotle, since it doesn't fit with their anatomy (any sperm they ate would just get digested); they're just describing some courtship behaviour. He doesn't say what fish do this, but my friend David Koutsogiannopoulos, who knows everything about Greek fishes, tells me it must be a wrasse, probably *Symphodus ocellatus*, and sent me a picture to prove it.

Fishermen's tales. Here are three that I heard from one who wanted to amaze me. First, that the monk seal that lives at the entrance to the Lagoon tracks the local fishermen and then plunders the fish from their nets. Second, that the seagulls of Vrachonisida Kalloni, a local islet, feed their chicks with olives instead of fish. Third, that the crows of Apothika drop walnuts in front of passing cars in the hope that they, the nuts, will be crushed beneath their wheels. Should the car miss, the crows retrieve their nuts and try again.

Amazed I duly was and said so. But, as Aristotle says, the problem is that fishermen don't really observe nature carefully, since they don't seek knowledge for its own sake. Popular lore may be a good place to start, but investigation of the natural world requires expertise, not only a general kind of expertise of the sort that enables us to evaluate rational arguments, but also expertise specific to a given subject. Experts, he says, will spot things easily missed by other people – for example, the shrivelled sperm-ducts of out-of-season dogfish. And, reports of tool-use in New Caledonian crows notwithstanding, I'd like to hear from a behavioural ecologist with a season in the field behind her before I believe that the crows of Apothika really are that smart. Aristotle's scepticism is the first stirring of scientific authority – the authority that has grown rampant in our day. He would surely marvel to see how in our day there is no topic, however arcane, that doesn't have its own caste of experts, authorized by PhDs and university posts, and primed with statistics, ready to trump popular opinion. He would relish it.

XIX

ARISTOTLE'S COYNESS ABOUT his sources extends to his own research. He never says, 'I have seen this – *that's* why it's true,' so it's hard to know which of his myriad facts on, say, reproductive behaviour come from personal observation. Yet, reading between the lines, it's clear that he did much empirical research. This, for example, has the stamp of personal authority:

> The appearance of the chameleon's body is, in general, like that of a lizard, though its ribs descend and converge towards the underbelly like that of a fish: its spine also sticks upwards like a fish's. Its face is very like a 'pig-ape's', but its tail is very long, descends to a point and is usually coiled up like a leather strap. It stands higher off the ground than a lizard but its legs are bent like a lizard's. Each foot is divided into two

parts whose relative position (*thesis*) resembles the opposition (*antithesis*) of thumbs in humans to the rest of the hand. Each part [foot] immediately divides into toe-like structures: the inside of the front feet is divided into three; the outside into two while the inside of the back feet is divided into two and the outside three. The feet have claws, as on a bird of prey. The whole body is rough, like a crocodile's. The eyes, very large and round, are covered in skin like the rest of the body and located in a cavity: in the centre is a small hole through which it sees and which is never covered by skin. Instead, the chameleon twists its eyes round, changes its line of sight in any direction and views whatever it wishes. Its change in colour occurs when puffed up, when its colour is actually black, not unlike a crocodile, or green like a lizard with black spots like a leopard. The same change occurs throughout the body including the eyes and tail. In movement the chameleon is dreadfully sluggish, like a tortoise. And when it is dying it turns green, keeping this colour after death. The oesophagus and windpipe are located as in a lizard, with no flesh anywhere except near the head and jaws and around the very base of the tail. Blood is located only round the heart, the eyes, the spot just above the heart, and, fanning out from them, the veins: the amount of blood in these is minuscule. The brain is linked to the eyes but located a little above them. In the eyes, after the external skin is drawn aside, something like a thin glinting copper ring is visible. Extending through most of its body, more than in other animals, are many strong membranes. Even after it has been cut open completely, the chameleon continues to breathe for a long time and a tiny motion remains around the heart. Though it is in the area of the ribs that the greatest contraction is visible, this occurs also in other parts of the body. There is no sign of a spleen. It hibernates, like a lizard.

It seems that he dissected, indeed vivisected, the chameleon, that beautiful and amiable creature that still lives in the olive groves of Samos.

XX

IN HIS ZOOLOGICAL WORKS Aristotle mentions the following mammals: *ailouros* (cat), *alōpēx* (fox), *arktos* (bear), *aspalax* (Mediterranean mole), *arouraios mys* (field mouse), *bous/tauros* (oxen), *dasypous/lagos* (hare), *ekhinos* (hedgehog), *elaphos/prox* (deer), *eleios* (dormouse), *enydris* (otter), *galē* (beech marten), *ginnos* (ginny), *hinnos* (hinny), *hippos* (horse), *hys* (pig), *hystrix* (porcupine), *iktis* (weasel), *kapros* (boar), *kastōr* (beaver), *kyōn* (dog), *leōn* (Asian lion), *lykos* (wolf), *lynx* (lynx), *mys* (mouse), *mygalē* (water shrew), *nykteris* (bat), *oïs/krios/probaton* (sheep), *onos* (ass), *oreus* (mule), *phōkē* (seal), *thōs* (jackal), *tragos/aïx/khimera* (goat).

All of these species are, or were, native to Greece and Asia Minor, so it is natural that he should do so. More surprisingly, the number of species that he mentions, but that are native to the Nile delta, the Libyan desert and the plains of Central Asia, is not much smaller: *alōpēx* (here the Egyptian fruit bat), *boubalis* (hartebeest), *bonassos* (European bison), *dorkas* (gazelle), *elephas* (elephant), *hyaina/trokhos/glanos* (striped hyena), *hippelaphos* (nilgai), *hippos-potamios* (hippopotamus), *ichneumōn* (mongoose), *kēbos* (monkey), *kynokephalos* (baboon), *onos agrios/hēmionos* (wild ass or onager), *onos Indikos* (Indian rhinoceros), *oryx* (oryx), *panthēr/pardalis* (leopard), *pardion/hippardion* (giraffe?), *pithēkos* (barbary ape), *kamēlos Arabia* (dromedary), *kamēlos Baktrianē* (Bactrian camel) – to which we can add creatures such as the *ibis* (sacred ibis), *strouthos Libykos* (ostrich), *krokodeilos potamios* (crocodile) and various African snakes. 'Always something new from Libya', says Aristotle – and, to judge by this list, the East too.

Where does Aristotle's exotic zoology come from? He was hardly ever out of sight of the Aegean Sea, so he could not have collected it himself. The Roman encyclopaedist Pliny the Elder gave an answer. As so often with Pliny's assertions, it has a fantastical air. He said that Alexander the Great supplied it.

King Alexander the Great, inflamed with a desire for discovering the natures of animals, entrusted this task to Aristotle, a man outstanding in every department of knowledge. Several thousand men in the whole region of Asia and Greece were put under his command – all those who made their living from hunting, bird catching and fishing as well as those who had in their care animal collections, herds of cattle, beehives, fish-ponds, aviaries. The idea was that nothing anywhere in the world might be overlooked by him. It was as a result of his thorough questioning of these men that he composed some fifty famous and distinguished volumes about animals.

In 343, while still in Lesbos, Aristotle was summoned to the Macedonian court. He had reason enough to go. Macedon was, after all, home, and it was no longer the backwater that he had left behind nearly a quarter of a century previously. Amyntas was long dead; Philip II had succeeded to the Macedonian throne, had raised an army and was flexing his military muscles. In Athens, Demosthenes warned the citizenry, in ever more apocalyptic tones, of the danger brewing on their doorstep. They ignored him – to their cost.

Philip wanted a tutor for his son: someone to rub the rough edges off the boy and give him the philosophical education befitting a prince. Did Aristotle make the boy into the man he would become? Or did he try to temper his natural powers? We long to know, but do not. For Aristotle's teenage pupil was not just any spoilt princeling, but Alexander himself, future King of the *oíkoumenē*, the Known World.

It is one of the most remarkable conjunctions in history: one of history's greatest thinkers has, for a few years, the whip hand over one of her greatest military leaders – and then unleashes him on the world. (Pierre-Simon Laplace merely examined Napoleon for admission to the École Militaire.) Writing four centuries later, Plutarch sets the scene:

> For their study and leisure Philip gave them the Nymphaion at Mieza: even today people point out to you the stone seats and shaded walks of Aristotle. It is likely that Alexander did not study just ethics and politics here but also those secret and more profound teachings (those so-called private lectures and special mysteries were not published or shared with the masses).

Those same shady walks and stone seats can still be seen.

In 336 Philip was murdered. Alexander became king. He began by reducing Thebes, second among Greece's cities, to rubble. In a letter Aristotle counsels him to be a leader to the Greeks and to look after them as if they were 'friends or relatives', but Alexander sold Thebes' citizens into slavery.

He later crucified all of Gaza's men. That was a bit more Aristotelian: in the same letter he tells Alexander to be a despot to barbarians and to 'treat them as if they were beasts or plants'. As the young general rampaged across the known world, he carried with him a copy of the *Iliad* in Aristotle's edition. In 335 Aristotle, his work done, returned to Athens, now under Macedonian hegemony, where he established the Lyceum. It is also where, if Pliny is to be believed, he dissected Alexander's zoological largesse.

XXI

P LINY'S STORY IS charming. Alexander, no mere kohl-eyed sensualist or megalomaniacal conqueror, loves plants and animals too, and, recalling his old tutor's interests, affectionately lays the biological booty of an empire at his feet. Writing a century or two later, Athenaeus says that Alexander gave Aristotle 800 talents for his research, and so turns the King into a Macedonian National Science Foundation. There is a whiff of romance about these tales. Eight hundred talents was several times Macedon's annual GDP; and in his biological works Aristotle says nothing about subsidies, a zoo nor even Alexander himself.

It is also clear that Aristotle got some of his exotic zoology from travel books. Ctesias of Cnidus, a fifth-century Greek physician to the Persian court, wrote several books about Persia and India that Aristotle felt he could neither ignore nor trust.

> None of these kinds [*genē*] of animals [live-bearing tetrapods, i.e. mammals] has a double row of teeth. Well, there is one, if Ctesias is to be believed. He claims that a beast that the Indians call the *martikhōras* has a triple row of teeth, resembles a lion in size, is just as shaggy and has the same sort of feet. It has a face and ears like a man's, blue eyes, vermilion colouring and a tail like a scorpion's. It has a sting in the tail, shoots spines like arrows, and has a voice halfway between a shepherd's flute and a trumpet. It runs as fast as a deer, is savage and a man-eater.

Behind the thicket of fable that is Ctesias' *martikhōras* lurks a tiger (the Persian is *martijaqāra*, literally 'man-eater'). Elsewhere, 'What Ctesias has written about the elephant's sperm [that it is as hard as amber] is false.'

'And in India, so Ctesias claims, there are no wild or tame pigs, but the bloodless and scaly animals are all large.' This is a reference to Ctesias' 'Indian worm' that lives in trees and devours domestic animals and is obviously a large python.

The wretched Ctesias is also the source of one of the classic problems in Aristotelian zoology. Aristotle refers to two kinds of animals that have a single horn. One, the *onos Indíkos* (literally 'Indian ass'), has a single hoof (i.e. is a Perissodactyl, specifically, a horse), the other, the *oryx*, has a cloven hoof (i.e. is an Artiodactyl, probably an antelope). He's cautious about the *onos Indíkos* and rightly so. Since at least the nineteenth century, scholars have supposed that it is a garbled description of the Indian rhinoceros, and that the *oryx* is the Arabian oryx glimpsed side on and far away. But of course that's far too late: sceptical though he was, Aristotle could not stop unicorns creeping into his books.

If Aristotle always suspects Ctesias of making things up, he's much more inclined to believe Herodotus (*fl.* 450 BC), borrowing from him often and with confidence. After all, Herodotus himself claimed he preferred to believe things that he'd seen for himself. *Historia animalium* is full of unattributed Herodotean facts: that the menopausal priestesses of Caria (Anatolia) grow beards; that camels fight horses; that in all of Europe lions are only found between the rivers Acheloos and Nestos (Macedonia); that in autumn cranes migrate from Scythia (Central Asia) to the marshlands south of Egypt where the Nile has its source; that Egyptian animals are larger than their Greek congeners, and so on. Sometimes, when the facts strike Aristotle as dubious, he will preface them with 'there are said to be', as in 'there are said to be certain flying serpents in Ethiopia'. Flying serpents may strike us as fantastical, but Herodotus claims to have *seen* their skeletons in Arabia, reports their vicious mating rituals and adds that each year they invade Egypt only to be beaten back by flocks of sacred ibises. Given this, Aristotle's tentative comment is admirably restrained. He simply ignores Herodotus' talk of gold-digging ants and griffins and refutes, without naming him, his belief that each hind leg of a camel has four knees. Indeed the only time that Aristotle names the historian – and you can hear the exasperation – is when he catches him saying something truly absurd: 'Herodotus is wrong when he says that the Ethiopians ejaculate black sperm.'

Since Ctesias and Herodotus account for only a small part of what Aristotle knew about Asia's and Africa's fauna, he must have raided the reports of other travellers as well. But the most puzzling aspect of his exotic zoology is how he manages to combine exact knowledge with profound ignorance. For example, Aristotle often refers to the elephant. Now he could have learnt something of the elephant's general appearance and habits – that it is big, has a trunk, tusks – from someone like Ctesias. But how did he know that the elephant does not have a gall bladder, that its liver is about four times the size of an ox's, that its spleen is rather less, and that its internal testicles are lodged near its kidneys?

Anatomical data like this are hardly the stuff of fourth-century travelogues. They are the sorts of surprising facts that have kept the tale of Alexander's largesse alive. Perhaps, then, Alexander captured one of Darius III's war elephants when he defeated the Persians in 331 at Gaugamela and dispatched it to Athens, a journey of about two thousand kilometres, where Aristotle dissected it in the shade of the Lyceum's Peripatos. L. Sprague de Camp, a minor science-fiction writer, wrote a curious novel, *An Elephant for Aristotle* (1958), based on just this premise and some scholars have not thought it absurd either. But even if we postulate this – prodigiously peripatetic – pachyderm, we may still wonder why, if Aristotle saw, and cut up, an elephant, does he say that its hind limbs are much shorter than its forelimbs?*

The rest of Aristotle's exotic zoology is equally erratic. Summarizing Aristotle on the Asian lion, William Ogle, one of the philosopher's most sympathetic translators and an expert zoologist himself, tartly observes: 'It is plain that Aristotle was not himself acquainted with the lion; for nearly all his statements about its structures are erroneous.' He's thinking, in particular, of Aristotle's claim that the lion has only one bone in its neck (it hasn't; like all mammals it has seven cervical vertebrae). The error is all the more peculiar since Aristotle could have seen lions without venturing far; in his day the Asiatic lion still skulked in Macedon's remoter valleys.† He gives a good description of the European bison, but then says that it fires caustic dung at its pursuers.‡ In the same way, his description of the ostrich

---

* The answer may be simple: the upper part of an elephant's hind limbs are covered by low-hanging folds of skin so that, to the casual observer, they look shorter than the forelimbs. This misapprehension surely could not have survived a dissection.

† The Asian lion, *Panthera leo persica*, was probably extinct in Europe by the first century A D. It now survives only in India's Gir Forest.

‡ An exaggerated story that has its origin in the bovine habit of arching their tails and squirting liquid faeces when threatened. The story may be an un-Aristotelian interpolation. It is repeated almost verbatim in *On Marvellous Things Heard*, a compilation of amazing stories that forms part of the *Corpus Aristotelicum* but that was written by one of Aristotle's successors.

is convincing except that he mistakes its (admittedly impressive) claws for hooves. He does better with the camel for he knows that it has a ruminant's multi-chambered stomach, that it has cloven feet and, surprisingly, that the cleft of the hind feet is deeper than that of the front. And he gives a very good description of the hyena's genitals.

XXII

IN *THE GENERATION OF ANIMALS*, Aristotle says that one Herodorus claimed that the *hyaina* has both male and female sexual organs, and that they take turns mounting each other in alternate years; that it is, in short, a hermaphrodite. Herodorus came from Heraclea, a Black Sea port, about which he wrote a *History* and where he fathered Bryson the Sophist, who tried to square the circle. His *hyaina* must be the striped hyena, *Hyaena hyaena*, for it is the only member of the family found there or anywhere else in the Hellenic world. Aristotle says that Herodorus is talking nonsense. The hyena isn't a hermaphrodite – but it does have an odd-looking undercarriage.

In *Historia animalium* Aristotle tells us more. When following his account, one must know that hyenas of both sexes have large glands that form a pouch around the anus; they explain the description he gives. I interpolate the modern terms:

> The hyena is wolf-like in colour but is more shaggy and has a mane along the whole of its spine. The claim that it has both male and female genitals is false. That of the male [the penis] is like a dog's or wolf's. That which resembles a female's [anal gland] is underneath the tail and, though its structure is similar to that of a female, it has no passage. What lies below it is the passage for residual matter [anus]. The female does indeed have what resembles what is claimed to be the female's genital organ [anal gland], but, like the male, it has it below the tail and no passage. After it comes the passage for residual matter [anus], and below it is the real genital organ [vagina]. The female hyena has a uterus, just like other female animals of that type. It's rare that one gets hold of a female hyena. A huntsman told me out of eleven hyenas that he had caught only one female.

A diagram shows the cause of the confusion: the invaginations formed by anal glands could easily be mistaken for vaginas. Aristotle, however, gets it right. But he doesn't say that *he's* seen all this; he says that *'it has been observed'*. Someone else evidently looked between a hyena's legs to see what he could find.

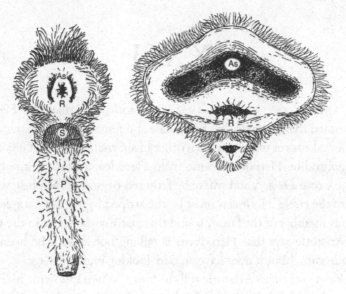

HYAINA — STRIPED HYENA — *HYAENA HYAENA*
LEFT: MALE GENITALIA. *RIGHT:* FEMALE GENITALIA
AS — ANAL SAC; R — RECTUM; S — SCROTUM; P — PENIS; V — VAGINA

Indeed, it does not seem likely that Aristotle saw any of the exotica that he describes. His accounts of their anatomy and habits simply lack the comprehensiveness, detail and accuracy that we would expect if he had — and that he gives when reporting on the anatomy of, say, a cuttlefish. The tale of Alexander's largesse is almost certainly a late invention designed to soften the conqueror's image — or boost the philosopher's. Instead, Aristotle seems to begin with travellers' tales — the various early *Histories* — which he vets as best he can, discarding the implausible, attaching cautionary phrases to the possible and keeping the probable. He then interweaves this material with fragmentary, but more scientifically sophisticated, reports sent by someone else. There is an unknown collaborator at work: someone who travelled, who knew anatomy and who sent Aristotle information about what he saw.

There are several candidates for the Unknown Collaborator. The most plausible of them is Aristotle's great-nephew, Callisthenes of Olynthos. The two men were not only kin, for Callisthenes was a student at Plato's Academy in Athens when Aristotle taught there. It is also likely that, when Aristotle left the Academy in 346/7 to go to Hermias' court at Assos, Callisthenes followed. When Hermias was tortured and executed by the Persians, he wrote, as Aristotle did, a hymn in the Tyrant's praise. Further tradition has it that Callisthenes followed Aristotle to Lesbos, and then, a few years later, to Macedon. Although a few years older than Alexander they may have been students together at Mieza. What is certain is that by the time Alexander came to power, Callisthenes had already made his reputation as an historian, having written the *Hellenica*, a ten-book history of Greece; and that, when Alexander crossed the Hellespont in 334 to conquer the East, Callisthenes went with him to record the campaign.

And to send reports to Athens of the army's progress. Alexander, still untested, just one petty monarch among many, wanted to make sure that the Athenians knew of his triumphs. But Callisthenes was no mere propagandist. He was also a natural philosopher capable of explaining the cause of the Nile's annual flood as the result of moisture-laden clouds hitting the Ethiopian massif. This was doubtless inspired by Alexander's swift traverse through Egypt in 332–1; Alexander may even have sent him south, towards the Sudan, to search for the great river's sources. Callisthenes also recorded Babylonian astronomical lore and proposed a theory of the causes of earthquakes. A fragment says that he sent information to Aristotle, though what about we do not know.

Callisthenes followed Alexander's battle train for seven years. He was present at the sack of Tyre and of Gaza, the entry into Oasis Siwa, the battles of the Granicus, Issus and Gaugamela and the epic pursuit of Darius across the deserts of Central Asia. He traversed Anatolia, Syria, Egypt, Mesopotamia, Babylon, Persia, Media, Hyrcania and Parthia. He skirted the Caspian Sea, the Kir Desert and the Sistan Marshes, climbed the Rock of Aornus and crossed the Hindu Kush. All of this is rich zoological territory, so we may wonder why Aristotle, drawing on all that Callisthenes saw, does not tell us more about the East than he does. That question is, however, easily answered: Aristotle never saw his nephew again. Somewhere in Bactria, modern-day Afghanistan, Alexander had the historian arrested and executed. The ancient sources disagree about why Callisthenes was killed and how, but concur that his death was a nasty one.

Alexander died in 323. Many said that he had been poisoned by Antipater, his Viceroy at Pella and Aristotle's friend. In his writings, which are entirely devoid of political and personal passions, Aristotle says nothing about his nephew's fate, but Theophrastus, the plant collector, mourned Callisthenes and wrote a dialogue in his name.

THE
ANATOMIES

ESTHIOMENON EKHINOS — EDIBLE SEA URCHIN
— *PARACENTROTUS LIVIDUS*

XXIII

ARISTOTLE REFERS TO the internal anatomy of about 110 different kinds of animals. For about thirty-five of them his information is so extensive or accurate that he must have dissected them himself. The quality of his work, at its best, can be judged by what he says about the anatomy of the cuttlefish. With one in hand his account is easily followed.

We place our cuttlefish – flaccid, pale, glutinous – on the table. We begin, as he does, with the external parts: the mouth, its two sharp jaws, the eight arms, two tentacles, mantle sac and the fins. We then have to get inside the thing. Aristotle doesn't tell us how. He may have just grasped the tentacles in one hand, and the mantle in the other, and ripped it apart – that's what a Greek housewife would do. We should not credit him with the skill, patience and fine instruments of a modern anatomist, but he was surely more careful than that. Elsewhere he describes cutting away the skin of a mole's face in order to reveal the stunted eyes beneath.

That being so, we slit the mantle lengthwise from tentacles to tail. A ventral incision reveals the reproductive organs; a dorsal one reveals the cuttlebone and, beneath that, a large, red, structure that he calls the *mytis* and the digestive system. We won't follow his anatomy in all its details, but merely note two remarkable things that he does.

First, that between the eyes with their iridescent argentae and black-slitted pupils, there is a cartilage. Shave it carefully away to reveal two small, soft, yellowish bulges: they're the cuttlefish's brain. It is very easy to miss or immolate, but he finds it. Once seen, the texture of neural tissue is unmistakable.

Second, we follow the alimentary tract. We start at the mouth, follow the oesophagus through the brain and through the *mytis* to the stomach that Aristotle aptly compares to a bird's crop. Then there's another sack, the spiral caecum, that he says looks like a trumpet snail's shell. The intestine emerges from the caecum but, where in most animals it runs posteriorly, here it doesn't. Instead it loops forward so that the rectum exits by the funnel. He's noticed one of the strangest features of cephalopod anatomy: that they defecate on their heads.

Aristotle gets some things wrong. He thinks that the *mytis* —a large, central organ — is the cuttlefish equivalent of a heart. It isn't: it's the cuttlefish equivalent of a liver. In the seventeenth century, Swammerdam found the true hearts — all three of them. He also notices 'feathery growths' in the mantle cavity but fails to identify them as gills even though they look very much like a fish's. He's oblivious to muscles and nerves.

Mistakes are to be expected. But something important is missing; not from the cuttlefish but from the book. *Historia animalium* lacks what any modern zoology text has: diagrams. Anatomy can't really be learnt, or taught, without them. It is only by abstraction and visualization that the logical structure of animal form becomes clear. As any anatomist knows, you don't really *see* until you *draw*. And, just as you're wondering how Aristotle got by without them, you come across this:

> For details of the arrangement of these parts, the diagrams of the *Anatomies* should be consulted.

There was a whole book of them. Eight, in fact, or so Diogenes Laertius says. Philosophers regret the loss of Aristotle's *Protrepticus*, an early summary of his philosophy. But the former, at least, can be reconstructed from people who quoted it. I mourn *The Anatomies* for they are lost complete.

What did a fourth-century BC anatomical diagram look like? Perhaps a bit like the fish paintings on Apulian pottery. But surely sketchier — Aristotle wasn't a professional artist — and he had a pedagogic point to make. An outline then, in swift, black brushstrokes, with alphabetical labels (A, B, Γ, Δ) for the various parts — he sometimes refers explicitly to them. We can try to reconstruct his diagrams but, in truth, can only guess. Nondescript ancient texts have been discovered on papyri wrapped around, or stuffed inside, Egyptian mummies. An Aristotelian diagram of the human heart might, then, yet exist in the eviscerated thoracic cavity of a Hellenistic corpse, but a papyrologist friend has told me that the chances of finding such a thing are comparable to those of finding a living dinosaur in the Congo. Even so, if I thought that a copy of *The Anatomies* lay buried in Egypt's sands, I should dig until I found it, until I could see what he saw, how he saw it.

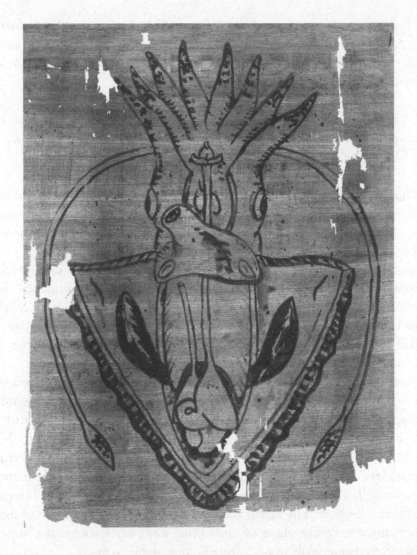

SĒPIA – CUTTLEFISH – *SEPIA OFFICINALIS*
AFTER *HISTORIA ANIMALIUM*, BOOK IV

XXIV

ALL ANIMALS INTEREST Aristotle, but none more than humans. They – we – are his ultimate model organism. The term is no anachronism, for *Historia animalium* begins with an account of human anatomy:

> First the parts of man [*anthrōpos*] must be grasped. People judge curren-
> cies, like everything else, by what is most familiar to them; and human
> beings are, necessarily, the animals most familiar to us.

Humans are not, he grants, very typical. He often mentions our peculiarities: that we alone have a face, eyelashes on both lids, variously coloured
eyes, are toothless at birth, are erect, have breasts in front and have hands.
Nevertheless we are the obvious place to start.

Did Aristotle ever dissect a man? It is much disputed. Denying it, one
sour scholar, Lewes, appealed to Sophocles who depicted Antigone – her
sweet, fierce loyalty, her virginal beauty, her vaulting courage – as she fought
to bury her brother. This, Lewes says, shows the respect that the Greeks
had for their dead, an attitude that would have ensured that Aristotle
would never have got his prurient anatomist's hands on a corpse.

It is not a strong argument. There were lots of slaves about in fourth-
century Greece; one imagines that their unloved, un-Greek cadavers were
always at surplus in Athens. Besides, in the next century Erasistratus of
Ceos and Herophilus of Chalcedon did, apparently, dissect humans, albeit
at liberal-minded Alexandria. The ancient sources even talk of vivisecting
prisoners. But we do not need sociological arguments to settle the matter.
Aristotle himself is fairly clear that he did not. Turning to our internal
anatomy he says: 'The fact is that the inner parts of man are extremely
unfamiliar to us: therefore we must bring along and examine the [inner]
parts of other animals whose nature is comparable to man's.'

Indeed, extrapolation accounts for some of the inaccuracies that litter
his account of our internal organs. He says that humans have a 'double
uterus' – a good guess since the uteri of most mammals are, to varying
degrees, bifurcate, and it's just too bad that ours aren't. He says that we

have 'lobed' kidneys – we don't, but an ox does. Some inaccuracies are inexplicable. He says that we have eight pairs of ribs – did he never see a skeleton? He records examining spontaneously aborted human foetuses. He does not say that he dissected one, but some of his apparent errors may be accurate descriptions of foetal anatomy.

No organ system interests Aristotle so much as the heart and its vessels. His discussion opens with the state of play. Syennesis of Cyprus, Polybus of Cos and Diogenes of Apollonia – two Hippocratic doctors and a *physiologos* – get anything from a paragraph to a few pages. Plato isn't mentioned at all. Perhaps this is because his model of the cardiovascular system, as given in *The Timaeus*, is only five lines long.

The two Hippocratics were hopeless. They started the blood vessels in the head and left the heart out. Diogenes was better and, in what is one of the longest fragments we have from any Pre-Socratic philosopher, Aristotle quotes him at length. Diogenes had the wit to attach the blood vessels to the heart and described the course of some of them in sufficient detail that they can be identified today. All three held that the vascular system is built on a left/right plan: one set of vessels feeds the left testicle, kidney, arm and ear; another, quite separate set feeds their cognates on the right. This, although neat, is wrong.

Aristotle's own account, by contrast, is a bravura bit of anatomical research. Where the Hippocratics seem to have traced the vessels visible through the skin or else simply guessed, Aristotle dissected:

As noted previously, the problem with visual examination is that it is possible to make an investigation effective only if the animals killed by strangulation have previously lost weight.

And:

The pointed end of the heart faces forward, but a shift in position during dissection can often cause one to miss this.

And:

A detailed and accurate study of the relative positions of the blood vessels should make use of the *Anatomies* and *Enquiries into Animals* [*Historia animalium*].

HUMAN VASCULAR SYSTEM
AFTER *HISTORIA ANIMALIUM*, BOOK III

Do not, he appears to warn, *think* to dispute my results without first mastering my techniques.

Those techniques gave him a coherent, detailed account of the heart's structure, the body's major blood vessels and their relationships and ramifications. Reading it, the thought even occurs that he did, after all, dissect a human; but, looking closer, it's clear that there's nothing in it that he couldn't have got from a goat. He places the heart at the centre of the entire system and orients the geometry of the major blood vessels so that the aorta lies 'behind' (dorsal) the 'great blood vessel' – the vena cava – as, near the heart, it does. We follow his account of the 'great blood vessel' and its tributaries:

The vena cava runs through the largest of the heart's three chambers (right atrium + ventricle). The superior vena cava runs towards the upper thorax and then divides to form the innominate veins which then merge into the subclavian veins that run to the arms and the two pairs of jugulars that run to the head. The jugulars give rise to the facial veins and many other small vessels in the head. The inferior vena cava runs through the diaphragm, where it branches into the hepatic vein that invests the liver, and then the renal veins that invest the kidneys, and then continues until it divides into the iliac veins that run down the legs to the toes. The veins of the stomach, pancreas and mesenteries, of which there are many, unite to form a single large vessel. A branch of the 'great blood vessel' (the pulmonary artery) divides and then branches and then branches again into ever smaller vessels that invest the lungs.

The terminology is modern, for Aristotle does not name any vessels except the 'great vessel' and the aorta whose tributaries he traces in much the same way. Yet his account is so good that we know what he means even if here his prose, always viscous, clots; it's so good that we can follow it with modern diagrams in hand; it's so good that its errors are immediately apparent.*

* Which include his belief that our heart has only three chambers instead of the four it does – he either fails to distinguish the right atrium from the right ventricle or mistakes the right atrium for part of the vena cava. Related to this, he confuses the connection of the pulmonary arteries – they enter the right ventricle not the vena cava. He also supposes that the digestive system's veins unite and join the inferior vena cava instead of running into the liver (that is, he misses the hepatic portal system); that the cephalic vein branches from the jugular near the ear (it doesn't; it is a tributary of the subclavian); and that the brain is devoid of blood. He also invents a pair of veins that run from the inferior vena cava to the arms (this may be a Hippocratic hangover). I have distinguished arteries from veins; he does not. Of course he does not know that blood circulates.

But dissection is hard. Open a corpse and you do not see organs neatly arrayed, logically connected and conveniently labelled in contrasting colours, but a morass of dimly discernible tubes and sacs and membranes swimming in pools of bodily fluids. What you see in that morass is deeply influenced by what you expect to see, for in dissection, as in all investigations, expectation and practical difficulties conspire to hide the truth. Expectations and difficulties can, however, sometimes be overcome. Aristotle wonders where the blood goes. He looks and describes, possibly for the first time, how blood vessels branch, and then branch again, until they become tiny vessels, the capillaries, and disappear into the flesh.

XXV

WHICH RAISES THE question: just how good is his biology as a whole? Never mind the theory – how many of his simple, descriptive claims are true? This question, one that will occur to any working scientist opening a volume of Aristotle's biological works and seeing the empirical claims roll by page after page, has never been answered.

It's not for want of trying. Over the centuries, many commentators have attempted to assess the truth of Aristotle's assertions. They have all been defeated by the immensity of the task. Consider the following passage:

> All live-bearing tetrapods have kidneys and a bladder. Some of the egg-laying animals (such as birds and fish) do not: of those that are tetrapods, only the turtle does, with a size proportionate to its other parts. In the turtle the kidney resembles those of cattle. An ox's kidney looks like a single organ composed of a number of small ones.

Only three sentences long, it contains six empirical claims: that (i) all mammals have kidneys – true; (ii) all mammals have a urinary bladder – true; (iii) no fish or bird has a kidney – false; (iv) no fish or bird has a urinary bladder – true; (v) among amphibians and reptiles, only turtles have kidneys – false; (vi) the turtle's kidney, like that of an ox, has a modular structure – true. Aristotle seems, then, to have missed the kidneys of fish and birds. Expectation surely played a part in that since the fish and bird kidneys are not kidney-shaped, but are instead long and thin. In fact, in another book, Aristotle says that fish and birds have 'kidney-like' parts.

But grading Aristotle on his knowledge of the excretory system is easy, requiring no more than a passing acquaintance with vertebrate anatomy. What, however, is one to make of his claim (to pick another) that there is a kind of woodpecker, of intermediate size, that nests in olive groves? Filios Akreotis, Greece's pre-eminent ornithologist, tells me that indeed there is – the middle spotted woodpecker, *Dendrocopus medius* – but that it does so only in Lesbos.

And then there are difficulties with the texts. In the *euripos Pyrrhaiōn*, Aristotle says, you can find the *esthiomenon ekhinos*, the edible sea urchin. He also says that you can tell this sea urchin (*Paracentrotus lividus*) from its inedible relations by the seaweed with which it decorates its spines. So one summer day we drove our scooters to the Lagoon's mouth and snorkelled for the garlanded urchins, cracked their tests or shells open on the rocks, and ate their gonads, the *ricci di mare* so beloved by Sicilians, raw. Among the debris of our lunch were the urchins' mouthparts: tiny, intricate devices made of bone-white calcite. In 1734 the Prussian polymath Jacob Theodore Klein described this structure in his *Naturalis dispositio echinodermatum*; or, rather, he redescribed it, for he noted that Aristotle had also seen it and so, adopting his predecessor's simile, called the structure 'Aristotle's Lantern'.

It is an iconic bit of anatomy. A zoologist may know nothing about Aristotle but will know of the sea urchin's mouthparts by Klein's name. Actually, it turns out that Klein, and pretty much everyone since, misread the texts and that when Aristotle compared the sea urchin to a 'lantern' he didn't mean just its mouth parts at all. An ancient lantern recently dug up from a necropolis in Lethe makes this entirely obvious, for it looks exactly like the sea urchin's test. The problem lies in the manuscripts: some say *sōma* (body), others *stoma* (mouth), and his interpreters have had to choose.

This is a cautionary tale. To determine the veracity of Aristotle's observations would take a squadron of zoologists, deeply versed in his thought and able to read ancient Greek, many years. Today such zoologists are rare. A few centuries ago, however, they weren't. Many could, and did, read Aristotle in the original. They loved what they found. Cuvier set the tone: 'In Aristotle everything amazes, everything is prodigious, everything is colossal. He lived but sixty-two years, and he was able to make thousands of observations of extreme delicacy, the accuracy of which the most rigorous criticism has never been able to impeach.' Cuvier, the author of *Leçons d'anatomie comparée* (5 volumes, 1800–5), *Le règne animal* (4 volumes, 1817) and

Histoire naturelle des poissons (with Valenciennes, 22 volumes, 1828–49) among other monolithic works, was by general estimation, not least his own, the greatest anatomist of his day. He thought that Aristotle could not be faulted – and he should have known.

He should also have known better. Instead he led the chorus: 'A master . . . who extends the limits of all sciences and penetrates to their very depths' – thus Geoffroy Saint-Hilaire *fils*; 'His plan was vast and luminous . . . he laid the basis of science which will never perish' – so de Blainville. That seems excessive. But Owen, Agassiz, Müller, von Siebold and Kölliker, masters of the scalpel in an age when all the animal kingdom came under the knife, all honoured Aristotle. They did so because he founded their science, but also because he knew things that they did not. They loved him, in particular, for having spotted three things that they had to rediscover: the catfish's paternal habits, the octopus' penis-arm and the placental dogfish.

XXVI

I N THE COOL RIVERS and lakes of Macedon lives a catfish of tender habits:

A river fish, the male *glanis*, takes great care of its young. The female abandons them on giving birth, but the male stays and does egg-guarding duty, wherever most spawn has collected. Its only useful service is to prevent other small fish from stealing the offspring during the 40 or 50 days it takes for the offspring to develop, making escape from other fish possible. Fishermen identify where it is on egg-guarding duty by the murmuring sound it makes while giving protection against other tiny fish. It is so affectionate and proprietorial in remaining close to the eggs that it does not abandon the offspring even when eggs attached to deep roots are moved by fishermen into the shallows. Here it can be swiftly caught in the act of grabbing the little fish as they approach. Experienced hook-eaters will not, even so, abandon the offspring but destroy the hooks instead by biting on them with their toughest teeth.

It is a lovely image. The male catfish, abandoned by his feckless mate, stands his ground muttering belligerently at all comers as his hapless fry huddle beneath his fins. It could be a vignette from a fable. That wouldn't be completely un-Aristotelian. He describes various animals as being 'good-tempered', 'sluggish', 'intelligent', 'timid', 'treacherous' and, in the case of one, 'noble and courageous and high-bred' – the lion, of course – all of which has an Aesopean ring.

In 1839 Georges Cuvier and Achille Valenciennes identified Aristotle's *glanis* as the wels, *Silurus glanis*. Too careful to dismiss Aristotle's account of the fish's paternal instincts outright, they nevertheless said that it 'borders on the marvellous', which it does. In 1856 Louis Agassiz, Professor of Zoology at Harvard University, considered the *glanis* again. Agassiz was much more inclined to credit Aristotle. Parental care had recently been documented in fish. He himself had seen an American catfish make nests and care for its young, so why shouldn't a Macedonian one? On the other hand, having grown up in Switzerland, Agassiz knew the habits of *S. glanis* intimately and had never seen it guard its young.

The problem was resolved when Agassiz received some Greek fish from one Dr Roeser, physician to the Greek king. In this collection 'were half a dozen specimens labelled *Glanidia*, caught in the Acheloos, the chief river in Acarnania, from which Aristotle had himself derived his information about the *Glanis*. The identity of the name and the place leave no doubt that I am in possession of the true *Glanis* of the Greek philosopher: that this *Glanis* is a genuine Siluroid, but not the Silurus *Glanis* of the systematic writers.' In 1890, his assistant Samuel Garman described the Macedonian catfish as a new species, *Silurus aristotelis*, differing from *S. glanis* chiefly by having four barbels on its chin rather than six.

Aristotle's description of *S. aristotelis*' breeding habits is exact. At least it is insofar as we know them. In another passage he describes the fish's courtship, the external fertilization, the 'sheath' (egg-envelope) that develops after fertilization, the embryonic eyes that develop a few days later and the unusually slow growth of the larvae. All of this is so detailed that Aristotle may well have studied the fish himself; he lived in Macedon as man and boy. His description of the *S. aristotelis*' parental care is also true to life. The females, after having deposited their eggs, do swim off leaving the male to stand guard. And the males do make a 'muttering' sound to scare off other fish (by beating their thoraxes with their pelvic fins). There is one puzzling feature of his account. Aristotle claims that the male stands guard for fifty days. This seems a very long time, for the eggs hatch in about a week or so. I have asked experts on this species whether the males also look after the growing fry as Aristotle says they do, but they say they do not know.

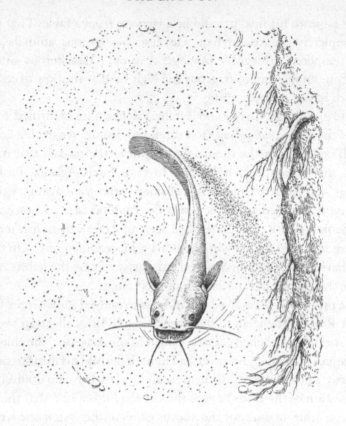

GLANIS – ARISTOTLE'S CATFISH – *SILURIS ARISTOTELIS*

Someone should investigate, for Aristotle may yet have things to tell us about this fish. They should do so soon: the IUCN (International Union for Conservation of Nature) lists *S. aristotelis* as 'endangered'.

XXVII

THE PAPER NAUTILUS, *Argonauta argo*, is a creature rather like an octopus. The animal itself is unimpressive, but its shell is beautiful. As thin and white as eggshell, it has a perfect planispiral geometry. And, although the paper nautilus is pelagic, living far out to sea, it is often washed ashore. After storms they can be found by the hundreds, dying on the beach.

In 1828 Delle Chiaje, an otherwise obscure Italian anatomist, studying paper nautili in the Bay of Naples, found that they appeared to be infested with a parasitic worm. He called his worm *Trichocephalus acetabularis* or 'hairy-headed sucker'. A year later Cuvier discovered a similar worm on an octopus at Nice. He called his worm *Hectocotylus octopodis* or 'cups that hold on [to] an octopus'.

There was nothing very remarkable about the discovery of a new parasitic worm. Marine animals are infested with them. *Hectocotylus* was, however, a strange sort of parasite for it oddly resembled its host. Its suckers looked very cephalopod-like. Suspicion grew that it was not a worm at all. In 1851 Heinrich Müller and Jean Baptiste Vérany independently showed that *Hectocotylus*, far from being a parasite, was in fact the paper nautilus' spouse, more precisely, its spouse's penis. All the paper nautili in the world that have ever been seen are, it seems, female; the male is an obscure and dwarfish creature that does not make a shell at all. One of his tentacles is a highly modified intromittent organ that, during copulation, snaps off in the female's mantle cavity, leaving the male with no penis, or one less tentacle, but in any event one less appendage than he started out with.

The paper nautilus' disposable penis-tentacle was an anatomical wonder of the nineteenth century. One that Aristotle, remarkably, had known all about. Or so, in 1853, claimed an enthusiastic von Siebold: 'Vérany and Müller, who have produced a new phase in the history of *hectocotylus*, will learn with astonishment that Aristotle may fair contest with them for priority of the relation between the male octopus and the hectocotylus arm.'

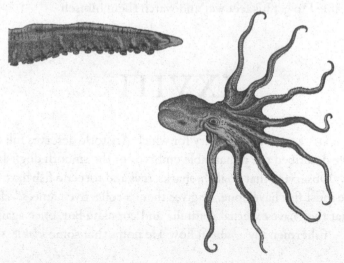

POLYPODON MEGISTON GENOS – COMMON OCTOPUS – *OCTOPUS VULGARIS*
ABOVE: COPULATORY ARM OR HECTOCOTYLUS. *BELOW:* ADULT

Could he? Aristotle certainly knew *Argonauta argo*. He calls it the *nautílos polypous* or 'sailor', describes it clearly and believes ('though knowledge from observation is not yet satisfactory') that it is not as firmly attached to its shell as other shelled animals such as snails and clams are. This is true, but he also repeats a story, that it uses the web between its tentacles as a kind of sail, and that is not. On its amours, however, Aristotle is as silent as the creature itself.

Yet Aristotle did see something. Describing the mating habits of the octopus, he says that one of the male's tentacles looks different from the rest, being pale, more pointed, with larger suckers at its base and a crease at its tip. During courtship, he continues, the male inserts this tentacle into the funnel of the female. In 1857 Steenstrup confirmed that *Octopus vulgaris* also has a penis-tentacle. It is a less outré version of *Argonauta*'s for when octopus male is done, he retrieves it intact from his mate's orifices – but it is just as Aristotle describes it.

Von Siebold exaggerated Aristotle's anatomical prowess. Aristotle certainly spotted the subtle specialization of the octopus' penis-tentacle, but he was unsure what it was for. In some passages he suggests that the tentacular probings are coitus itself; in others he says no, that's a fisherman's tale, the octopi are merely bracing themselves for sex. He cannot understand how semen can be transferred via a tentacle and doubts the whole business on *a priori* grounds. This approach, sound enough when considering fellating fish, led him astray on amorous octopi. But both passages tell us something about the way he thought. And, perhaps, that he wasn't inclined to get his feet wet and watch them himself.

XXVIII

BUT THERE IS one discovery for which Aristotle deserves full credit. He described the remarkable embryos of the smooth dogfish. Observing that dogfish, sharks, rays and torpedo fish have cartilage where most fish have bone, he gives them a collective name, *selakhē*.* He knows that they have external genitalia and copulate but, once again, he's cautious – 'fishermen say' – about how. He notes that some *selakhē*, such as

---

* Aristotle's *selakhē* is not equivalent to our Order Selachii which includes only sharks, but roughly equivalent to our Class Chondrichthyes which includes sharks, rays and skates.

the *batides* (rays or skates) and *skylion* (spotted dogfish), lay eggs with hard shells and tendrils – the 'mermaid's purses' that can be found washed up on beaches – but that most give birth to live young. Moreover, he knows that if you cut a female *akanthias galeos* (spiny dogfish) open the foetuses can be seen still enclosed in their egg cases; they are, as we would say, ovoviviparous.* This was probably common knowledge. Nowadays the infants, which are known as *koytabakia* or puppies, are eaten with garlic sauce.†

The *selakhē* are clearly strange fishes. But one selachian, the *leios galeos*, is stranger yet. Here:

> the animals develop with the umbilical cord attached to the uterus, with the result that as the eggs are used up the embryo resembles that of a tetrapod. A long umbilical cord is attached to the lower part of the uterus, each one fixed by a kind of sucker. The embryo is fixed to the cord in the middle, at the liver. The nourishment, in a dissected embryo, is egg-like, even when it no longer has the egg. A chorion and membranes grow round each of the embryos, as in tetrapods. When young the embryo has its head upwards, downwards when they are well grown and complete . . .

LEIOS GALEOS – SMOOTH DOGFISH – *MUSTELUS MUSTELUS*
Placentation

---

* Aristotle also says that the *batrakhos* (frogfish, *Lophius piscatorius*) is a cartilaginous fish that lays hard-shelled eggs in a mass by the shore. Although he knows the fish well, here he's all at sea. First, *Lophius* is not a cartilaginous fish; second, although *Lophius* is oviparous, his description of its egg masses does not tally with reality since, as Alexander Agassiz showed in 1882, the frogfish lays millions of eggs in enormous, gelatinous, pelagic 'veils'. I think that either Aristotle, his informants or subsequent scribes have partially confused the *batrakhos* with the *batos* (rays, skates).
† *Skylion* is derived from the Attic Greek for 'puppies'.

This could not be clearer. Aristotle is describing the fact that the pups of the smooth dogfish, *Mustelus mustelus*, are linked to their mother's uterus by an umbilical cord and a kind of placenta. He even notices that this remarkable arrangement is otherwise found only in live-bearing tetrapods – that is, mammals.

In the 1550s Pierre Belon and Guillaume Rondelet confirmed the smooth dogfish's peculiar reproductive structures. The latter even figured an infant dangling by an umbilicus from its mother's belly. In 1675 the Danish naturalist Niels Stensen (Steno) dissected one and showed how the umbilical cord feeds into its guts. After that the smooth dogfish was forgotten for nearly two centuries. Cuvier and Valenciennes do not mention it. Johannes Müller discovered it again in 1839. In a masterpiece of dissection, he showed that the placenta of the smooth dogfish is in fact the yolk sac that has become attached to the mother's uterine wall and has a structure as elaborate as that of any mammalian placenta. In homage to his master he titled his monograph *Über den glatten Hai des Aristoteles* (*On Aristotle's Smooth Shark*).

Many zoologists have praised Aristotle, for they have seen him as one of their own. Some, in their enthusiasm, have ignored his defects; they have attributed to him their own insights and obsession with accuracy by way of compliment. However, one scholar and zoologist's assessment seems to me particularly beautiful and just:

> Now I take it that in regard to biology Aristotle did much the same thing as Boyle, breaking through a similar tradition; and herein one of the greatest of his great services is to be found. There was a wealth of natural history before his time; but it belonged to the farmer, the huntsman and the fisherman – with something over (doubtless) for the schoolboy, the idler and the poet. But Aristotle made it a science, and won a place for it in Philosophy.

Thus D'Arcy Thompson.

NATURES

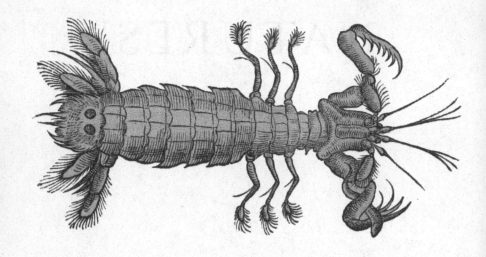

KRANGŌN – MANTIS SHRIMP – *SQUILLA MANTIS*

XXIX

SCHILLER SAID THAT the Greeks saw nature without sentiment; Humboldt that they did not portray her for her own sake. I think that both are wrong.

> Cicadas cry softly under high leaves, and pour down
> shrill song incessantly from under their wings
> The artichoke blooms, and women are warm and wanton –
> but men turn lean and limp for the burning Dog-Star
> parches their brains and knees.

Alcaeus' lovely fragment may be about Lesbos, for that is where he came from. He wrote it in the sixth century BC and may have been Sappho's lover. Perfectly capable of comparing a loved one's face to a squadron of cavalry or the ranked oars of a battle-fleet, Sappho too wrote of golden broom by the seashore, the dew on wild roses and thyme, and how light pours on to the sea. And, if one reads further in the *Greek Anthology*, that compilation of broken odes and bittersweet epigraphs, it is clear that for the thousand-odd years of its span, nature was always near to the Greeks and filled with meaning.

Yet there is sense, a narrow sense, in which Schiller was right. The Greeks may have also celebrated the swallow's spring return but their 'nature' is not the Romantics' nature, a catch-all for everything wild and inhuman. For the *physiologoi* it sometimes meant 'creation'; at least Xenophanes, Heraclitus, Empedocles, Gorgias, Democritus and, later, Epicurus all wrote works titled *peri physeos* – *On Nature* – that contain cosmologies. Aristotle, too, wrote a work with that title (the first four books of his *Physics*), but it isn't a cosmology at all. Rather, it is an analysis of change.

Rocks fall, hot air rises, animals move, grow, copulate and die; the heavens rotate – all is in motion. We take it for granted that the causes of change are various. Steam rises from a cooking pot towards the sky and so does a plant growing in a garden, yet these phenomena are obviously *so* different that they must have different causes. Aristotle sees that too – though not in quite the same way that we do – but he also sees that change itself is the thing that needs to be explained, and so he identifies it with *physis* – 'nature'. He says it would be absurd to try to prove that nature in this sense exists. Lots of things have natures – that's just self-evident. The scientist's job is to discover how nature works and what it is.

He did not invent this conception of *physis* for it can, perhaps, be found in Homer: 'So saying, Hermes gave me the herb, drawing it from the ground, and showed me its nature.' It's certainly very close to Democritus: 'Nature and teaching are similar, because teaching changes a man's shape, and nature acts by changing shape.' By descent it is also close to our use of 'nature' to describe the innate causes of something, as in this piece of doggerel by Isaac Watts (1674–1748), 'Let dogs delight to bark and bite / For God hath made them so; / Let bears and lions growl and fight, / For 'tis their nature too,' or Hobbes' 'Nature (the art whereby God hath made and governes the world)'.

But Aristotle isn't an eighteenth-century deist, and dragging God into the causal chain risks obscuring what he means. His nature is an *internal* principle of change and rest. That's the fundamental difference between natural objects and artefacts: the former move and stop by themselves; the latter don't and can't. And, although he thinks that inanimate natural things such as elements also move of their own accord, it's plain that this definition of 'nature' is really built for biologists. Its purpose is to pin down the mysterious way in which creatures do all that they do – and do it by themselves. No one cranks the clockwork, no one points the little machine in the right direction – nature does.

XXX

IN DEFINING NATURE AS an 'internal principle of change and rest', Aristotle merely delimits the scope of natural science. The question – the great question that motivates his entire scientific enterprise – is what are the *causes* of change?

To answer this question Aristotle began to read. By the time he arrived at the Academy in 367 BC, the intellectual tone was anti-scientific and the great line of *physiologoi* extinct. But their books – papyrus scrolls – were still around. I won't venture to say how or when Aristotle retrieved them from the library shelves; I note only that when he left the Academy he was thirty-seven, so he'd had plenty of time to read, take notes and think.

Among the works that he read were Democritus'. Only Plato looms larger in Aristotle's intellectual hinterland than he. The natural philosopher's natural philosopher, Plato is said to have loathed him and wished his books burnt. Later philosophers evidently read them, so we know that he did not destroy them; yet posterity has granted Plato his ignoble wish for not one now exists. Aristotle's physical theory is constructed largely *contra* Democritus', but much of what we know about the latter comes from the former for, unlike Plato, Aristotle paid his opponents the compliment of preserving their words.

As Aristotle tells it, Democritus held that the world was ultimately composed of entities that were invisible, solid, indestructible, indivisible, immutable, infinite in number and variety and perpetually in motion – in short, atoms. He called his atoms *onta* – 'things'. He learnt the theory from his teacher Leucippus. Today he and Leucippus are jointly celebrated as the fathers of atomic theory and all that it entails, for the theoretical thread that connects them to Dalton and Rutherford is, however etiolated, real.

Democritus developed his atomic theory into a cosmology. The theory, which is sketchy – though whether due to Democritus' failures or to history's vicissitudes we do not know – proposed that atoms floating in the void collide with, and adhere to, each other and so form larger entities, ultimately the planets and stars. He also apparently explained sex determination, sensation and movement in animals by appealing to the shapes and motions of atoms. He may have elaborated a whole reductionist theory of life – the doxographers list three books on the *Causes of the Animals* – but we do not know since they are lost. Even so, the general thrust of his theories is clear. When Democritus sought to explain the nature of things, why they change, he appealed exclusively to matter – the stuff of which they were made. He was not the first to do so; materialism is one of the great threads of Neo-Ionian speculation, but his account was the most sophisticated. Aristotle would spend much of his life trying to show why it is wrong; in a way his scientific works are one long argument against the materialists. We have arrived at one of the great turning points of scientific thought. It's often been judged a wrong turn.

The problem with Democritus' cosmology, Aristotle argues, is that it has the universe arising *spontaneously* from atomic collision. To explain why this is improbable, Aristotle analyses the meaning of 'spontaneous'. Suppose, he says, we see a tripod standing on its three feet, we would naturally suppose that someone had deliberately placed it so. But that need not be the case; perhaps the tripod fell from a roof and just happened to land on its feet – the Greek is *automaton* from which our 'automatic' comes. Democritus supposes that the cosmos is like a tripod standing on its feet, one that wasn't put there deliberately, but just happened to land so.

It seems like a peculiar argument. Why shouldn't the cosmos just happen to have landed on its feet? But Aristotle's point is that spontaneous events are those that appear to have a purpose but in fact don't. And that is the nub of the matter: Aristotle thinks that the cosmos – the stars, the planets, the earth, the living things it contains, the elements themselves – obviously have a purpose; they show the hallmarks of design. And although purposeful things *can* arise spontaneously it just seems implausible to him that a cosmos that is so exquisitely ordered could spontaneously self-assemble.

Most modern cosmological theories suppose that the universe does not have a purpose but just is. Only a child would ask 'what are the stars *for?*' But that's not a childish question to Aristotle. His sense of purpose embraces almost everything. Perhaps this will seem less strange if we see him as a kind of cosmic biologist. We may think that he's on uncertain ground when looking at the stars, but he's obviously right to argue that the random collisions of atoms can't explain the regular and purposeful features of life on earth (or anywhere else).

Aristotle's biologist's vision of the world is explicit when tackling another of the *physiologoi*. Whenever he discusses Democritus, Empedocles is usually close by. For Aristotle, they are both materialists, albeit of different stripe. Empedocles thought that the world was composed of four basic elements – earth, water, air, fire – which can be read as matter in its solid, liquid and gaseous phases, with fire as an extra. These elements combine in particular proportions to give all the different kinds of stuff – stone, iron, bone, blood – that we see.

Existing things have no nature – only a mixing and a separating of what has been mixed. Nature is a name given by human beings.

'Nature' is just mixology. Empedocles' verses explain how a conflict between Love and Strife brings about the cyclical creation and destruction of the world and, with it, the periodic creation of living things. In the first phases of each cycle Love forms tissues, each to a particular chemical recipe, and from these tissues strange creatures emerge composed mostly of single organs: 'eyes without faces', 'heads without necks' and 'single limbs'. Love waxes, Strife wanes, the cycle turns, body-part creatures fuse together in random combinations to produce creatures that have two faces, two chests, or else are part male and part female, or else are hybrid 'man-faced ox calves' or 'ox-faced men' – a teratological bestiary complete with a Minotaur. And it may seem that Empedocles is far from being able to produce the animals that we actually see except that he has a brilliant solution. Simplicius, a sixth-century AD commentator on Aristotle's *Physics*, tells us what it is:

> Thus Empedocles says that under the rule of love parts of animals first came into being at random – heads, hands, feet and so on – and then came into combination: 'There sprang up ox progeny, man-limbed, and the reverse [obviously meaning human progeny with oxen limbs, i.e. combinations of ox and man]. And those which combined in a way which enabled them to preserve themselves became animals, and survived because they [the parts] fulfilled each other's needs – the teeth cutting and grinding the food, the stomach digesting it, the liver converting it into blood. And the human head, by combining with a human body, brings about the preservation of the whole, but by combining with the ox's body fails to cohere with it and perishes. For those which did not combine on proper principles perished. And things still happen the same way nowadays . . .

Most of the recombinants were unfit and perished, so we see only the survivors today. Many early natural philosophers, Simplicius remarks, had this idea. That, if true, is remarkable, for it suggests that in Aristotle's time the idea of selection as a source of order was a commonplace. Certainly Epicurus, a generation younger than Aristotle, gave an even more elaborate selection-based cosmogeny than did Empedocles – at least he did if Lucretius' Epicurean verses are to be relied on.

One might expect Aristotle to like Empedocles' model. The Sicilian has – at least as Simplicius tells it – a perfectly reasonable mechanism capable of producing complex, functional creatures out of chaos. Surely Aristotle, seeking an explanation for purpose in nature, would see, and seize upon, this? He certainly sees the force of the logic. He picks a lovely bit of biological design: teeth. In infants, the front teeth – incisors – come up sharp fitted for tearing food while the molars emerge broad and useful for grinding food. Why, he asks, shouldn't we view this as the product of a process in which things that are fittingly ordered survive and those that aren't don't? Why aren't teeth 'spontaneous'?

Aristotle can think of several reasons why not. But, to understand them, it's necessary to have Aristotle's version of selection clearly in mind. It probably isn't Empedocles', for the Sicilian's extant verses tell only of recombination-selection events that occurred in some remote historical past; ever since then the forms of the survivors – the plants and animals we see – have been fixed. Aristotle, by contrast, supposes that selection is working today. But Aristotelian selection isn't Darwin's either; it's much more radical. Darwinian natural selection supposes that organisms have a system of inheritance that transmits their features more or less intact from one generation to the next, that the inherited material varies slightly nevertheless, and that this subtle variation is the substrate for natural selection. Aristotelian–Empedoclean selection, however, assumes that every individual forms itself *de novo* by a variational-selection mechanism. The womb contains, as it were, a formless soup from which selection produces a child complete with teeth. Aristotle, in short, turns a cosmological model into an embryological one.

Which he demolishes with ease. His arguments are fascinating, for some of them have been used against the theory of evolution by natural selection. (1) Spontaneous events are rare, but the signature of genuinely purposeful events is that they are common: teeth always come up in exactly the same way. This is a probabilistic argument for the existence of a purposeful agent and, like all such arguments, it is wrong, for selection can regularly produce order from disorder.* Admittedly, Empedocles helps Aristotle to this conclusion by making his cosmogonies indeterminate: '[sometimes] it may happen to run one way, but often it ran otherwise' – the line is quoted by Aristotle. (2) It's not just the *end* of development that has the appearance of purpose; it's also the process. Every step in development is obviously directed towards a final goal, rather like each step in the construction of a house. These steps must be the product of an intelligence

---

* This same argument was used by the astronomer Fred Hoyle in a radio interview in 1982: 'The probability of life originating on Earth [by natural selection] is no greater than the chance that a hurricane, sweeping through a scrapyard, would have the luck to assemble a Boeing 747.' The argument, known as the 'Boeing 747 Gambit', is similar to Aristotle's insofar as both hold that, since chance alone can't bring about the regular production of some complex structure (a child's teeth, an aeroplane), some purposeful agent must do so. Both fail to see that selection is not 'chance' but a determinate, creative process.

that has the final product in mind. (3) Although development is very regular, mistakes do happen (in *The Generation of Animals* he has a lot to say about conjoined twins and dwarfs), but they are *mistakes* – deviations from some existing, purposeful programme that must be already in place. Indeed, even Empedocles' original recombinant animals could not have come from nothing; they must have sprung from 'some corruption of some principle corresponding to what is now the seed'. (4) Besides, we simply don't see that much variation. Granted, monstrous progeny sometimes appear, some perhaps even as monstrous as Empedocles' man-headed calf, but why don't we see the same thing in plants, an olive-headed vine sapling, say? 'An absurd suggestion' – and one wishes that one could have shown him a homeotic mutant flower. (5) Organisms inherit their forms from their parents. A given seed doesn't develop into *any* creature, but rather into a specific one: a cicada, a horse or a man. Selectionism can't do that. Aristotle's right – his version of it can't.

The heart of Aristotle's rejection of materialism is his conviction that the cosmos, and the creatures it contains, have order and purpose. His dismissal of Democritus' conviction that order can simply arise spontaneously is, perhaps, understandable. His objection to Empedocles is less certain, for selection – even non-Darwinian selection – *can* bring about order from disorder; it is, indeed, the only known naturalistic explanation for it. Aristotle seems to have painted himself into a corner. Where, then, *does* order come from? And what is its purpose?

XXXI

SPEAKING OF THE *physiologoi*, Aristotle concedes that one of them wasn't completely clueless. 'Whoever said that, in nature as in animals, mind was present as the cause of all order and arrangement appeared like a sober man compared to the random utterances of those before him.' The object of this backhanded compliment was Anaxagoras of Clazomenae (*c.* 500–428 BC). In Anaxagoras' cosmology, for he too had one, the cosmos began with a mixture of various kinds of eternally existing matter. This mixture was set in motion by the action of *nous* – 'Intelligence' – so that a partial separation of ingredients occurred to give the various kinds of matter that we see. The fragments do not tell us what the ingredients were, the recipes for any existing matter or the source of the Intelligence; but it seems that Anaxagoras' Intelligence was not so much a designer as the cosmological mixer's power source.

In *Phaedo* Socrates expresses his disappointment with this. Back in the day, he says, when he was still interested in natural science, hearing that Anaxagoras had made Intelligence the order and cause of things, he had hoped that he, Anaxagoras, would also explain *why* things were arranged as they were, why indeed they were arranged for the best. But then he bought Anaxagoras' books and found that 'the fellow made no use of Intelligence and assigned to it no causality for the order of the world, but adduced to it causes like air and *aither* and water and many other absurdities'.

That's just the sort of reaction we expect from Socrates. Unexpectedly, Aristotle has much the same complaint. For, a few pages after complimenting Anaxagoras on his invocation of Intelligence, Aristotle retracts and accuses him of using Intelligence as a *deus ex machina*, dragging it in only when he's at a loss, and generally explaining events by appealing to all sorts of other causes. The problem is not that Anaxagoras invokes Intelligence; the problem is that he doesn't give it full rein. When we find Socrates–Plato and Aristotle, the first so hostile to science, the other so utterly committed to it, united in their disparagement of a third (or is it a fourth?) philosopher, we can be sure that we have found a deep connection between them. And so it proves. For Aristotle's conviction that the cosmos must be explained in terms of goals or ends is one that he learnt at Plato's knee.

Explanations that appeal to goals, or purposes, or final causes are known as 'teleological' explanations. The word is derived from *telos* – 'end' – and was coined in 1728 by the German philosopher Christian Wolff. The perennial fascination of teleological explanations is that, in attributing purpose to the world, they appear to demand the existence of purposeful agents; indeed, the phenomena they explain become evidence that such agents exist. It was for that reason that William Paley in his *Natural Theology* (1802) limned the functional perfection of the eyelid:

> Of the superficial parts of the animal frame, I know none which, in its office and structure, is more deserving of attention than the eyelid. It defends the eye; it wipes it; it closes it in sleep. Are there, in any work of art whatever, purposes more evident than those which this organ fulfils? or an apparatus for executing those purposes more intelligible, more appropriate, or more mechanical?

And for that same reason Socrates did too:

> And besides all this, do you not think this looks like a matter of fore-sight, this closing of the delicate orbs of sight with eyelids as with folding doors, which, when there is need to use them for any purpose, can be thrown wide open and firmly closed again in sleep?

Socrates goes on to argue that the foresight and purpose manifest in the eyelid comes from God, 'a wise artificer full of love for all living things'. It is the first appearance in history of the Argument from Design, the argument upon which Paley's *Natural Theology* and *The Bridgewater Treatises* (1833–40) depend. It is a gesture towards the sort of account that Socrates sought from Anaxagoras and the other *physiologoi*, one that would lead him from the phenomenal world to the beautiful and the good and the Divine. It is almost certainly Socrates' argument, for it appears in Xenophon's *Memorabilia* rather than in Plato's works. But if Socrates merely gestured towards an account of the world, Plato wrote one – or something that looks like one.

The Timaeus may be a 'myth', but it is a myth written by Plato and so, between the jokes and the moralizing, pullulates with ideas. Of course Genesis and the *Rig Veda* contain ideas too, and those in *The Timaeus* would be as irrelevant to the history of science as theirs are, were it not that Aristotle read *The Timaeus* and transmuted its conceptual lead into the gold of scientific explanation.

The myth that Plato tells is one of intelligent design. The cosmos and its creatures exist and are beautiful because a divine craftsman, the *Dēmiourgos*, made them so. Plato, no zoologist, mentions only six kinds of living things: the heavenly gods (a.k.a. stars and planets), humans, land animals, birds, fish and shellfish. Even so, he has a good deal to say about how and why the *Dēmiourgos* made them as he did.

His account of our digestive tract reveals the *Dēmiourgos'* design priorities. Our intestines are, Plato says, looped into coils to ensure that nourishment does not pass through them too quickly. The coils, therefore, restrict how much food we can eat. That's good, because when we eat we become 'deaf to the command of the divinest part of our nature' – we literally stuff ourselves silly – and so cannot think, and that's bad. Philosophy, it seems, begins in our bowels.

The *Dēmiourgos* is also remarkably farsighted. Plato explains that we have fingernails 'for our framers knew that some day men would pass into women and also into beasts, and that many creatures would need nails (claws and hoofs) for many purposes; hence they designed the rudiments of this growth from the very birth of mankind'. One is tempted to suppose that Plato is thinking of evolution here, and that fingernails are pre-adaptations for claws. That would make this a weird but interesting passage; in fact, it's weird but boring. It's just another of his transmutationist bizarreries, rather like the one that has astronomers becoming birds.

It's not that there aren't some interesting ideas in *The Timaeus*. Aristotle uses many of them in his zoology. But Plato, characteristically, does not think that we should accept his divine teleology on its scientific merits. In *Laws* he explains that materialism – the materialism of Empedocles and Democritus – is malignant, for, dispensing with divine purpose, it leads to atheism and so social disorder. There's a moral sting in every Platonic tale.

XXXII

U PON PLATO'S UNNATURAL teleology Aristotle built a functional biology. When Aristotle invokes a teleological explanation, he often uses the phrase *to hou heneka* – 'that for the sake of which' – or a grammatical variant thereof. He gives a crisp definition of the term in *The Parts of Animals*: 'We all say x is *for the sake of* something when some movement is unimpeded in its progress towards an apparent goal.' He identifies this teleological impulse with natures, the internal principle of change, and then goes on to give a concrete example, the development of a horse. Thus, he is saying: when we see processes that, by their nature, are directed towards an end (for example, the development of a horse from its parent's seed into a foal and finally an adult), then we should explain that process in term of 'this is for the sake of that', where 'this' is some feature of the animal, and 'that' is the adult animal itself.

Aristotle was deeply impressed by the resemblance between organisms and artefacts, particularly machines. In passage after passage he draws on axes, beds, houses and, more mysteriously, *automata* to explain and elucidate various features of animal life. Sometimes they provide mechanical models for explaining how animals work. In *The Movement of Animals* he compares the workings of a limb to those of a puppet. But Aristotle's real interest in comparing organisms and artefacts is that both of them 'come to be': they grow or are made. And both bear the stamp of design.

This talk of artefacts is all very Platonic. And it may seem that Aristotle, too, is working his way towards an intelligent designer. Yet repeatedly and decisively he denies that there is a divine craftsman who made it all. There's no room for a *Demiourgos* in Aristotle's cosmos because it was not made; it's always been there. Besides, a craftsman isn't needed. Consider, he says, the apparently purposeful actions of animals: the way a spider weaves its web or a swallow makes its nest. Some people suppose that this ability must make them as intelligent as human craftsmen. But that clearly isn't so for even plants, devoid of intelligence, show purpose in how they grow. In the same way the various parts of organisms may look *as if* they have been designed by an ingenious external mind, but they have not: each animal and plant is the result of its very own nature; each living thing crafts and maintains itself, like a doctor doctoring himself.

Aristotle denies that Plato ever used 'that for the sake of which'-type explanations. This is odd. *The Timaeus* seems to be full of them and Plato even used the phrase. Perhaps Aristotle thought that his kind of teleology was very different from Plato's. It is. In *The Timaeus* Plato gave a teleological explanation for the coils of the intestinal tract and in *The Parts of Animals* Aristotle gives one too; the explanations are related, for both argue that intestinal morphology regulates appetite; but where Plato explains that human gut is designed just so by a divine craftsman to make sure that we philosophize, this is what Aristotle has to say:

> When feeding, some animals require greater moderation (i.e. they do not have a space in the lower stomach, nor a straight gut, but many spirals). Space creates desire for bulk. Straightness speeds up desire. Such animals end up gluttonous in either speed or quantity.

No divine philosophy-loving craftsman there; just comparative digestive physiology.

Such examples could be multiplied – *The Parts of Animals* is full of them. 'Every part of the body is *for* some action: so what the body as a composite whole is *for* is a multifaceted action.' And although Aristotle's explorations of these profound truths are delightfully detailed and endlessly ingenious it appears that he has impaled himself on the horns of a dilemma. He sees, as Socrates and Plato did before him, the evidence of purpose writ across the face of the world; sees too that material forces alone cannot explain it, but refuses to yield to the expedient of a cosmic designer. So the question remains: from where do plan and purpose in nature come? Aristotle's answer to this question is brilliantly subversive. He appropriates another of Plato's doctrines, one that underpins his, Plato's, entire ontology and epistemology, one that is indeed the very mainspring of his contempt for the perceptible world, destroys it, rebuilds it and turns it to the service of science. There is much to dislike in Plato: his anti-science, his totalitarianism and the seductive charm of his prose; but to his credit let this be said: he taught Aristotle.

XXXIII

COMING ON THE heels of Democritus' atomic cosmology, Plato's creationism may seem like a throwback to the naive natural theology of Hesiod's *Theogony*. And so it would be had not Plato underpinned it with a whole new ontology. Seeking a source of stability in a shifting, mutable world, Plato argued that the physical entities that we see are but imperfect copies of abstract, immaterial entities that he called Forms. It is an obscure doctrine, but if we think of Forms as blueprints in the mind of God we will approach, perhaps, what he had in mind. The entire cosmos is but a copy of a Form. In *The Timaeus*, Plato calls its original the 'Intelligible Living Creature', a title that reflects his conviction that the cosmos is alive. This ultimate Form contains within it countless subordinate Forms, the blueprints for all the objects that the cosmos contains. Beds, birds and men are all but hazy reflections of unseen ideals.

Plato's theory of Forms is the ancestor of all species of idealism. Modern scientists are generally realists and so will find it incomprehensible or bizarre. So did Aristotle. He wanted to explain the features of the physical world. But, if Forms are eternal and static how then, he asks, can they actually *do* anything? And what does it mean to say that the physical world 'participates' in the world of Forms? And if a Form is merely a mental conception, then does not any one physical object have as many Forms as ways that you can think about it? And if a Form exists for any given physical object, say Socrates, then why should there not be two, three or an infinite number of copies of Socrates walking about? Platonic Forms, he concludes, are merely empty words and poetical metaphors. They annihilate the study of nature.

It is all the more remarkable, then, that this unpromising theory was the source of one of Aristotle's deepest ideas. For Aristotle believes that the nature of a living thing, or at least the most important part of it, is in fact its form – if not its Form. The term he uses for 'form' is the term that Plato used, *eidos*. It is one of the most vital organs of his thought.

Aristotle holds that any sensible object is a compound of form (*eidos*) and matter (*hylē*). One can speak of 'form' and 'matter' in the abstract, but in practice they're actually inseparable. To explain what he means Aristotle appeals to various metaphors. If wax is *hylē*, then *eidos* is the impression made in it by a signet ring. In its most general sense, *eidos* is the way in which matter is structured to make the things we see. That seems fairly clear. However, when he applies the term to the world of living things, he uses it in several distinct, but related, senses.

The first biological sense in which Aristotle uses *eidos* is close to the English meaning of 'form' – as the appearance of an animal. His word for a taxon of animals is *genos* (pl. *genē*) – which I translate as 'kind'. Some *genē* are small – the sparrow kind; others are large – the bird kind. So when he wants to describe the features that make a sparrow a sparrow rather than a crane, or a bird a bird rather than a fish, he speaks of its *eidos*.

When using *eidos* in this sense Aristotle usually speaks of forms *within* a kind: 'There are many *eidē* of fishes and birds.' Which brings us to the second sense of *eidos* – as the fundamental unit of biodiversity, that is, close to what we mean by 'species'. Indeed, the traditional Latin translation of *eidos* is precisely *species*, just as *genus* is of *genos*.* One could, then, rewrite the above passage as: 'There are many species of fishes and birds.'

The ambiguity is problematic. To say that there are many different species of birds or fishes is a much richer claim than saying that they come in many forms, and it's often hard to know which of these Aristotle means. Older translations of the biological works often simply use 'species' for *eidos*. Read William Ogle's *de Partibus animalium* (1882) or D'Arcy Thompson's *Historia animalium* (1910) and it's hard to resist the conclusion that Aristotle's sense of the reality of species isn't that different from Linnaeus'. These days most scholars agree that Aristotle rarely uses *eidos* in this second sense. Sometimes he'll refer to an *atomon eidos* – an 'indivisible form' – as when he says that Callias and Socrates share an *atomon eidos*. He obviously doesn't

---

* The Platonic–Aristotelian terminology of division – *eidos/species* and *genos/genus* – filters through Roman encyclopaedists, Neoplatonic commentors, medieval schoolmen and Renaissance naturalists to Linnaeus, from whom we got it.

mean that they are identical, but that they have the same essential features. That seems to correspond to our 'species'. But he names very few indivisible forms, among them, humans and horses, sparrows and cranes.

The problem is that here, as so often, Aristotle's technical vocabulary is underdetermined. He's very reluctant to coin new terms even when he badly needs them. He's not entirely oblivious to the problem. He'll often say that a given term is used in several different senses, and even tell us what they are – but then, as often as not, leaves us guessing which one he means.

Indeed, there's a third sense in which Aristotle uses *eídos*. It's related to the other two, but goes much deeper, and is much more surprising. It's the appearance of an organism, but – if this is not too paradoxical – its appearance when it cannot yet be seen. It is the 'information' or the '*formula*' which was transmitted to it by its parents, from which it built itself in the egg or womb, and which it will, in turn, transmit to its progeny. It is in this sense that Aristotle thinks that the nature of a thing resides primarily in its form.

To speak of *eídos* as 'information' risks anachronism. Aristotle certainly does not conceive of information in the general sense that we do. Yet this interpretation is supported by passages in which he draws a parallel between the transmission of animal form and the transmission of knowledge. In *The Parts of Animals* Aristotle considers how a woodcarver might explain his art. He clearly wouldn't just talk about the wood – that's merely the matter out of which it's built. Nor would he just talk about his axe and auger – they're merely tools. Nor would he just talk about the strokes that he makes – that's mere technique. No, if he is really to convey the origin of the thing he's making, he has to talk about the idea that he had when he began his work – the process by which it will unfold in his hands, its final design and ultimate purpose – he must talk about its *eídos*. In the same way, when a scientist seeks to explain why living things have the features they do, he must talk about their *eíde*. It's just that the forms of living things are not, as Plato held, located in the mind of some divine craftsman, but rather located in their parents' seeds.

There's a passage in the *Metaphysics* where Aristotle gives another metaphor for the relationship between material and formal natures. Rather marvellously, he compares the components of a body to a symbolic system. Some things, he says, are compounds. The syllable *ab* is a compound of the letters *a* and *b*. But putting *a* and *b* together is not enough to give you that particular syllable; you need something else: you need to specify the *order* of the letters (lest you get *ba* instead) or, as we would now say, you need information. In the same way, flesh is a compound of fire and earth and *something else*: the way in which they are ordered. And that order is the form and nature of flesh.

Aristotle's belief that we should attend less to the matter than to the informational structure of living things makes him seem like a molecular geneticist *avant la lettre*. He did not somehow miraculously anticipate the discovery of DNA: it's mere coincidence that he used an ordered alphabetical sequence – *ab v. ba* – to describe forms as we describe nucleotides. Yet, in retrieving forms from the Platonic realm-beyond-the-senses, Aristotle answered, and answered correctly, the question: what is the immediate source of the design that we see in living things? It is the information that they inherit from their parents.

XXXIV

FOR ALL THE severity with which he handled his predecessors (and Aristotle *never* minced his words), he nevertheless drew on them all. Democritus and Empedocles showed him the power of matter; Anaxagoras, Socrates and Plato the prevalence of purpose; Plato the origin of order. His own explanatory scheme contains all these elements.

It had to. The whole problem was that none of his predecessors saw that natures could be, *should* be, understood in several different ways. Our hearts beat – but not just because of chemistry; nor just to keep us alive; nor just because one grew in our embryonic torsos; nor even just because our parents had hearts too; rather, our hearts beat because of all of these things. All these kinds of causes complement each other, indeed, are deeply intertwined. Or so Aristotle argues in a famous methodological dictum known as the 'four causes'. But 'cause' isn't quite right: 'four questions' or 'four kinds of causal explanation' capture his meaning better:

> There are four basic causal explanations: first, what something is for (i.e. what its goal is); second, the formal cause or 'definition of the essence' (these first two should be treated as pretty well the same thing); third, its material basis and, fourth, its efficient cause or origin of movement.

I take them in reverse order. The efficient (or moving) cause is an account of the mechanics of movement and change. It is now the domain of developmental biology and neurophysiology. The material cause is an account of the matter – the stuff – of which animals are made, and their properties. It is now the domain of modern biochemistry and physiology. The formal cause is an account of the information transmitted that any creature received from its parents, and that is responsible for the features that it shares with other members of its species – that is, the subject matter of genetics. The final cause is teleology, the analysis of the parts of animals in terms of their functions. It is now the part of evolutionary biology that studies adaptation. To the degree that function moulds the stuff of which animals are made, the way they develop, how they maintain themselves in the face of the world's vicissitudes, and how they reproduce and die, the final cause embraces, as Aristotle says, the other three. He gives us the structure of our thought even when we do not know it.

Its fault-lines too. Since its revival in the seventeenth century, biology has often been roiled by great conflicts. Many were merely arguments over how to explain. In the 1950s the battle was between the formal-materialistic molecular biologists *v.* teleologically minded organismal biologists. Scientists still live who bear the scars. The zoologists Ernst Mayr and Niko Tinbergen tried to broker a peace – or at least check molecular triumphalism – by arguing for the equal recognition of 'four causes' or 'problems'. Their lists of causes were not quite identical and not quite Aristotle's (they were, after all, evolutionists) but the recognition that living things *must* be explained in several different ways certainly is. These days, in most universities, each kind of explanation gets a department of its own.

Is it Aristotle's thought that has so influenced us? Some scholars, pointing to the sources of Aristotle's system, have suggested that he was only a very industrious magpie. Karl Popper, in a fit of *lèse-majesté*, judged him 'a thinker of no great originality' (though he also conceded, with no apparent sense of contradiction, that Aristotle had invented formal logic). Plato's fans – he still has some – are particularly prone to viewing Aristotle as his teacher's epigone. That can be accomplished only by studiously ignoring how Aristotle transformed Plato's ideas. As a student Darwin read, and enjoyed, Paley's *Natural Theology* and may have even acquired from it his keen sense of the design displayed by living things. Yet who would call Darwin a Paleyite? Calling Aristotle a Platonist is like that.

For Aristotle not only produced a new system of explanation, but also applied it. His predecessors viewed the world as if from Olympus. It lay far below them blurred by distance or obscured entirely by mist, and speculation filled in what they could not see. Aristotle, however, went down to the shore. He observed, applied his causes to his observations and wove them together in the books that make up his Great Course in Zoology: *The Parts of Animals, The Length and Shortness of Life, Youth & Old Age, Life & Death, The Soul, The Generation of Animals, The Movement of Animals* and *The Progression of Animals*. By the time he was done matter, form, purpose and change were no longer the playthings of speculative philosophy but a research programme.

THE
DOLPHIN'S
SNORE

DELPHIS – Atlantic bottlenose dolphin – *Tursiops truncatus*

XXXV

THE BIRD HALL of London's Natural History Museum has four old cabinets. They show three different visions of nature. The first is a walnut cabinet dating from the early 1800s filled with, perhaps, a thousand hummingbirds (but they are hard to count). Collected from throughout the New World, they are mounted in crowded flights so as to suggest an impossible avian Garden of Eden or else the Heathrow Approaches. Here, the cabinet declares, is the Trochilidae in all its glory. Observe the variety and brilliance of their plumage (now dimmed by time), the varied lengths of their bills and the protean forms of their tails; observe the endless variations upon a common theme fashioned by a Creator but ordered by men. It is a very eighteenth-century vision for it represents the science of Linnaeus and Banks, their delight in the creatures of the New World and their desire to pin them down.

The second cabinet, in the centre of the Hall, is made of oak, dates from 1881 and contains not species nor even individuals but rather body parts. The birds have been dismembered so that a duck's webbed foot is juxtaposed against a raptor's talons and the hook of a parrot's bill against the slenderness of a hoopoe's. It is an essay in functionalism. Filled with minutely printed labels, it explains with fussy didacticism why birds have the various bills, feet and feathers that they do. It must have once seemed very modern.

The third and fourth are at the back of the Hall. The birds are mounted among branches and leaves, whole, in pairs, with their nests and their progeny. They belong to group called 'British Birds – Nesting Series'. In one cabinet a pair of petrels crouches against Hebridean boulders; in another a blackbird peers from a hawthorn hedge as his mate stands guard over their ivory eggs. The youngest of the exhibits, they show nature as the Romantics hymned it and as we should like, but struggle to, believe it still exists, filled with animals at home in an untroubled, timeless world going about their business which is to pair and raise young. They are also a vision of England, the England of *Selborne*, *The Hay Wain*, *Adlestrop* and *The Lark Ascending*, caught on the wing and preserved in a vitrine. Upon reading the labels we learn, without surprise, that where now there are two cases once there were 159, the rest having been destroyed by the Luftwaffe in the summer of 1944.*

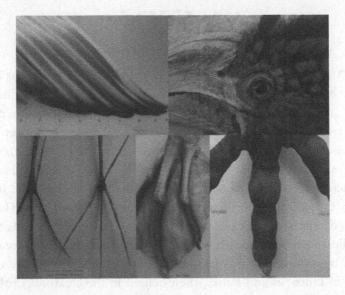

BIRD PARTS. NATURAL HISTORY MUSEUM, LONDON, MAY 2010

The beauty of living things arises from their endless variety, the sense that they give of unity within diversity, and the intricacy of their relations. In the face of nature's munificence it's all too easy to surrender to an inarticulate sense of the numinous connectedness of things, or else let it all surge by in a kaleidoscopic blur. Haeckel looked down upon the coral gardens of Al-Tur and babbled of the Magical Hesperides; Darwin entered the *Mata Atlântica* at Rio and discovered devotion – faced with a rainforest, anyone can go weak at the knees. If, however, we wish to *understand* the natural world, we must focus on, isolate and name its components. We must dissect, parse and label. But, as the Bird Hall shows, there are many ways to carve reality, and each cut reveals a different facet. The question that confronts us, then, is this: where did Aristotle make his cuts? What sort of science did he invent?

---

* And yet the British Birds – Nesting Series have a very German pedigree for they were made by Albrecht Günther (1830–1914), the Tübingen-born Keeper of Zoology at the British Museum (Natural History), who took his inspiration from a taxidermy display that he saw at the Crystal Palace in Southwark that had been originally made for the Great Exhibition of 1851 by the naturalist Hermann Plouquet, another German. Besides the petrel and the blackbird, still on display, a few of the original nests were salvaged and are now in the research collection.

XXXVI

THE NATURAL PHILOSOPHERS of the Renaissance looked at the world with curiosity, discovered that they knew almost nothing about it and turned, quite naturally, to Aristotle as one who did. For them Aristotle was primarily a naturalist who sought to give a comprehensive account of all the creatures that he knew, but who had unaccountably failed to order his data properly.

In 1473 Theodore of Gaza presented his translation, into Ciceronian Latin, of Aristotle's zoological works to his patron Pope Sixtus IV. In a preface Theodore described what the books were all about:

> The rational inquiry of nature proceeds in orderly fashion through all the distinctions that nature has made so that all its living beings are diversified one from another: it groups the major genera and expounds singly the remaining aspects: it distributes genera in species and describes them one by one (and these books contain about 500 of them); it continues by explaining in which way each one reproduces (both terrestrial and aquatic species), of which limbs it is constituted, by which aliment it feeds, by what it is injured, what are its customs, how long the time it is allowed to live, how big its body is, which one is the largest and which the smaller, and the shape, the colour, the voice, the character and submissiveness; in short, it does not neglect any animal that nature generates, feeds, grows and protects.

The false advertising is blatant. Aristotle does name some 500 'species',* and does have much to say about many of them, but certainly does not describe the 'species one by one'. Consider Aristotle on the elephant. Given that he never saw one he has much to say about it. But to find out what you must have recourse to the index, first, of *Historia animalium* where you find the elephant's parts and habits dismembered and distributed throughout the length of the book:

---

* Gaza is already imposing an un-Aristotelian use of 'genera' and 'species' here.

Elephant: age of 586a3–13; 630b19–31; mammae of 498a1; 500a17; breeding of 540a20; 546b7; 579a18–25; capture of 610a15–34; diet of 596a3; disease of 604a11; 605a23b5; driver of 497b27; 610a27; feet of 497b23; 517a31; gall of 506b1; genitals of 500b6–19; 509b11; habits of 630b19–31; hair of 499a9; limbs of 497b22; sleep of 498a9; skull of 507b35; sperm of 523a27; temper of 488a28; teeth of 501b30; 502a2; trunk of 492b17; 497b23–31; voice of 536b22 . . .

and then to *The Parts of Animals*:

Elephant: aquatic habits of 659a30; trunk and its multiple offices 658b30; 661a25; 682b35; forefeet of 659a25; foot has toes 659a25; mammae of 688b15; protected by bulk 663a5 . . .

where its dissolution continues.* Theodore, perhaps trying to make zoology palatable to his Papal patron, disingenuously elided this arrangement and presented Aristotle as encyclopaedist. He was, in fact, promoting Aristotle as a Greek Pliny.

In the first century AD Pliny the Elder wrote and published his *Naturalis historia* which, essay by essay, covered most of what he claimed it did – everything. It was a genuine natural history, probably the first. Pliny borrowed his zoology from anywhere he could and arranged it by species. He said that he valued first-hand reports, but didn't mean it. He may well have seen elephants in Roman triumphs, circuses and battles, but this abundant source of data was for naught. A few quotes give the flavour:

[The elephant] is pleased by affection and by marks of honour, nay more it possesses virtues rare even in man, honesty, wisdom, justice, and also respect for the stars and reverence for the sun and moon.

[O]ne elephant fell in love with a girl who was selling flowers and (that nobody may think it was a vulgar choice) who was a remarkable favourite of the very celebrated scholar Aristophanes.

* Aristotle's manuscript didn't have an index. In the absence of one, I find it hard to conceive how he retrieved, from the hundreds of scrolls in his library, his earlier thoughts on a given topic. Actually, it seems that he often didn't bother, for he has a vexing habit of contradicting himself over trivial matters of fact, as if he's forgotten what he wrote earlier. Indeed, as we'll see, he does so on the elephant.

[B]ut the biggest [elephants are produced by] India, as well as serpents
that keep up a continual feud and warfare with them, the serpents also
being of so large a size that they easily encircle the elephants and fetter
them with a twisted knot.

This is the authentic voice of ancient natural history: gossipy, credulous
and insistent that what the author has to tell is marvellous, as indeed it
would be were any of it true. If Pliny had a precursor it was surely Herodotus
with his stories of gold-digging ants, griffins and one-eyed Arimaspi, and
even Herodotus finds the last too outlandish for his taste.

Yet it was Pliny rather than Aristotle who provided the model for
Renaissance natural history even if it was Aristotle, happily, who provided
most of the substance. In 1551 the Swiss physician and scholar Konrad
Gesner published the first volume of his *Historia animalium*, a compendium
of all that was known about the animal world. He filleted Aristotle's works
and, like Pliny, arranged his material encyclopaedia-wise. Unlike Pliny,
Gesner was primarily interested in the biology of the creatures he wrote
about, was commendably cautious and sought to confirm the data of antiquity; it is in his hands that the modern natural history manual takes
shape. After him all that is required to give us the British Birds – Nesting
Series is the perception that nature is not merely marvellous but filled with
terror, beauty and pathos.

XXXVII

ODERN BIOLOGICAL TAXONOMY – the science of classification – began in 1758–9 with the publication of the tenth
edition of Carolus Linnaeus' *Systema naturae*. It set the agenda
for one of the great projects of nineteenth-century science: the discovery,
classification and cataloguing of all life on earth, a project that his successors prosecuted by publishing vast multi-volume monographs depicting
nature in chromolithographed glory. Cuvier and Valenciennes' *Histoire
naturelle des poissons*, Voet's *Catalogus systematicus coleopterorum* (2 vols, 1804–6),
Esper and Charpentier's *Die Schmetterlinge* (7 vols, 1829–39), Agassiz's
Recherches sur les poissons fossiles (5 vols, 1833–43), the Sowerbys' *Thesaurus
conchyliorum* (5 vols, 1847–87), Gould's *Monographs of the Trochilidae, or Family of*

Humming-birds (1849–61), Darwin's *Living Cirripedia* and *Fossil Cirripedia* (4 vols, 1851–4), Bell's *Tortoises, Terrapins and Turtles* (1872) – to name but a handful among hundreds – bow library shelves with their weight even now.

The taxonomists, too, remade Aristotle in their own image. Their Aristotle was no mere natural historian, but the founder of their own particular science. Aristotle too, they felt, must have had the classificatory impulse, now said to be an attribute of the mildly autistic, that drives a boy to arrange and rearrange his shell collection searching for the unique organizing principle that will unify their disparate forms. He too must have felt the triumph that comes with the discovery of some creature that no one had ever noticed before, a species (delectable words) *new to science*, and have allowed himself the delight of giving it a name. *Historia animalium* must be too – though, granted, it isn't plain to see – a catalogue.

They saw him as a proto-Linnaeus constructing classifications by the Aegean shore. A gifted one, too. Cuvier, characteristically, was fulsome in his praise:

> Aristotle, right from the beginning, also presents a zoological classifica-
> tion that has left very little to do for the centuries after him. His great
> divisions and sub-divisions of the animal kingdom are astonishingly
> precise, and have almost all resisted subsequent additions by science.

This, of course, is mere hyperbole. Cuvier himself constructed a classification of the animals that was vastly superior to Aristotle's, one in which the Greek's major divisions were added to, subtracted from or simply abandoned so that hardly one survived intact. Still, hagiography aside, the view that Aristotle's project was fundamentally a taxonomic one has progressivist appeal. After all, a science can hardly get off the ground unless its objects have first been pinned down and named. As biology needed Linnaeus' system, so astronomy needed Johan Bayer's star atlas, crystallography the Abbé Haüy's geometries, and chemistry Dimitri Mendeleev's Periodic Table. But why talk only of science? No sooner did He make the animals than He gave them to Adam to name – even God sees things that way.

Most of Aristotle's kinds – *genē* – correspond roughly to our species. *Erythrinos, perkē, skorpaina, sparos, kephalos* can all be matched to one or a few modern species of fishes. Sometimes, however, his kinds designate our breeds or varieties: 'there are several kinds of dogs . . .' – the Laconian and the Molossian hounds. His animal names were not, as far as we can tell, invented technical names of the sort that Linnaeus devised for his species.

Rather, they were the vernacular zoology of his day: he got them from the fishermen, hunters and farmers that he talked to. 'Near Phoinike [Lebanon], there are crabs that are known as *hippos* – horse – because they run so fast that they're hard to catch' – thus the ghost crab whose Latin binomial, *Ocypode cursor*, means 'swift-footed runner'. 'There is a rock-bird called *kyanos* ["blue"]. It is most common in Skyros and it spends time on rocks. It is smaller in size than a *kottyphos* [blackbird], a little bigger than a *spiza* [chaffinch]; it has large feet and climbs on the face of the rocks; is entirely deep-blue; its beak is thin and long, its legs short, and like those of the *hippos* [woodpecker]' – probably the rock nuthatch. The fact that, for Aristotle, a *hippos* can be a crab, a bird and a horse doesn't make his zoology any easier to read.*

There is a widespread, and rather romantic, notion that the fishermen and hunters of traditional societies are extraordinarily skilled taxonomists, able to distinguish at a glance species that mere scientists struggle to tell apart. New Guinea Highlanders are said to identify unerringly 136 different kinds of birds. Maybe they can. Modern Greek fishermen, however, appear to be much less gifted when it comes to identifying their fish. There is no reason to suppose they were ever any more so.

We were at Skamanoudi, a tiny port on the eastern shore, where, we were told, the remains of ancient Pyrrha's harbour could be seen just off shore when the light is right. But the wind was up, the Cape exposed and whitecaps obscured the view, so we sat down and ordered some ouzo and a plate of salted fish. Someone said how good the papallinas were. David K., the expedition ichthyologist, demurred. You mean sardellas, he said. The sardella is *Sardina pilcharus*, the papallina is *Sprattus sprattus*, he continued – and, to prove the point, he produced *Ta psara tis Helladas* (*The Fishes of Greece*) of which he is authorially proud and rarely without, so that we could see his beautiful gouaches of two practically identical fishes.

We asked the proprietor. They're papallinas, he said. But, we pointed out, the menu says sardellas. Of course – a papallina is a sardella inside the Lagoon and a sardella is a papallina outside, and these ones came from inside which is why they're so good. An adjacent table of fishermen intervened. The proprietor had not spoken truthfully or at least not comprehensively. The sardella and the papallina were indeed one and the same species, but the essential difference lay not in their geographic origins,

---

* *Hippos* as woodpecker may be a copyist error from *pipō*, Aristotle's usual name for the bird. So maybe this isn't Aristotle's fault.

but in their age or perhaps simply their food – but whether one or all of these factors was salient, none could agree. Some contrarians took the scientific view. The sardella and papallina were, they said, quite different species, just as the *kyrios* said: anyone could taste the difference.

The diversity of their views on the relationship between the two fishes, and which of them we were eating, was puzzling. Kalloni exports thousands of tons of sardellas or papallinas or at least little silvery fish annually; there's no Greek supermarket in which they cannot be bought, and you'd have thought that the men who fish them daily would have arrived at a taxonomic consensus. They've had a very long time in which to work one out. Does Aristotle recognize the inherent ambiguity of vernacular names? Perhaps. His faith in the zoological prowess of fishermen is limited, and he certainly sees that folk-taxonomies do not capture the diversity of life: 'the other kinds [of *karkinoi*, crabs] are smaller in size and don't tend to have special names'. But he never remedies the deficiency.

Even so, many of Aristotle's kinds can be convincingly identified with modern species, among them: dogs *in toto*, horses, two cicadas, four woodpeckers, six sea urchins and humans. Unsurprisingly, he's very good on cephalopods, naming the *polypdōn megiston genos* (common octopus), the *heledōne/bolitaina/ozolis* (musky octopus), the *sēpia* (cuttlefish), the *teuthos* (sagittal squid), the *teuthis* (European squid) and the *nautilos polypous* (paper nautilus). He also speaks of another shelled cephalopod 'that lives in its shell like a snail, sometimes protruding its tentacles'. The identity of this creature has been much disputed. It would be a lovely description of that exquisite mollusc, the chambered nautilus, were it not for the fact that *Nautilus pompilius* lives in the Indo-Pacific west of the Andaman archipelago – *very* far outside Aristotle's range. Some scholars have suggested that he took his description from a specimen seen by someone who had accompanied Alexander the Great to India; others that he is referring to the male of a pelagic octopus, *Ocythoe tuberculata*, that makes its home inside a salp's test or else to a pelagic snail, *Janthina janthina*, that doesn't look like a cephalopod at all. None of these seems particularly plausible and the identity of the ninth cephalopod remains unknown.

Aristotle also recognizes larger groups that resemble modern higher taxa such as Genera, Families, Orders, Classes and Phyla. He calls them *megista genē* – 'greatest kinds'. Some of their names are obviously vernacular too: *ornis* (bird), *ikthys* (fish). But others were apparently invented as part of a technical vocabulary. Aristotle saw that folk-taxonomies aren't much good at classifying animals into larger groups, especially when the animals in question are the sort that most people ignore. The names of his greatest kinds often have a descriptive air: *malakostraka* ('soft-shells' = most crustacea), *ostrakoderma** ('hard-shells' = most echinoderms + gastropods + bivalves + barnacles + ascidians), *entoma* ('divisibles' = insects + myriapods + chelicerates), *malakia* ('soft-bodies' = cephalopods), *kētōdeis* ('monster-like' = cetaceans), *zōotoka tetrapoda* ('live-bearing tetrapods' = most mammals), *ōiotoka tetrapoda* ('egg-laying tetrapods' = most reptiles + amphibia), *anhaima* (bloodless animals = invertebrates), *enhaima* (blooded animals = vertebrates).

Aristotle seems to have believed that, in a good classification, kinds are subordinate to kinds and that each kind has a unique, defined position relative to all others — in other words, that they should be arranged as a nested hierarchy. 'The most important kinds of blooded animals are egg-laying tetrapods, live-bearing tetrapods, birds, fish, cetaceans and any that are unnamed because the group does not exist, merely the simple form in each individual case.' 'We must now speak of bloodless animals. There are several kinds' — which he then lists. 'There are four greatest kinds of soft-shells: they are called *astakoi*, *karaboi*, *karides* and *karkinoi*' — which tells us that lobsters, crayfish, prawns and crabs are subordinate greatest kinds within an even greater kind, the soft-shells. But some of his hierarchies are very shallow: man is a blooded animal but otherwise stands alone.

It is now obvious that animal relationships should be described as a nested hierarchy. It is the only way to describe the topology of a tree graph using words; and a tree graph is the only way to describe descent by modification from a common ancestor using a picture. But if it's obvious to us, we may wonder why it was so to Aristotle, who, after all, never read the passage in the *Origin* where Darwin explains it. ('The affinities of all the beings of the same class have sometimes been represented by a great tree. I believe this simile largely speaks the truth . . .') There are, after all, logical alternatives. Aristotle could have built a classification from taxa that were quite independent of each other. In his delightfully disingenuous description of the *Heavenly Emporium of Benevolent Knowledge*, a Chinese encyclopaedia, Borges tells of one in which each taxon is defined by features such as belonging to the emperor, being embalmed, being a mermaid, being a stray dog or resembling, from a distance, a fly. Aristotle could also have built one from purely orthogonal, rather than nested, taxa. In *Politics* III, 7 he classifies forms of government in this way on the basis of two features, the degree of concentration of power and their quality:

* Literally 'pottery-shard-skinned'.

	high	medium	low
good	monarchy	aristocracy	constitutional
bad	tyranny	oligarchy	democracy

In the event he did not apply this structure to animals.

Perhaps it is just obvious to anyone who studies the diversity of life with care that it should be arranged hierarchically. Linnaeus didn't need Darwin to tell him to put his animals into genera, orders and classes. Aristotle's term for 'taxon', *genos*, is also inherently hierarchical for it originally meant 'family', by which the Greeks meant a patrilineal clan. But it's also true that nested hierarchies emerge naturally from his classification method.

Aristotle's classification of the animals was probably the first.* But classification is very close to definition, and definition was an Academic obsession. Plato thought that to define something was to understand it. His method of definition entailed successive dichotomous division of the thing's features. Investigating the nature of monarchy in the *Statesman*, he began with 'all human knowledge', which he then divided into successive specialist branches of knowledge until he could show that a king is a kind of herdsman. But what does a king herd? To find out Plato divided the animals successively by their various features and concludes that a king is a herder of the tame, hornless, featherless bipeds more commonly known as humans. Plato acknowledged that any given class of activities, people or animals can be divided in many different ways yielding many possible definitions (he gave eight or so pedigrees for the sophists, most of them aimed at defining them as unsavoury, mercenary corruptors of youth). Even so, he argued that one can 'pull together the threads' of the various definitions and so discern the nature of the beast. His later dialogues show signs of definitional mania.

In the *Metaphysics* and *Posterior Analytics* Aristotle tweaks Platonic division a bit; in *Historia animalium* and *The Parts of Animals* he transforms it. He widens its object to embrace classification and subjects its method to a withering assault. He gives many reasons why dichotomous division won't work, but the most telling is the arbitrary nature of the results. Plato divided animals into 'water-dwellers *v.* land-dwellers' and 'gregarious *v.* solitary' and 'tame *v.* wild', which is all very well except that, whichever of those you choose, birds will end up in both sub-groups and that doesn't seem right. Living things, Aristotle saw,

---

* Athenaeus, though, says that Speusippus wrote a book called *Resemblances* in which he claimed that trumpet shells, murexes, snails and clams are similar. On what grounds he did this, and to what ends, we do not know.

have a deep, natural order that a good classification should reflect; when dividing, he says, 'one should avoid tearing each kind apart'. Actually, Plato
expressed the same thought more stylishly: 'we shouldn't cut across the joints
like a clumsy butcher' – a wise precept that he invariably ignored. It is also one
that prompts the question: how do we find nature's joints?

XXVIII

THE WHOLE PROBLEM is that they can be hard to see. Aristotle has
much to say against Plato's approach but less about his own.
Nevertheless, his practice and various programmatic passages
speak of a sophisticated method of division, one that rests upon two important insights.

The first of these is his recognition that animal features vary at different scales of nature's hierarchy. The *diaphorai* – differences – between kinds
within a given greatest kind, between, say, a sparrow and a crane, are relatively subtle. They share the same basic body parts, differing only in shape
and size. His term for such variation is 'the more and the less':

> Differentiation between bird [kinds] involves excess and deficiency in
> their parts and is a matter of the more and the less. Some are long- or
> short-legged, have broad or narrow tongues and so on for the other parts.

Much of his descriptive biology, then, is about how beaks, bladders, bowels
and brains vary in size and proportion.

The differences between the greatest kinds, say, birds and fishes, are
much more radical. They lie in the kinds of parts that animals have and
their arrangement. They are architectural. Modern zoologists speak of a
'body plan' from the German *Bauplan*; Aristotle doesn't have an equivalent
term, but he uses the concept. The relative position of hard and soft parts
and the number of legs are particularly important. Some soft-bodies
(cephalopods) have a hard internal structure (a squid's quill and cuttlefish's
cuttlebone),* but soft-shells (crustaceans) and hard-shells (snails, clams,

---

* When making this contrast he seems to forget about the shells of the paper nautilus and
the mysterious ninth cephalopod.

sea urchins) have an external hard part or, as we would say, an exoskeleton.
Fish have no legs, humans and birds have two, tetrapods have four, divisibles and soft-bodies many.

Greatest kinds also differ in their geometries. For Aristotle an animal
has three axes with six poles: *above—below, before—behind* and *left—right*.* Above
is the pole of an animal that takes in nutrition, *below* is that which expels
it; *before* is the pole towards which an animal's sense organs face and the
direction in which it moves, *behind* is its opposite; his *right* and *left* are the
same as ours. This geometry is based on humans and distinguishes them
from tetrapods. In a tetrapod, *above* (location of the mouth) and *before*
(orientation of the sense organs) are the same pole, and *below* (location of
the anus) and *behind* (opposite to the sense organs) are too. This is one
reason that Aristotle doesn't classify us with the live-bearing tetrapods
(mammals).

Modern zoologists will find his way of geometrizing bodies rather odd.†
But there's no reason why he should do things as we do. And it does give
Aristotle a genuine insight into the weird geometry of cephalopods. Since
their feet are arrayed around their mouths and their guts are twisted into a
U, Aristotle asserts that the cuttlefish has the geometry of a tetrapod that
has been bent double so that its *above* and *below* and *before* and *after* all meet
in the same place.

That's rather brilliant.‡ But his geometry also leads him into some less
perspicacious claims. Ignorant of photosynthesis, he imposes an animal
model of nutrition on to plants. Plants, he thinks, get their nutrition
through their roots which, therefore, are analogues of animal mouths. They

---

* Since all of Aristotle's translators use different terms to describe these poles, I give the
Greek originals here: the before (*to emprosthen*); the behind (*to opisthen*); the above (*to anō*);
the below (*to katō*); the right (*to dexion*); the left (*to aristeron*).
† For us, humans and tetrapods have the same axes: anterior—posterior, dorsal—ventral,
left—right. This is because we ignore the fact that humans are upright, while, for Aristotle,
it's fundamental. To put it another way, where Aristotle bases his axes on functional analogy, we base ours, at least in vertebrates, on structural homology. Yet the difference
between his approach and ours isn't quite as deep as it seems. Looking beyond the vertebrates our axes aren't really defined by structural homology either. By convention, a
fruit-fly's belly is ventral and its back is dorsal, but molecular genetic data suggest that
insects are inverted relative to us, so that our dorsal is homologous to a fly's ventral and
our ventral to a fly's dorsal. In that light, 'dorsal/ventral' is also now merely a statement of
functional analogy.
‡ Equally brilliantly, he notices that gastropods have the same twisted geometry. In both
cephalopods and gastropods, this is the result of a process called 'torsion' during which, as
embryos, their bodies become twisted about.

must also excrete something at the opposite end – namely, fruit. These analogies lead him to the conclusion that the *above* end of a plant is buried in the soil, while its *below* end bends in the breeze.

But Aristotle doesn't just delineate his greatest kinds by their body plans. He also asks whether or not they share the same kind of body part. He appropriates an existing term, *analogon* – analogue. He never defines it but his examples suggest that he means something like 'a part in one kind of animal that has the same function or position as a part in another animal yet is different in some fundamental way'. The term is mathematical in origin: 'as A is to Y, so B is to Z'. In his zoology he applies it metaphorically: 'as feather is to bird, so scale is to fish'. If two creatures have analogous parts then they belong to different greatest kinds. Analogues differ in their fine structure or physical properties. Crabs and snails both have hard external parts, but step on a crab and its carapace is crushed; step on a snail and its shell shatters. Carapaces and shells must, then, differ in some fundamental way* and so, too, must the kinds that possess them.

Aristotle identifies quite a few analogues. Live-bearing tetrapods, humans and dolphins have skeletons made of bone, but fishes, sharks, cuttlefish and squid have bone-analogues: 'fish spine', cartilage, cuttlebone and quills. All of these structures have the same function: to preserve and support the soft tissue. Bird feathers and fish scales are both obviously coverings. Blooded animals have hearts, but bloodless animals – particularly cephalopods – have something analogous to blood and something analogous to a heart to handle it. Lungs are analogues of gills. Sometimes he seems unsure whether two parts are analogous or really just variants of the same thing. In one passage he says that the cephalopods have only a 'brain-analogue'; in others that they have 'a brain' which seems to imply an identity with the tetrapod brain.† Aristotle did not invent an antonym of *analogon* for parts that 'are the same', but only speaks, with considerable ambiguity, of parts that are the same 'without qualification'. It was only in 1843 that Richard Owen filled the terminological gap with 'homologue'. Aristotle probably thinks that most of the internal organs of vertebrates are homologous in this pre-evolutionary sense; at least he speaks of hearts,

* As indeed they do, one being mostly made of chitin, the other of calcium carbonate crystals.
† This sounds trivial, but it's a startling discrepancy. In effect, it transfers the cephalopods from the blooded to the bloodless animals.

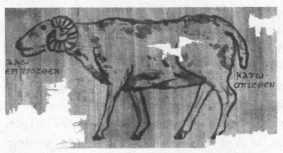

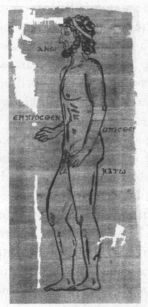

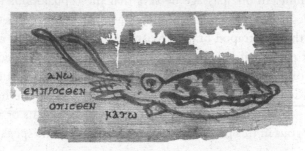

THE GEOMETRY OF LIVING THINGS
AFTER *THE PROGRESSION OF ANIMALS* 4

stomachs, livers, gall bladders, etc. in egg-laying tetrapods, live-bearing tetrapods, birds and fishes without qualification.

Smaller kinds – breeds, species – are, then, distinguished by variety in the size and shape of the same parts; greatest kinds – higher taxa – are distinguished by variety in body plan and analogy of parts. To put it more abstractly, Aristotle adjusts the weight of his features to the scale of his classification. This logic is still the basis of modern systematics. Yet his sense of unity beneath diversity is often acute. He recognizes that his terms 'analogy' and 'the more and the less' are ambiguous. When, after all, does a quantitative difference become so pronounced that it becomes a qualitative one? Compare the skeleton of, say, a cow with that of a sardine and the distinction between true bone and fish spine seems clear (at least to him): they are analogues. But, as Aristotle notes, while birds and snakes generally have bone, small birds and snakes have something more resembling fish spine. In such animals he observes 'nature makes a transition by small steps'. He recognizes that the boundaries between his greatest kinds are not sharp, but shade into one another. Speaking of snakes and lizards he says, 'Serpents as a kind have parts comparable to lizards (if you increase their length and take away their feet),' and he even calls them *syngennis* – kin. And seals may live in water yet their flippers are just odd limbs; they are, he says, 'imperfect' or 'crippled' tetrapods.

XXXIX

ARISTOTLE'S SECOND METHODOLOGICAL insight is his solution to one of the great problems of biological classification – namely, the vexed tendency of organisms to display a mix of apparently incongruous features. Nature's hierarchy isn't neat; in fact, it's a mess. Classify animals by their mode of reproduction (egg-layers *v.* live-bearers) and you will place them into two groups; classify them by their appendages (legs *v.* fins) and you will also place them into two groups – but quite different ones. There are, as taxonomists say, conflicts in the data, and either solution, as Aristotle says, risks tearing kinds apart. This is the problem that Plato's method does not solve. In Platonic division each feature is considered sequentially with inevitably arbitrary results. Aristotle, however, has a far better feel for the order of nature. Here he is deciding how to divide up some land animals:

Wingless tetrapods all have blood, but the live-bearing ones have hair, the egg-laying ones have scales, as an equivalent of fish scales. The snakes are a kind of animal that has blood, can move on land, but naturally lacks feet and has horny scales. Snakes are generally egg laying but the *ekhidna* (Ottoman viper) is exceptional in being live bearing. But not all live-bearing animals have hair, some fishes are also live bearing too.

The trick, it seems, is to consider several features simultaneously – feet (four *v.* none), reproduction (live-bearing *v.* egg-laying) and covering (hair *v.* scales) – and in combination. Three features with two states each yields eight possible combinations, eight possible kinds of animals. But only four actually exist:

(1) hairy, live-bearing tetrapods
(2) scaly, egg-laying tetrapods
(3) scaly, egg-laying apods
(4) scaly, live-bearing apods

The first three of these are greatest kinds: *zōotoka tetrapoda*, *ōiotoka tetrapoda*, *opheis* (snakes). The fourth combination, the exceptional viper, is in all respects a snake except that it gives birth to live young. So how should it be classified? Had a Platonic taxonomist considered the viper he would have defined it as a 'scaly, live-bearing apod' and so sundered it from the rest of the snakes. Aristotle is subtler. For him a kind is a group of similar creatures, but one with fuzzy boundaries.* Quite sensibly, then, he's clear that, although the viper may be viviparous, it is nevertheless a snake. This pragmatism is very Aristotelian. He's always talking about things that are 'for the most part' true, as if the organic world is filled with exceptions that one should note but not fuss too much about.

This sounds casual, but in fact the Platonic and Aristotelian approaches to division represent two very different ways of carving up the world. In 'monothetic' classifications the presence of a feature state (e.g. live-bearing) is necessary and sufficient for that object to be included in a class (of live-bearing animals); in 'polythetic' classifications classes are identified

* By this I do not mean that kinds have *overlapping* boundaries. For Aristotle, an animal cannot belong to two kinds at the same level of the hierarchy simultaneously. A viper may (implausibly) be a live-bearing tetrapod that has scales and no legs *or* it may be a snake that is live bearing *or* it may be something else entirely – but it can't be a live-bearing tetrapod/snake.

by the central tendency of *all* the features and possession of no single feature state is either necessary or sufficient for class membership.* When delineating *genê* Aristotle takes an implicitly probabilistic stance, analyses the feature matrix and clusters. He needs no computer to do this. When classifying, humans naturally attend to many features and look for associations among them. It is in this spirit that Aristotle says we should begin with the *genê* that most people use (birds, fish) – at least we should when they've got it right.

The viper is not the only troublesome creature in Aristotle's bestiary. Ostriches, apes, bats, seals and dolphins are also hard to classify. Most of these animals have features that point to divergent affinities. The origin of the problem is as plain to us as it was obscure to Aristotle: the vagaries of evolutionary history. Closely related species tend to have many features in common due to their descent from a common ancestor. Distantly related species may, however, also share features due to convergent evolution – birds and bats may both have wings, but that does not mean that they are related. Animals may also be confusing mosaics of ancestral and derived features – witness the egg-laying, hair-coated, milk-secreting, duck-billed platypus. The history of systematics can be written as the search for a solution to such confusions. Aristotle may not have understood the cause but he saw and dealt with the consequences. He used the word *epamphoterízein* – to dualize – of animals whose bodies pointed two ways.

Aristotle classifies some dualizers as he does the viper, by blurring the boundaries of an existing greatest kind. The *strouthos Líbykos* (ostrich, literally 'Libyan sparrow') appears, on balance, to be a bird. He avoids the implications of the Barbary macaque. He says that it has some human features (face, teeth, eyelashes, limbs, hands, chest, female genitals, no tail), some tetrapod features (hair, hips, general proportions, male genitals) and some unique features (hind feet that resemble hands), but not where in his classification it should go. On the dolphin, however, he is decisively radical.

---

* Aristotle's procedure bears some resemblance to phenetic classification methods developed in the 1970s insofar as it results in polythetic taxa. However, pheneticists traditionally insisted on overall similarity (use of all assayable characters with equal weight), which Aristotle does not.

XL

IN THE MIDDLE OF relating Greece's battle-scarred dynastic history, Herodotus, inconsequentially, tells the story of Arion, a musician from Lesbos. The beauty of Arion's music, Herodotus says, was second to none; he invented the dithyramb, the wild measure of the Dionysiac hymns. Arion had long lived at Corinth. This is in the time of the Tyranny of Periander which puts it in the mid- to late seventh century BC. Then Arion moved to Sicily, where he played his harp and grew rich. But after a while he yearned for rocky Corinth and so hired a ship and crew at Tarentum in Apulia to take him back. The crew were Corinthians, so decent fellows all, except that they weren't. As Italy disappeared, spotting his cash, they made to pitch Arion overboard. Not so fast! said Arion, let me sing for you first. Why not? said the crew. And so Arion put on his finery, plucked his harp, sang his song and then threw himself over the side where a friendly dolphin picked him up, inquired where he was bound and gave him a lift all the way back to Corinth. Of course no one there believed this tale, but then the crew showed up, were duly shocked at finding Arion alive and so confirmed their guilt. And there is still, concludes Herodotus, at Taenarum a shrine with a small bronze statue of a man perched upon a dolphin's back.

That Herodotus has Arion leaving Italy via Tarentum (Taranto) is no coincidence, for a youth riding a dolphin was entangled in the city's foundation myth and stamped on its coins. Pausanias, Aelian, Pliny, Oppian, Ovid and a dozen other ancient writers besides tell of Arion or other dolphin riders, but Aristotle, seeking the plausible core of the myth, only says: 'On sea-animals: much evidence attests to the mildness and gentleness of dolphins and the passion of their love for boys in the regions of Taras, Caria and elsewhere.' If that sounds paedophilic, it does in the Greek too. He then goes on to tell how dolphins protect their own, particularly their young, but he's mostly interested in their anatomy.

Dolphins, Aristotle says, are supremely swift swimmers and voracious hunters. He says that they copulate and give birth to one or two live young that they suckle via ventral slits. They have internal testicles near the belly and no gall bladders. They have proper bone. They breathe air, have a windpipe and lungs, and a blowhole through which they spout water. When hunting they will plunge into the deep, calculate how long they can stay down, and then shoot to the surface like arrows, flying out into the air, sometimes clearing the masts of boats. As such, they're just like divers bolting for the surface. When caught underwater in nets they drown, but conversely survive for a long time on land. If taken from the sea they moan but cannot articulate for their tongues are immobile and they don't have lips. Sleeping dolphins do, however, snore – or so it's said. They live in male and female pairs for up to thirty years. We know this because fishermen nick their tails and then release them again – which seems to be an account of history's first mark–recapture study. Sometimes they strand themselves for no obvious reason at all.

Most of this is accurate. That dolphins snore is dubious, but perhaps we'll let it go since they do, apparently, vocalize in their sleep. Some scholars think that Aristotle must have dissected a dolphin. I don't for he also makes some serious mistakes. He says – and says twice – that the dolphin's mouth is slung under its head rather like a shark's. That's an error made by someone who never saw a dolphin close up. (Pliny, amplifying Aristotle's error, says that dolphins have their mouths on their bellies, which makes you think that, where the Greek is sometimes wrong, the Roman is often a fool.)* Aristotle also thinks that the blowhole is connected to the mouth since he says it expels water taken in during feeding, but it isn't and doesn't. It's clear that he got his anatomy from some fisherman who butchered a dolphin on the beach. It's often said that the Greeks cherished dolphins as sacred animals. Oppian, who loved dolphins, said that to hunt them is immoral, as loathsome as homicide, and described, in terms that would do a Greenpeace activist proud, how the beastly Thracians harpooned them. But dolphin hunts must have been widespread for Aristotle describes another technique. He says that nets are set in utter silence and then, when the dolphins are encircled, the hunters make a racket which stupefies and entraps them. There's no hint of censure: he's interested only in the fact that dolphins can evidently hear even though they don't have ears.

Whether first hand or not, Aristotle put his dolphin anatomy to good use. Although it is in many ways like a fish, he recognizes that its moans and snores, lungs and bones, internal testicles and live-born-and-milk-suckling

---

* The error may be based on popular iconography which, as in the coinage of Tarentum, often depicted dolphins with underslung mandibles. To his credit Pliny did, however, get the function of the blowhole right.

offspring are typically tetrapod features. It also has a feature of its own, the blowhole. In *The Parts of Animals* he seems unsure what to do with the dolphin, but in *Historia animalium*, parts of which were probably written later, he assigns it, along with the porpoise and the whale, to a new greatest kind, the *kētōdeis* from whence derives our Cetacea. He was probably led to erect a new taxon by the fact that *several* kinds of animals shared this distinctive combination of features; he is a taxonomic pragmatist. Aristotle did not label the cetaceans as mammals since 'mammal' was a concept that he did not understand. For him the cetaceans were just one of the great kinds of blooded animals of equal rank to the birds, fish and viviparous tetrapods. Still, he did a lot better than his successors who for two thousand years just called them 'fish'.

I have never seen dolphins in Kalloni, but they are sometimes there. A fisherman told me that, in the summer of 2011, a large pod of bottlenoses entered the Lagoon to hunt. Some fishermen, other fishermen he implied, not him, though this was not completely clear, rounded them up and killed them. He explained that the youngsters damage the nets that cost three thousand euros each and that, of the fifty bottlenoses, three got away.

XLI

IN CONGRATULATING ARISTOTLE on his success in ordering the animals I have, however, elided the problem that I began with – namely, whether or not his project was, at heart, a taxonomic one. The zoologists of the eighteenth and nineteenth centuries thought it was. We should not take them at their word. They sought an illustrious predecessor. There are reasons to doubt whether he really was anything of the sort.

Contra Cuvier, Aristotle never produces anything resembling a coherent, comprehensive classification in which every animal has its place. Further, though he may have perceived nature's hierarchy, he does not name its levels: from Race to Kingdom, *genos* suffices for all. He also never tells us how to distinguish one kind from another, and he's terribly casual about names. In Aristotle's day salted *Sardina pilchardus* and *Sprattus sprattus* were Aegean staples, but he mentions neither the 'sardella' nor the 'papallina' for both are Roman names. Instead he speaks of the

membras, the *khalkis*, the *trikhis*, the *trikhias* and the *thritta*, all of which seem
to be clupeids, but whether sprats, sardines, shads or pilchards (to intro-
duce the equally underdetermined English names) is hard to say, for he
gives us few clues to their identities. His higher taxa are feeble. He sees
that snakes and lizards are somehow kin but doesn't bother to give them
a family name. He forgets to tell us whether bats are birds or tetrapods
or something else again. He never gives the diagnostic features that
might define a kind in a usable way; never says 'a fish is an animal that
has gills + scales + fins . . . etc.', but instead just says 'fish are a kind' and
assumes that everyone knows what one is. Compare the relentless lists
and tables of names and definitions in *Systema naturae* with the narrative
discursions of *Historia animalium* and it's plain that very different scientific agendas are at work.

Aristotle seems to name and classify only when some other purpose
demands it. He even says as much. When describing animals we could, he
says, talk about a sparrow or crane individually, but 'insofar as this will
result in speaking many times about the same property because it belongs
in common to many things, in this respect speaking separately about each
one is somewhat silly and tedious'. It's simply much easier to discuss larger
groups composed of animals that have a lot in common.

But if Aristotle's descriptive biology is neither Pliniesque natural
history nor Linnaean taxonomy what, exactly, is it all about? A hint comes
from the structure of *Historia animalium* itself. At the start of the book he
considers how to reduce his data to some semblance of order. The problem
that he faces is that faced by any zoologist: should he order it by taxon (e.g.
reptile, fish, bird) or by feature (e.g. reproductive system, digestive system,
behaviour, ecology)? His solution, a sensible one, is to compromise:
'Animals differ from each other in their mode of subsistence, in their
habits, and in their parts. Concerning these differences we shall first speak
in broad and general terms and subsequently we shall treat them with close
reference to each particular kind.'

He begins with a general synopsis, paying particular attention to
humans, his model. He then treats the gross anatomy of blooded animals:
limbs, skin, secondary sexual characteristics, alimentary system, respiratory
system, excretory system. Then he considers the bloodless animals system
by system, returns to the blooded animals for a look at their sensory
systems, the sounds they make and how they sleep. Then come two books
on reproductive organs and behaviours again ordered by blooded *v.* bloodless animals, a book on habits and habitats, one on behaviour and, finally, a

book on human reproduction. By the end it's apparent that he has constructed a comparative zoology – the first.

He looks at feet and describes how some of the blooded live-bearing tetrapods (mammals) have many digits (man, lion, dog, leopard) while others (sheep, goat, deer, pig) have bifurcate feet with hooves instead of nails and others (horses) have a single solid hoof. Elsewhere he examines fish guts. Besides the usual stomach and intestines many fish have pyloric caecae, appendages that increase the absorptive surface of the intestine. He describes how they vary in number and position. Elsewhere he considers distribution of the sense of smell and so on. All this is a precursor not to the great systematic monographs such as Cuvier's *Poissons* but rather to his *Anatomie comparée* or Owen's *Vertebrate Zoology* (1866) in which the animals are cut into bits. You can read Aristotle on tetrapod feet or fish guts, find the cognate sections in Owen and illustrate Aristotle with his plates. We are in the Bird Hall looking at the Cabinet of Parts – he has a section on bird bills and feet too.

But it isn't easy to work out Aristotle's aims. Like all of his extant works, *Historia animalium* is poor in structure and rich in redundant, inconsistent, misplaced and barely assimilated data. The reader itches to edit it. It was never a polished piece of work, but always a thing in flux; he seems to have composed it piecemeal, adding new information as it came, or else revising in the light of theory worked out elsewhere. It's also been messed about with – though by whom and how much is hard to say.

Even so, modern scholars generally agree that it does have a clear purpose. Beneath the disorder, it provides the materials for a data trawl. Aristotle is searching for patterns – patterns of a very subtle sort. He isn't interested merely in how parts vary, but also in how they *covary*. This is how he describes the famously intricate four-chambered stomach of a ruminant (modern terms interpolated):

> Live-bearing horned tetrapods which have unequal numbers of teeth in the upper and lower jaw (also called ruminants) have four chambers. The *stomakhos* [oesophagus], starting from the mouth, goes down past the lung from the midriff to the *megale koilia* [rumen]. This is rough on the inside and partitioned. And attached to it near the entrance to the oesophagus is the *kekryphalos* [reticulum], so called because it looks like a stomach outside but inside is like those woven hair-caps. The reticulum is much smaller than the stomach. Connected to this is the *ekhinos* [omasum], rough and laminated on the inside and similar in size to the

reticulum. After this is what is called the *enhystron* [abomasum], greater in size than the omasum and more elongated in shape. It has many large smooth internal folds. Straight after this comes the gut.

The description is detailed and true, but its real interest lies in how he introduces this strange stomach as a property of live-bearing tetrapods that are horned and do not have the same number of teeth in both jaws (he's thinking of the incisors and canines missing in the upper jaws of many ruminants). It is from such associations that Aristotle constructs his greatest kinds, but it's the associations themselves that he's after. You can pull together his data and present them as a data matrix that includes, say, six classes of features (tooth number, stomach type, foot type, etc.) and twelve kinds of animals (cattle, pig, horse, lion, etc.).* It shows how the various features go (imperfectly) together. He never constructed such a table – everything is explained, laboriously, in words. But that he had something like this in mind is clear from *Historia animalium*'s sequel, *The Parts of Animals*, in which he summarizes the patterns of variation and covariation that he has discovered and explains why they exist. He pulls his data together and weaves a vast causal web that has a single purpose: to discover the true natures of living things.

* For the full matrix see Appendix I.

THE
INSTRUMENTS

ELAPHOS — RED DEER — *CERVUS ELEPHAS*

XLII

EVERY SCIENTIST HAS a conception of what constitutes 'good science'. It is a sense, as firmly held as it is poorly articulated, of which causal claims are sound and which aren't. It's not, of course, that scientists necessarily agree on the soundness of any given claim. If you have ever contemplated the reviews of a manuscript, submitted with such hope to *Nature* or *Science* (for at the gates to these journals hope truly does spring eternal), you will know that your peers' notions of what constitute sound causal claims are often very different from your own and really quite confused.

Aristotle also faced the problem of securing causal knowledge from observation, but he faced it alone. Behind him lay generations of speculative theories about the causes of the natural world; at his feet stretched the world itself. He saw, and saw as no one before him had, the need for a way to connect them. So he developed one.

In Book I of *Historia animalium* he alludes to it. First, he says, we have to get the facts about the different features of animals, then we have to work out their causes. Doing things in that order, he continues, will make the subject and target of our demonstrations clear. It seems like a rather banal introductory statement. It isn't. For, when Aristotle talks of 'demonstration', he means an intellectual structure of daedal complexity whose foundations are sunk in metaphysical bedrock and whose pillars are constructed of steely formal logic. He means his scientific method.

XLIII

ORGANON MEANS 'TOOL' or 'instrument'.* It's the title often given to six of Aristotle's books. It's an apt one for they are tools for the production of knowledge. One of these books, the *Posterior Analytics*, contains his scientific method.

Aristotle distinguishes the rules for debating opinions from the rules for constructing scientific explanations. The first he called 'dialectic', the latter 'demonstration' (the Greek is *apodeixis*). By 'demonstration' he means exactly what a modern scientist means when he says, 'we have demonstrated that A is the cause of B' – that is, he and his collaborators have shown that the presence of A is a necessary and sufficient condition for B. He had a high notion of the power of scientific demonstrations: he thought that they delivered truth. That's because they are the products of logical operations. Aristotle invented the theory of inferential logic known as his syllogistic. It was his greatest technical achievement and dominated the subject for millennia even if it was incomplete and, in parts, wrong. His syllogistic aimed to deduce new conclusions from established premises where the premises are propositions that contain a subject and a predicate, e.g. 'All octopuses [subject] are eight legged [predicate].' To analyse such statements he invented a formalism that substituted letters for the terms, e.g. 'All A are B.' This formalism allowed him to speak generally of all propositions of a given form, manipulate them and derive the many results that he did.

For Aristotle, a scientific demonstration rests upon a syllogism. But to qualify as a demonstration a syllogism must meet certain conditions. First, the premises of the syllogism must obviously be true. Second, the premises of the syllogism must be more immediate, more empirically apparent, than its conclusion (at least in natural science as distinct from geometry). Third, it must concern universals rather than particulars. In fact, Aristotle thinks that it's impossible to have scientific knowledge of individuals. To say that *this* octopus has eight legs gets us nowhere; scientific knowledge can only

* The general usage is almost relict in English, but lives on in Greek. During the Colonels' regime policemen were called Organa because they were the instruments of authority.

begin once we've established that *all* octopuses have eight legs – or at least that all normal octopuses do. Finally only universal, assertive and assertoric propositions can form the basis of demonstrations: 'All A are B; all B are C; therefore all A are C.' Logicians refer to such syllogisms as being in the 'mood' of 'Barbara'.

Such logical strictures may seem remote from the modern scientific method and so, in a way, they are. But Aristotle's reason for grounding scientific knowledge in his syllogistic is, I believe, one familiar to any modern scientist. I suggested that, far from being a natural history or a taxonomy, *Historia animalium* is a search for associations among the traits that animals possess; that it is, in fact, a data trawl. His syllogistic, then, provides a powerful way of securing those associations – of showing that they are true. Secure associations, in turn, demand causal explanations – which his syllogistic also identifies.

A lovely bit of modern biology can be pressed to illustrate this strategy. In the bays and estuaries of Northern Europe and America there lives a small fish called the three-spined stickleback, *Gasterosteus aculeatus*. The binomial translates as 'bony stomach with spines', which is apt since it has a pelvic girdle with spines on its belly. Although the stickleback usually lives in the sea, it's a versatile fish and, in the last ten thousand years, has invaded freshwater lakes many times. The lake fish have evolved rapidly and have lost their pelvic girdles and spines. Recently, several beautiful studies have shown that the lake sticklebacks carry a mutation in an enhancer of a gene called *Pitx1*, a mutation that their marine relations do not have. Had he known them, Aristotle would have surely wondered about the connection between these facts, but before investigating it, he might have proved its existence, so:

All lake sticklebacks lack pelvic spines;
All sticklebacks that lack pelvic spines have a *Pitx1* mutation, therefore,
All lake sticklebacks have a *Pitx1* mutation.

The truth of this syllogism guarantees a connection between several stickleback predicates: living in a lake, a lack of pelvic spines, and the presence of the *Pitx1* mutation. A demonstrative syllogism implies not just a logical connection, however, but a *causal* one that can be expressed as a 'definition'. We usually think of a 'definition' as a description of a word – that is, a nominal definition: 'a lake stickleback is one that lacks pelvic spines'. Aristotle, however, would point to the middle term of the syllogism – the

Pitx1 mutation – as the causal link and give a definition of the following sort: 'a lake stickleback is one that lacks pelvic spines *because* it has a *Pitx1* mutation'. That's demonstration, he would say; that's science. Such definitions are the *logos* – the 'essence' or 'formula' of the things he studied. So his scientific method turns out to be a way of expressing the fundamental causal identities of things shorn of all incidental, and hence scientifically uninteresting, features.

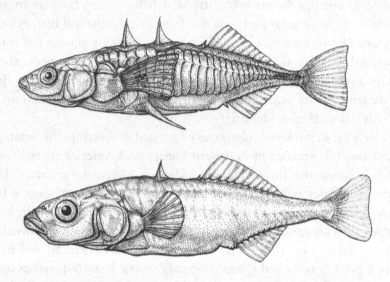

STICKLEBACKS – *Gasterosteus aculeatus*
Above: MARINE (ANADROMOUS) MORPH FROM CALIFORNIA
Below: LAKE (BENTHIC) MORPH FROM PAXTON LAKE, BRITISH COLUMBIA

I have been speaking of Aristotle's 'theory of demonstration' as if there is just one of them. In the *Posterior Analytics* he certainly devotes most space to the method I have sketched. But he also allows that there are other modes of demonstration – though he's quite vague about how they work. In *The Parts of Animals* he says that the methods of demonstration in natural science are actually different from those in 'theoretical sciences' such as geometry. In biology, he suggests that we should start with the *end* – the teleological purpose – of an animal and work our way deductively back to infer how the animal's various parts serve that purpose. Such demonstrations can also, with some twisting, be couched in syllogistic terms.

Although demonstration is the beating heart of his scientific method, Aristotle acknowledges that science rests on various indemonstrable statements. These include the axioms of his syllogistic as well as various primary definitions. For example, geometry requires a definition of 'spatial magnitude' and arithmetic a definition of 'unit'. The axioms and primary definitions of biology are less obvious but include statements such as 'nature does nothing in vain' – an apophthegm that he puts to hard use. Aristotle isn't clear how such ideas can be justified, and suggests that their truth is just apparent by induction (*epagōgē*), but argues nonetheless that they're needed if science is to get off the ground.

This is certainly right. In our day there are people who think, all evidence to the contrary, that science is just one system of beliefs among many. Aristotle had to contend with them too. Some people, he says, claim that scientific knowledge is impossible because any inference we make must rely on some previous inference, and that must rely on another, and so on to infinity so that, ultimately, we can know nothing. Other people, he continues, claim that anything can be demonstrated: everything is true hence nothing is true.

Aristotle recognizes that both thoughts are lethal to the possibility of science, and he deals with them briskly. No, there isn't an infinite regress of inferences, nor is it true that everything can be demonstrated, because our arguments ultimately begin with axioms and our perception of the empirical world. His language is combative. It has to be. He has to show against his opponents, not just Plato but the sophists with their razor-sharp dialectic, that it is possible to extract real knowledge from the sensible world. We may wonder whether he succeeded. Modern science rests on fundamental axioms no less than Aristotelian science, and scientists mostly justify them by the fact that they work. But Aristotle could hardly defend his assumptions, as a modern scientist can, by flicking on a light.*

* The operative word is 'scientists'. It is certainly the business of philosophers to worry over epistemological issues, but never was there a scientist who lost an hour's sleep over, say, foundationalist *v.* constructivist justifications of truth. It's not clear that Aristotle did either.

XLIV

ARISTOTLE'S THEORY OF demonstration isn't without its problems. Every science undergraduate learns that 'correlation does not equal causation'. Nor does it: which is why we do experiments. To their get their work published in *Science*, Chan *et al.* not only had to show that the *Pitx1* mutation is coextensive with a missing pelvic spine; they also had to show experimentally that the *Pitx1* mutation really does cause spine loss – and (rather heroically) made a transgenic stickleback to do so. Aristotle, who never did controlled experiments, is much less cautious. Having identified a coextensive set of features, he tends to jump to the causal relationship. Perhaps it can be shown syllogistically that having horns, an incomplete set of teeth and many stomachs is completely coextensive (i.e. that ruminants and only ruminants share these features), but are they really causally related in the direct way that Aristotle says they are? In the absence of other evidence, we may be inclined to doubt it.

Another problem lies in the direction of causation. 'Horn-bearing animals have many stomachs *because* they do not have a complete set of teeth.' Maybe – but why not the other way around? Surely it's just as plausible that they have incomplete teeth because they have many stomachs? In the case of the sticklebacks we are confident that the arrow of causation runs: invade lakes → gain *Pitx1* mutation → lose pelvic spines because two theories, the theory of evolution by natural selection and the fundamental dogma of molecular biology, tell us that it must be so and not the other way around. Aristotle considers the problem in the *Posterior Analytics* but doesn't solve it. In practice the directions in which he aims his causal arrows also depend on all sorts of theoretical beliefs independent of the syllogisms upon which they are based.

Finally, Aristotle argues that all demonstrative claims can be stated in the form of a syllogism. Some certainly can, but all? Much modern science depends on mathematical models that posit quantitative phenomena and relations. Tests of such models require measurement and a probabilistic theory of inference. Aristotle's models, by contrast, are invariably qualitative and he seems never to have measured a thing.

Some scholars have suggested that when we open his actual scientific works, The Parts of Animals say, we should see Aristotle's scientific machine at work; that we should see axioms and syllogisms neatly arrayed as in a treatise of geometrical proofs. They are puzzled by the fact that we don't. All the treatises are a messy mixture of data, arguments and conclusions (a messiness that, given its pervasiveness, cannot be blamed on their having been scrambled in transmission). If we look hard enough, traces of Aristotle's machine can be found throughout his scientific works. They may not contain syllogisms, but they contain their results. His works are filled with causal definitions: 'The horn-bearing animals have many stomachs *because* they do not have a complete set of teeth'; 'selachians have skin that is rough *because* they have cartilaginous skeletons'; 'the ostrich has toes rather than hooves *because* it is large' – all these are quotes from The Parts of Animals. Still, he never spells out the syllogisms themselves. Why not?

Perhaps he felt he didn't need to. Or perhaps he felt that he'd only do so when he understood the causes of everything, once his work was complete, but he never did, it never was, and so he didn't. But I think that he didn't express his causal claims as syllogisms because he couldn't. In Aristotle's demonstrative logic the predicates of his syllogisms are typically coextensive, but in real animals they're not. Horns, multiple stomachs, cloven hooves and missing teeth merely tend to go together. The camel has all of these features except one: horns. The problem with syllogistic reasoning is the same problem that monothetic classification has: some creature or other always spoils the show.

It is the problem that our theory of probabilistic inference – statistics – addresses. When we search for associations among attributes, we demand that they be not fully coextensive but only correlated. Indeed, Chan *et al.* do not claim that living in lakes, lack of pelvic spines and possession of a *Pitx1* mutation are fully coextensive; they show (remarkably strong) statistical associations and warn that other genetic factors may be at work. Aristotle's solution is simply to say '*many* [italics mine] of the cloven-hoofed animals have horns' and then give a patently *ad hoc* explanation for why the camel doesn't. In fact, he'll often assert that some association or other is true 'for the most part'.

I think that the Posterior Analytics sets a gold standard for scientific knowledge. It establishes the conditions under which we really know a given causal relationship to be true. But, in practice, natural science – and by this I mean, as Aristotle did, the study of the natural world rather than mathematical or geometric objects – rarely admits of rigorous proofs. Most of it depends on much weaker forms of inference, the claim that *this* explanation is the best one around. Data are incomplete, results are tenta-tive, causes are complicated and inferential gaps yawn at every turn. It is so for us; it was for Aristotle too. The result is that his practice is more casual and dialectical or, to give it a positive spin, much more reasonable and probabilistic than the rigours outlined in the *Posterior Analytics*. In the *Nicomachean Ethics* (for ethics is an Aristotelian science as well, even if not a *natural* science) this ambiguity is explicit:

> Here, as in all other cases, we must set down the appearances [*phainom-ena*] and, first working through the puzzles, in this way go on to show, if possible, the truth of all the beliefs we hold about these experiences; and if this is not possible, the truth of the greatest number and most authoritative. For if the difficulties are resolved and the beliefs are left in place, we will have done enough demonstration.

Untangled, this amounts to the following: start with some ordered infor-mation about some part of the world, identify the problems that it presents, collect the best explanations for those problems and then demonstrate which of those explanations are coherent and which aren't. Those left standing are the answer.

Although this passage appears to be about 'demonstration' it actually suggests a quite different procedure from the theory of the *Posterior Analytics*. That's shown by Aristotle's use of the word *phainomena*. The syllogistic theory of demonstration requires that the premises of the argument be indisputably true. If they're not, then you can't prove anything. But *phain-omena* don't have that kind of epistemological certainty since, according to Aristotle, they include opinions – the opinions of 'wise' and 'reputable' people to be sure – but opinions nevertheless. We are in the realm of dialectic, which, it turns out, isn't that far from demonstration after all. Most of his biology lies in this twilight realm.

That is the consequence of the world's messiness. But Aristotle has another, deeper strategy for coping with a lack of coextension. He recog-nizes that if a group of individuals (or kinds) share some feature, but that this feature is differentially associated with other features, then multiple causes may be at work. In such cases, he suggests, we should divide our classes and search for the common cause, and keep doing so until we have identified a single cause for each.

Quite a lot of modern biomedical science follows exactly this recipe. Melanomas of the uvea – a part of the eye – afflict about one in 167,000

Americans. How shall we aid them? The answer – one upon which researchers have bet their working lives – is by searching for the cause of the disease, its definition, its formula, its essence. It's a cancer so it's probably caused by a particular mutation or combination of mutations. But there are at least two 'species' of uveal melanoma, each of which has its own mutational 'formula'. Class 2 tumours have mutations in a gene called *BAP1*, while Class 1 tumours do not. This has consequences since where Class 1 tumours can be treated, Class 2 tumours currently cannot, are aggressively malignant and nearly always kill. Even these two classes are heterogeneous and can be subdivided further by the presence and absence of other mutations. And so oncogeneticists, searching for the causes of this disease – or rather, of these several diseases – chase the formulas ever deeper, subdividing as they go. Actually, Aristotle alludes to just such a case:

> Every definition is always universal; the physician does not prescribe what is healthy for some particular eye, but rather for every eye *or for some determinate form of [afflicted] eye* [italics mine].

Yet there is a difference between Aristotle and the modern biologist. Aristotle is convinced that if you burrow deep enough you will actually be able to delineate stable classes of objects, indivisible forms – true species – that all share some unique, defining causal formula. We, too, are impressed by natural variety, but having seen so much more of it than he did, have surrendered to it completely. Our technology – DNA sequencing is just the latest – shows us that no two sticklebacks, no two cancers, no two people, not even 'identical' twins have exactly the same formula. This difference in view is profound, but in practice it doesn't matter that much. For still we burrow and divide and seek the formulae of things, just as Aristotle did. And if, in our heart of hearts, we know that we'll never reach that vein of pure causal ore, we also know that we'll strike it rich on the way down.

And that is the point. For all its limitations, Aristotle's theory of demonstration is a genuine scientific method. It is part of ours. Scientists may quarrel about methodology but they also agree about a lot. They understand the domain of science: the kinds of things it investigates. They understand how it delimits things and problems and investigates them piecemeal rather than trying to study and answer everything. They understand the reciprocal role of theory and evidence and the distinction between hypothesis and fact. They understand that science begins with induction to give generalizations from observations and then deduction to give firm causal claims from generalizations. They understand that a scientific claim must be a logical one – and can recognize a logical argument when they hear one. They understand that some causal claims are strong, others weak – and that the trick is to tell them apart. That they understand all this is because Aristotle told them it was so.

THE
BIRD
WINDS

EPOPS – EURASIAN HOOPOE – *UPAPA EPOPS*

XLV

SEVENTY DAYS AFTER the winter solstice, some time in early March, the *ornithiai anemoi*, the bird winds, begin to blow. That is when the migrants begin to arrive in Lesbos. In the marshes and pools between Skala and the mouth of the Vouváris, where the Lagoon melds softly into the land, they flutter among the reeds and wade in the shallows, while far above the raptors stream over from Africa. Birders follow them in from Arlanda, Schiphol and Gatwick. They track the birds through telephoto lenses, quarrel like stilts and update their websites with sceptical precision ('7 May: A common snipe, present at the salt works, and said to be present yesterday too, casts a shadow over yesterday's claimed great snipe from these pools'). The island swarms with Aristotle's birds. Here is his description of just one (though it is really a winter migrant): 'The *tyrannos* is just a little bigger than a locust, its crest is the colour of sun shining through mist and it is, in every way, a beautiful and graceful little bird.' That's the goldcrest, *Regulus cristatus*, which lives in Lesbos' pine forests.

Perhaps, on spring days, when the olive groves at Pyrrha were red with anemones and Olymbos' summit was swept clear, Aristotle went birding too. The beauty of birds lies in their lucidity. Fish lurk beneath the waves, mammals skulk in the woods, but the lives of birds are open to us. That, I believe, is why when Aristotle wants to explain 'the more and the less', the subtle variations in size and shape that kinds within greatest kinds show, it is of them that he often speaks.

He begins by sorting his birds into groups: carnivorous birds, water birds, marsh birds and the like. These are not taxonomic groups – *gene*/kinds – but functional classes like the guilds of modern ecology. He explains the features of each class in terms of how the birds make their living. Carnivorous birds (eagles, hawks) have to find and overpower prey – and so have large talons, powerful wings, short necks and very good eyesight. Water birds (ducks, grebes) have to swim, reach down into the water and tear off water plants – and so have short legs and webbed feet rather like oars, long necks and flat bills. Marsh birds (herons, cranes, stilts) live in swamps and catch fish – and so have long legs, long necks and spear-like beaks. Small birds (finches) collect seeds or grasp mites – and so have small, hollow beaks. Some birds are powerful fliers so that they can migrate to distant lands.

This is exactly where functional explanation in evolutionary biology begins. 'Seeing this gradation and diversity of structure in one small, intimately related group of birds, one might really fancy that from an original paucity of birds in this archipelago, one species had been taken and modified for different ends.' People who quote Darwin on the finches of the Galapagos archipelago are usually interested in the *one species taken and modified*. But forget about that and consider just the *gradation and diversity of structure* and then the *different ends*, the fact that one species of finch has a beak adapted to cracking tough, spiny-shelled seeds, another for delicately picking at tiny grains, another for poking holes in boobies to drink their blood, several for eating insects, that one has even taken to using cactus spines the better to winkle insects, woodpecker-like, from the bark of trees; consider all this and you have a thoroughly Aristotelian analysis.

In *The Parts of Animals* Aristotle says that 'nature makes instruments to fit the function, not the function to fit the instrument'. Of course this now seems trite. But it was he who first saw that a bird is not just a cabinet of parts, but a toolbox on the wing.

XLVI

W*HAT IS IT for the sake of?* To fully explain any natural phenomenon we must ask and answer four questions. But, Aristotle is clear, this question is the first. We should begin, as it were, with the end.

In the best case, Aristotle asserts, individual organisms would be immortal. But, in fact, all individual organisms die. So they do the next best thing: they reproduce. To achieve this they need, in turn, body parts with which to eat, breathe, copulate and so on. His term for such a functional body part is *organon* from which our 'organ'.

Calling body parts 'instruments' may suggest that Aristotle's much vaunted teleology is nothing more than naive functionalism of the Socratic/Platonic/Paleyite type: eyelids are 'doors for the eyes' – that sort of thing. He certainly gives plenty of explanations like that. He has the standard textbook list of basic animal capacities – nutrition, respiration, protection, locomotion, sensation – and he parcels out the organs among them, allowing that some do lots of things. Sometimes he finds this easy to do – a stomach is obviously for the sake of nutrition. Sometimes it's harder – he's uncertain what the spleen does or whether it does anything at all. Sometimes he thinks it's easy when it's not – he's sure that he knows what the heart and brain do, but he really hasn't a clue.

But, right or wrong, this kind of general teleology is only a beginning, for Aristotle's programme is much more searching and ambitious than anything that Socrates, Plato or Paley even dreamt of. He's a comparative biologist; his real interest is *specific* teleology; he wants to know not only why this animal has that feature, but also why others haven't. To answer this question, and the countless others like it, prompted by all the parts of all the animals in all the world, he devised a system of teleological principles and precepts. It's the core of a system that has been used ever since. *The Parts of Animals,* then, is about why some animals fly, some swim and others walk; it's about teeth and talons, jaws and claws, horns and hooves. It's about birds and their wings, legs and beaks. And it's also about the elephant's nose.

XLVII

'THE ELEPHANT'S NOSE is unique among animals because of its length and extraordinary versatility.' The elephant can use it like a hand. He can feed with it. He can defend himself with it. He can trumpet with it. He can even use it to uproot trees. Aristotle may never have seen an elephant, but he has a lot to say about its trunk.

When Aristotle wants to explain why some animal has the particular features it has, he sometimes appeals to its lifestyle – its habitat, diet and relations to other living things – in a word, its *bíos*. That is how he explains birds in all their beauty. Occasionally he manages to peer beneath the waves too. 'In marine animals you can also see many skilled activities [*technika*] in relation to their lifestyles, for the stories about the *batrakhos* and the *narkē* are true.' He then tells how the frogfish conceals itself in the mud, rears its lure and vacuums up the fish that are drawn to it, and how the torpedo ray narcotizes its prey.

In this spirit, when Aristotle explains the elephant's remarkable nose, he also begins with its lifestyle. He thinks, only slightly erroneously, for elephants *do* love water, that it lives in swamps:

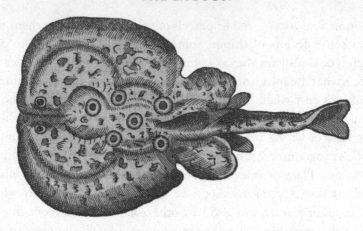

NARKĒ —TORPEDO RAY — *TORPEDO TORPEDO*

Divers, when they stay for a long time in water, provide themselves breathing equipment and use it to breathe in air from outside. Nature makes a similar mechanism for elephants in the length of their nostrils.

Did ancient divers and ancient elephants really snorkel? I frankly doubted both claims. But D. L. Johnson, in a paper titled 'Problems in the Land Vertebrate Zoogeography of Certain Islands and the Swimming Powers of Elephants', reports that African elephants swim the Zambezi, that Asian elephants swim between islets off Sri Lanka and that they do so with a sort of porpoising motion while holding their trunks aloft. Their maximum velocity and range is 1.5 knots and twenty-six nautical miles. He adds that one rarely sees an elephant swimming because they usually do so at night. To silence doubters he supplies a blurry photo.

The elephant's swampy habitat doesn't, however, entirely explain its trunk. Hippos, seals and crocodiles are at least as amphibious and they don't have one. The elephant must face some unique problem to which a trunk is the best solution. It does. It's not a single problem though, but rather a cluster of them. Not only is it swamp-bound but it also has to fulfil basic animal functions such as breathing, eating and protecting itself from predators, and it has to do these things while being constrained by other features. Aristotle's complete explanation for the elephant's trunk, then, begins with these functions and features and follows their consequences in intersecting causal chains.

The elephant requires protection. From what Aristotle doesn't say; presumably nothing less than a triple-tooth-rowed *martikhŏras* can take it down. It defends itself by sheer bulk. That has consequences. Because it's so big its legs must be thick. Thick legs make inflexible legs, and inflexible legs make the elephant rather slow. Perhaps that doesn't matter so much on land, but the elephant lives in swamps. It occasionally finds itself in deep water, but can't dash out to catch its breath, so there it is, wallowing in Indus mire, in mortal danger of drowning. At least it would be had not nature, providentially, given it a snorkel.*

Aristotle calls this kind of explanation 'conditional necessity'. It is a principle of the following sort: given some goal, X, and some instrument, Y, for the sake of X, then condition Z is a necessary for Y to fulfil X. The example he gives is banal. If the goal (X) is to cut wood, and the instrument (Y) is an axe, then the axe must be made of something hard (Z), e.g. bronze. But the principle is general: if your goal is to breathe, and you're a sluggish swamp-dwelling tetrapod, then you need a long nose. It's his way of expressing and investigating the truth that a living thing is an integrated whole whose every part is adjusted to all others so as to ensure its survival. If you were to shuffle parts randomly among forms, you would get monsters, and hopeless ones at that. That is why Empedocles' selectionist schemes are so absurd.

In *Historia animalium* Aristotle disassembles his creatures to establish the associations among their parts; in the *Posterior Analytics* he gives a method by which to demonstrate the causes of those associations; in *The Parts of Animals* he reassembles his creatures and puts the method into practice. The principle of conditional necessity therefore becomes the single most important teleological principle at work in that book. The causal chains multiply and ramify through the text so that it's hard to see where they start and stop. He even gives another conditional-necessity chain for the elephant's trunk that ends by explaining its use as a hand, but begins with its toes.

The thing about the elephant's toes is that there are many of them. As such, the elephant has a functional affinity with other multi-toed animals such as cats, dogs and humans. Multi-toed animals seize their food with their forelimbs. But the elephant can't because its legs are inflexible, because they're thick, because it's big.† So there's the elephant starving to

---

* Aristotle's semi-aquatic elephant is a bit absurd. But it was also an inspired guess. Recent studies of elephant embryology, fossils and molecular phylogenetics show that the elephant evolved from an aquatic mammal. The inference is that its trunk, whatever its manifold uses now, was originally a snorkel. Interestingly, Aristotle knows another piece of evidence for this claim, the fact that the elephant, like seals and dolphins, has internal testicles. But he doesn't make the connection.

† Aristotle vacillates on just how bendy the elephant's legs are. Five centuries later, Aelian, the Roman paradoxographer, thought it odd that elephants could dance even though they don't have joints. He may have picked up a distorted Aristotelian echo or perhaps he got it from someone else. In any event, the idea that elephants don't have knees and sleep

death in the teak forest. At least it would be had not nature, providentially, given it a kind of hand. Put this together with the snorkel argument and you really need a causal diagram to see that it all makes sense. It does, but you wonder how Aristotle got by, as he presumably did, without one.

XLVIII

ARISTOTLE'S ANALYSIS OF avian anatomy is so clear, so obvious, that you expect The Parts of Animals to be filled with more of the same. You expect to find a full-blown adaptationist programme demonstrating the exquisite – the adjective is rarely absent – fit of animals to their environments.

It isn't there. Yes, Aristotle does occasionally explicitly explain animal adaptations in terms of their lifestyles, but the principle of conditional necessity – the explanation of parts in terms of each other – dominates. One reason for this is his taxonomic focus. Some biologists paint the canvas of life in the broadest strokes, comparing the great phyla with each other, oblivious to the legions of species that they embrace; others construct intimate family portraits of just one group, tiger beetles, say; many study just one species, the mouse, the fly, the worm, ourselves. Aristotle does it all. His gaze wanders up and down his taxonomic hierarchy, sometimes coming to rest on humans, sometimes on all blooded animals. In The Parts of Animals, however, it is to the differences among the greatest kinds that he mostly attends.

It is the obvious thing to do. It is among the greatest kinds that most of life's variety can be found. The parts of the greatest kinds differ, not as those of birds do, by 'the more and the less', but in the entire order of their bodies. If their organs are similar, then they are so only by analogy. And yet it is also among the greatest kinds that specific teleological explanations are most elusive. Birds have beaks and tetrapods have teeth – why? Most animals have a mouth at one end of their bodies and a rectum at the other, but not cephalopods – why? Some animals have blood, others do not – again, why? Each of the greatest kinds is so different from every other, and embraces so much variety, that it is hard to correlate form with lifestyle unless trivially so. Yes,

standing up became firmly established in medieval bestiaries and survived to become the subject of a couplet by Shakespeare, a stanza by Donne and the subject of Sir Thomas Browne's withering scorn. Kinematic studies show that the elephant's legs are quite flexible.

fish have fins rather than legs, and gills rather than lungs, because they live in water and not on land – Aristotle does indeed tell us so. But, in general, when contemplating the diversity of his greatest kinds he doesn't even try.*

It may seem that Aristotle has come to the end of explanation. He has just begun. His approach is as follows. For each greatest kind he takes certain features as primitive (in the epistemological rather than evolutionary sense). They are givens, not necessarily to be explained. They are, however, starting points for explanations. Aristotle often indicates primitive features by saying that they are part of the 'definition [*logos*] of the entity [*ousía*]' of an animal. For birds, flying is a feature of this sort; for fishes, swimming is. For birds (and presumably others) having a lung is. For live-bearing tetrapods (= most mammals), egg-bearing tetrapods (= most reptiles and amphibia), birds and fishes, having blood is; for all animals, sensation is. Aristotle only explicitly identifies a few features as givens of this sort, but he acts as though many are.

Such as bird beaks. Aristotle never explains why birds have beaks rather than teeth; they just do. But he does chart the consequences. Because birds have beaks, rather than teeth, they cannot chew their food. To compensate for this deficiency they store and 'concoct' (digest) their food using a variety of other devices. Some birds (pigeons, pelicans, partridges) have a crop; others (crows) have a broad oesophagus or else (kestrels) an expanded part of the stomach (proventriculus). Most birds have a fleshy and hard stomach (the gizzard). Marsh birds have neither a crop nor a wide oesophagus

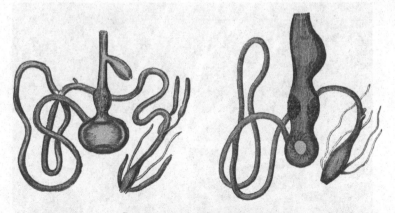

BIRD ALIMENTARY TRACTS
Left: ALEKTŌR – CHICKEN – *Gallus domesticus*
Right: AIETOS – EAGLE – *Aquila* SP.

* Nor, in general, do evolutionary biologists. Attempts to explain the characters of phyla or classes in adaptive terms are rather rare, unless it is to tell some story about the 'rise of the mammals' or the 'fall of the dinosaurs'.

because their food is easily ground up. All this anatomy is broadly correct. So is Aristotle's reasoning: they do so because they lack teeth.

Aristotle uses the same logic to explain why some grazing animals (horses, asses, onagers) have the same number of teeth in their upper and lower jaws and simple stomachs, but others (cows, goats, sheep) have an unequal number of teeth in their jaws and have complex stomachs. Or why some fish have single gills, but others double. Or why the ostrich cannot fly. Or why . . . but such examples could be multiplied, for nearly every line of *The Parts of Animals* is an explanation.

XLIX

IN THE PORT OF Mithymna, the pretty Turkish town on Lesbos' northern cape, I once found the desiccated remains of a sea-urchin species that I had never seen before. I found it on the quay, evidently discarded by a fisherman who had tied up there to sort and clean his nets. I recognized it as one described by Aristotle who says that it has a small body and large, hard spines and so is very different from the fat, fragile-spined sea urchins

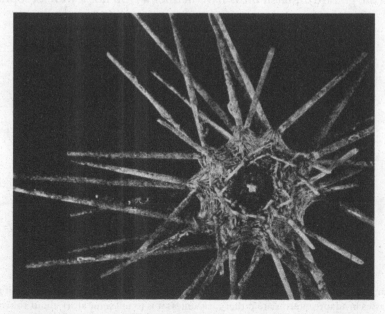

EKHINOS GENOS MIKRON — LONG-SPINED SEA URCHIN — *CIDARIS CIDARI*

of the Lagoon. Characteristically, he does not give it a name but he does say that it lives far off shore, a hundred metres ('60 *orguiai*') down – which fits with the location, for Mithymna looks out over the strait of Mytilene where the sea floor drops to three hundred metres or more.

In his *Greek Fishes* D'Arcy Thompson identified the nameless sea urchin of the deep as *Cidaris cidaris*; he was certainly correct, for that is what I held in my hand. Aristotle says that this sea urchin is used as a remedy for stranguary, the urgent need to urinate coupled to the painful inability to do so, but I don't think it's used like that any more. In any case, he's not very interested in its medicinal qualities, but is keen to explain its really long spines.

Aristotle knows that spines protect sea urchins, so you would expect him to argue that the sea urchin of the deep needs particularly long spines for some functional reason, perhaps because the fishes down there are especially fierce. But that's not the explanation he gives. Instead, he argues that they're of no particular benefit to the sea urchin at all, but are simply the result of its 'material nature'.

Though he may heap scorn upon the crass materialism of his predecessors, Aristotle believes in the power of matter. A form, after all, cannot actually exist without it. To be sure, considering form and matter in the abstract, form is the more important: a form (a sphere) that manifests in different kinds of matter (wood, iron) remains essentially a sphere. Yet a sphere is a very abstract example; living forms are much more dependent on the stuff from which they're made. You can make a wooden statue in the form of a man, but it obviously won't walk or talk.

Every Aristotelian animal is a triple-levelled hierarchy. At the bottom it's made of elements, at the top organs. In the middle are the 'uniform parts' – blood, semen, milk, fat, marrow, flesh, sinew, hair, cartilage, bone. I would call them 'tissues' except that the term slightly distorts Aristotle's meaning. We know that tissues are composed of cells, but Aristotle thinks that his uniform parts really are uniform – that is, totally devoid of microscopic structure. Every uniform part has its own particular 'material nature', a set of functional properties – soft, dry, moist, pliant, brittle – that depends on the particular mix of elements of which it is composed. And, although the uniform parts have functions in their own right (bone protects flesh), their real purpose is to be the stuff from which organs are made.

Aristotle notices that the uniform parts differ among animals. Animals vary in the heat of their blood, the hardness of their bones, flesh, fat and skin, and the quantity of their marrow. Many animals, of course, have no blood, flesh or bone at all but have some other, analogous, usually unnamed, uniform parts instead. Aristotle seems to think that the uniform parts of any given animal kind have some innate norm, but that their quality and composition vary with health, diet and season. They are the basic units of his physiology and so the link between an animal's environment and its body.

That opens up a whole different kind of explanation for animal variety. Aristotle does not think that all variation is teleologically explicable. Some of it is directly due to the effects of the environment. The deeps in which the nameless sea urchin lives are, he says, cold. For that reason it does not have the warmth that it needs to 'concoct' or digest its food properly. (Sea urchins are, in his physiological scheme, rather cold creatures anyway.) For want of warmth, then, the sea urchin has a lot of 'residual matter' left over from concoction that it diverts into spines, which is why they are so long. The cold causes the spiny material to petrify, which is why they are so hard.

This is all very mechanical. Long spines are just the result of 'necessity' – though here he means material rather than conditional necessity, for their length is not to be explained by any functional goal however remote but only by brute physiology. As it happens, sea urchins do grow different spines in response to the environment, a phenomenon that biologists call 'phenotypic plasticity', though that isn't the explanation for *Cidaris cidaris*' long spines for they are a feature of the species. For whatever reason, however, that is not how Aristotle sees it. He sees it as a case where matter has got the upper hand.

L

WHEN ARISTOTLE ANALYSES animal design he thinks like an architect or an engineer. His first thought is the purpose that a given organ serves. However, he's also deeply aware of the stuff of which it's made. These two kinds of explanation – conditional and material necessity – interact in subtle ways.

Aristotle supposes that an animal's organs are generally made of the right stuff, that the material natures of their uniform parts – the biophysics of their tissues – match the animal's functional needs, yet he also allows that animals don't always have the right stuff. The fact that the bodies of particular animals are made of certain kinds of matter limits the kinds of organs they have; it can even prevent them from having otherwise desirable organs. Conversely, animals produce material surplus to requirements – 'residues'. Sometimes these residues are put to good use to make organs that aren't, perhaps, vital but are desirable nevertheless. For Aristotle, as for an architect, function isn't omnipotent; function stands in chains.

His examples of how the properties of uniform parts match functional demands are a biophysicist's delight. Snakes need to see what's coming up behind them, but they can't easily turn their bodies around because they don't have feet, so instead they twist just their heads – which is why their vertebral columns are so flexible. When rays swim they undulate, so they, too, need a flexible skeleton – which is why they have cartilage rather than bones. The human oesophagus must be able to dilate to swallow food and yet resist being scraped as the food goes down – which is why it is both elastic and fleshy. The human penis must both droop and jaunt erect – which is why it's made of some stuff that can be both soft and hard. These are all examples of conditional necessity where the consequent is a property of a uniform part.

The epiglottis is a different story. Aristotle notices that the design of the neck is rather poor. Since the larynx and windpipe are located right in front of the oesophagus, animals easily choke on their own food. In mammals, nature has solved this problem by devising a lid for the larynx, the epiglottis, which closes during swallowing. But birds and reptiles don't have an epiglottis. Why not? His answer is that their flesh and skin are 'dry' and so they *can't* have one, since an epiglottis needs to be 'fleshy' to work. Nature has, therefore, devised some other device for them, contraction of the larynx.*

Aristotle is usually confident that he knows what a given organ does. The spleen is mysterious. He's reasonably sure that it's not a vital organ. He knows that many blooded animals have only a very small spleen and thinks that some don't have one at all, so he moots the possibility that it's there to counterbalance the liver (he likes his organs to come in bilaterally symmetrical pairs); it may even aid it in 'concocting' nutrition (i.e. digestion); and that it helps to anchor the blood vessels. But mostly he thinks that it's just a 'residue' – an excretion product that nature has put to various, not particularly essential, secondary uses.†

---

* Aristotle's zoology is broadly correct. Is his explanation for the presence of the epiglottis in mammals, and its general absence in reptiles and birds, correct too? Surely not. But why did mammals evolve an epiglottis when their ancestors had already solved the problem?
† The function of the spleen was mysterious until the last century. It filters the blood and, in doing so, removes red blood cells, maintains iron balance and is a centre for mounting

Of course, some bodily productions really are just useless 'residues'. Urine and faeces are obviously just excretory products, but Aristotle thought that bile is too. He was going against the grain of received opinion. The Greeks had forever probed the livers and gall bladders of sacrificed animals to predict the future. The more rationally minded *physiologoi* speculated that the gall bladder has a sensory function. The Hippocratics and Plato thought that bile was the product of disease. Aristotle rejects these ideas and, drawing once more on comparative data (some animals have gall bladders, others don't), argues that bile is a residual product of blood that is produced in the liver, is excreted into the intestine and is quite useless: 'Sometimes nature puts even the residues to some use, though that is no reason to seek a purpose for all of them. Actually, just because some have a purpose, many others are there necessarily.'*

Sea urchins, snakes and sharks; gall bladders, penises and spleens – as Aristotle dissects and analyses, he's picking his way along a precarious path. Above him soar the heights of Plato's heedless teleology, beneath him lies the abyss of the *physiologoi*'s relentless materialism. Aristotle, recognizing that neither cause can be ignored, considers every part in turn and assigns primacy now to functional goals, now to physiology, and often – and this is his great contribution – to the subtle interplay between the two. Yet, as we read *The Parts of Animals,* it becomes evident that beneath such explanations, which derive directly from his four causes, there is also a quite different set of principles at work – axioms that are neither directly teleological, nor material, but economic.

LI

ZEUS, POSEIDON AND Athena are having a contest to see who can make the best thing. Zeus makes a man, Athena makes a house, Poseidon makes a bull. They ask a colleague, Momus, to judge their creations, and Momus promptly derides them all. The man, he says, should

adaptive and innate immune responses.
* Actually, bile is probably an example of an excretory product that has been secondarily utilized. It's an excretion product of bilirubin, the product of defunct red blood cells harvested by the spleen, transported to the liver, gathered by the gall bladder and excreted into the intestine where it is used in fat digestion.

have a window into his heart so that we can see his plans; the house should
be a caravan so that it can shift location; the bull should have (extra?) eyes
under its horns so that it can see what it's goring. Irritated by this carping
– he had spent a lot of time on man – Zeus slings Momus off Olympus.

In *The Parts of Animals* Aristotle alludes to Aesop's fable, but his version
has nothing about horn-eyes; instead Momus suggests that the bull
should have had its horns on its shoulders since that's where they'd be the
most use. Aristotle retorts with plonking humour, 'Momus' criticisms
here are obtuse.' He should have done some research on the strength and
direction of blows; besides, if horns were on the shoulders (or anywhere
else) they'd impede the bull's movements; they're just where they should
be: on the head.

That's a straightforward conditional-necessity-type teleological argument: horns are for protection, so they're in the best possible position
subject to other functional constraints. He adds some other details: how
they're hard, how in deer they're solid, how in bulls they're hollow but are
strengthened by a bone at the base and so on. It's all as adaptationist as
can be. Again, you'd expect that when he turns to explaining why most
animals don't have horns, he'll demonstrate how the lifestyles of some
animals necessitate horns while those of others don't. He doesn't. Instead,
he invokes a set of auxiliary principles that depend on the economics of
the body.

In the *Politics* Aristotle argues that household management reduces to two
problems: command and control, and economics. It's about who rules whom
and about the acquisition and allocation of goods. He has a strong sense of
the natural order of things. There is a natural hierarchy: master, wife, children, slaves, animals. There is – or should be – a natural limit to the
acquisition of wealth. He's very severe on the getting of money beyond one's
needs: retail is unnatural, usury loathsome. He has an intellectual's contempt
for the bourgeois' money-obsession. His tone is that of a 1940s Cambridge
don (F. R. Leavis springs to mind): autocratic, moralistic, puritanical.

When writing of household economics he repeatedly refers to animals.
When writing of animals he repeatedly refers to household economics.
'Like a good housekeeper, nature is not in the habit of throwing away
anything from which anything useful can be made'; 'nature does nothing in
vain'; 'what nature takes from one place it adds to another'; 'nature does
not act out of cheapness'. Aristotle needs these principles to make his teleological explanations work, but he doesn't argue for them, they are axioms
whose truth is obvious. Heraclitus once said that 'Nature loves to hide.'

Not from Aristotle. He writes as though nature is living next door and running a taverna.

Aristotle is curiously ambivalent about the utility of horns and antlers. To be sure, animals have them for the sake of defence, but he also suggests that they're dispensable or even deleterious. He's impressed by the fact that stags shed their antlers annually. I also suspect that he never saw them in action. He speaks of their use against predators, the huntsman's point of view, but not of their use as sexual weapons. He can never have seen the clashing antlers of deer in rut.*

That horns aren't very useful is reflected in their physiological origins. As animals build themselves from the nutrition they assimilate, they first construct the most vital organs using the highest-quality nutrition and then, if they have something left over, make the less important ones. Thus his 'good housekeeper' who, we have to imagine, has some scraps left over from the family meal that she throws to a stray cat skulking around the kitchen door – a cat that's, frankly, a bit of a pest and you certainly wouldn't want it in the house, but the children like it and it does help keep the mice down. Horns are like Greek cats: low in both cost and marginal utility.

So why don't all animals have horns? Aristotle gives two reasons. Consistent with his 'good housekeeper' image, he thinks that animals are efficiently designed in that they generally lack functionally redundant organs. He notes that animals can protect themselves by being big, fast, horned, tusked or fanged. But if an animal has one means of protection it does not need another, for 'nature does nothing in vain or superfluous'.

There's another economic principle at work. In *Historia animalium* Aristotle has identified a web of associations among live-bearing tetrapods (mammals) including the fact that horned animals (ruminants) have an unequal number of teeth in their jaws (they're missing canines and incisors in their upper jaws) where hornless animals (e.g. horses) don't. Horns and teeth must be hard and so both have lots of earthen stuff in their mix. He accordingly argues that there is a trade-off between the making of horns and the making of teeth: an animal can make either horns or a full complement of teeth, but it can't make both because, as he often puts it, 'what nature takes from one place it adds to another'.† He wields this principle of

* Compare The Parts of Animals II, 2 with The Descent of Man & Sexual Selection II, 17.
† This same principle is still much used in evolutionary biology to explain apparent trade-offs between organs or other features. Recently, for example, it has been used to explain the evolution of horns in scarab beetles and the way they appear to trade off with other head structures.

resource allocation with great subtlety. He notices that the horns of large animals are disproportionately large compared to those of smaller ones, such as the gazelle, which is the smallest ruminant he knows, and explains this pattern by arguing that large ruminants have relatively more surplus earthy material to devote to their horns than smaller ones do. He's touching on one of the great patterns that living things show – one that we are still hard-pressed to explain.*

Although Aristotle's nature is generally parsimonious, sometimes parsimony can be taken too far. Many animal organs have multiple functions; the elephant's trunk is especially versatile. But he also observes that there are functional trade-offs. It's hard to do many things well, so he generally holds that it's better for parts to be specialized. As he puts it, nature doesn't act 'like a coppersmith who, out of cheapness, makes a turnspit and lamp holder in one', which curious device presumably doesn't work very well as either. He also supposes that more complex animals tend to have more specialized parts.

These auxiliary principles pervade his explanations of diversity. Nature's good housekeeping explains (or helps explain) the presence and absence of all sorts of weakly functional organs such as eyebrows, spleens and kidneys. That nature does nothing in vain explains *inter alia* why fish don't have eyelids, lungs or legs, why fanged animals don't have tusks, why only animals with molars grind their teeth from side to side, why our teeth last as long as they do and why males exist. The fact that nature can only give to one part what it's taken from another explains *inter alia* why sharks don't have bones, why bears don't have hairy tails, why birds don't have bladders, why lions have only two teats, why birds have either talons or spurs but not both and why the frogfish has its funny shape. It also explains much of life-history variation and why we die.

Collectively these auxiliary principles are a model of the body's economic design. Just as the master of a human household has a certain

---

* If, for some collection of related animals (say, mammals), we plot some feature (horn size, metabolic rate, longevity, etc.) against body size, we very often find that the feature does not scale with body size, but increases faster or slower. The relationship is, in modern jargon, *allometric* rather than *isometric*; indeed it is often best described by an exponential rather than a linear function on the arithmetic scale. The mathematics of allometry were first worked out by Julian Huxley in the 1920s, so Aristotle didn't use them, but he did notice the phenomenon and try to explain it. Many have followed. Stephen Jay Gould famously used allometry to explain the monstrous antlers of the giant Irish elk, but didn't credit Aristotle for having pioneered the area – I suppose he hadn't read him.

natural income out of which he must feed, shelter and clothe his charges, an animal has a certain nutritional income out of which it must build its parts and perform its functions. Some organs and functions are vital; others are useful but dispensable. Vital organs and reproduction have first call on nutritional income, and dispensable organs are made if there's something left over. In general, however, animals operate under fairly severe budgetary constraints and organs are expensive. This has two consequences. First, the manufacture of some organ must often be paid for by an inability to make another. Second, animals must make efficient use of their nutritional income and so tend not to make functionally redundant organs. Although all animals have to keep within their nutritional budgets, larger animals have a disproportionately greater surplus than smaller ones and so can afford to allocate more nutrition to non-vital organs. Finally, although multifunctional organs are cheap, and many animals have them, the virtues of functional specialization mean that it's best to have one organ perform a single task if possible.

Aristotle doesn't spell this model out. Nowhere does he speak of nutritional 'income', 'efficiency' or 'budgets'. This is a model of his model, but one that makes sense of much of his auxiliary teleology. Economics is woven into the theoretical structure of modern evolutionary science; it has been so since Darwin. Darwin lived in an age of *laissez-faire* capitalism, belonged to the rentier class and absorbed Adam Smith and Malthus through his pores. Aristotle did not, yet I believe that he understood and applied these simple but profound economic truths too.

THE
SOUL OF THE
CUTTLEFISH

CUTTLEFISH EGGS ON A BRANCH

LII

KALLONI'S FISHING BOATS are the small, double-ended kind called *trehantíri*, painted blue with trims of yellow and green. When we got to the harbour most of the fleet was still tied up. A pelican standing on the quay silently yawned and ruffled his feathers. The dawn Lagoon was very still, a reflected symmetry of grey-washed oranges, pinks and blues, bisected by a slash of white marking the western shore.

We were going for cuttlefish. In spring they migrate into the Aegean's shallow bays to mate, spawn and die. In his *Halieutica*, which was written in a town just up the Turkish coast, Oppian says that's when you can catch them in conical traps made of rush. The technique is still the same except that the traps are now made of plastic mesh. The first traps we brought up were empty, and for a while we suspected that someone else had lifted them (Kalloni's fishermen, a fractious brotherhood, are not above stealing each other's catch), but then a small octopus slithered bonelessly on to the deck. Oozing intellect, it headed straight for the scupper, but we caught it, stunned it and threw it into a bucket. An entangled mullet came up with its head chewed off – 'See, *soupia* did that' – and then a few kilos of cuttlefish, just enough to cover the fuel.

'At the salt mines of Salzburg,' wrote Stendhal, 'they throw a leafless wintry bough into one of the abandoned workings. Two or three months later they haul it out covered with a shining deposit of crystals' – thus the famous metaphor for the crystallization that happens if you leave a lover alone with his thoughts for twenty-four hours. Throw a branch into Kalloni during the spring and, within a day, it will be covered with berries resembling small, Greek grapes. They're cuttlefish eggs. 'The cuttlefish spawns close to land near seaweed or reeds or any debris such as brushwood, branches or stones; fishermen even put branches in position deliberately for them to lay on,' says Aristotle, and in Kalloni fishermen still do that. But cuttlefish will lay on anything hard and our traps were covered with their eggs. There must be cephalopod orgies down there.

A cuttlefish's eggs are solitary, rubbery and, when first laid, stained by their mother's ink an opaque violet-black. As they mature the egg case clears. I plucked one of these translucent berries from the net, held it up to the sun and saw within it a minute, twitching sketch of a cuttlefish, white with startlingly pink eyes, floating in its golden perivitteline fluid. Aristotle must have done the same:

> *The development of the young cuttlefish*: developing inside, from the moment the female spawns, is a sort of hailstone. It's out of this that the young cuttlefish develops, attached by the head: birds have a similar fastening by the belly. As yet there is no visual evidence of the exact nature of this umbilical attachment: just that, as the young cuttlefish grows, the white bit gets smaller and eventually, as with the yolk in birds, disappears. Its eyes, like those of other animals, appear very big to start with. In the diagram A represents the egg, B and Γ the eyes, and Δ the young cuttlefish itself. Pregnancy occurs in spring; laying within fifteen days. When the eggs have been laid, another fifteen days later something like a bunch of grapes develops: when these burst, the young cuttlefish come out.

As we drifted, a pair of copulating cuttlefish swim by. Aristotle: 'Soft-bodies, such as the octopus, the cuttlefish and the squid, all copulate in the same way, that is to say, they unite at the mouth, by an interlacing of their tentacles.' He fails to mention that both partners need not be alive. With necrophiliac ardour, the male was dragging about a very pale, and very dead, female. Females die once they've spawned and males grab anything with tentacles, twitching or not. As the cuttlefish came out of the traps they flushed dark-red with irritation, squirted jets of black ink and hissed like small, but very angry, kittens. We headed back, common terns transecting our wake.

CUTTLEFISH EMBRYO
AFTER *HISTORIA ANIMALIUM*, BOOK V

LIII

WHAT IS LIFE? It is Erwin Schrödinger's question. His answer was that life is a system that feeds on negative entropy. Herbert Spencer defined life as 'the definite combination of heterogeneous changes, both simultaneous and successive'. Jacques Loeb thought that living things were chemical machines 'consisting essentially of colloidal material, which possess the peculiarities of automatically developing, preserving, and reproducing themselves'. Hermann Muller thought that any entity that had the properties of multiplication, variation and heredity was alive. For the authors of any biology textbook that you care to pick up it is a rather arbitrary list of properties: metabolism, nutrition, reproduction and so on; for most biologists it is a question best ignored.

Aristotle asked Schrödinger's question, and answered it. At first he gives the conventional list of properties: 'By life we mean the capacity for self-nourishment, growth and decay.' But that doesn't really capture the terms in which he analyses the problem. He's after a much more abstract description of what it is that separates the living from the dead. His deeper answer is that living things, uniquely, have a soul.

LIV

THE TRADITIONAL GREEK conception of soul was Homer's. Patroclus falls at Troy and his disembodied soul takes wing for the House of Hades. Perhaps this explains why the Greek name for a butterfly is the same as that for 'soul' – *psychē* – for, as the soul flees a corpse at death, so a butterfly clambers from its chrysalis.

In *The Phaedo* Plato elaborates the traditional theory. The soul is no longer merely something lost when we die; it reasons and regulates the body's desires while we are alive. Now it is Socrates who is dying. His soul, too, will leave the body in which it is trapped and travel to Hades, but, where Patroclus' soul will live, at best, a kind of feeble afterlife, Socrates' soul can look forward to the prospect of perpetual reincarnation – or so he optimistically argues. In *The Republic* the soul gains further complexities. It becomes the seat of moral virtue. Plato describes the human soul as being mutilated by evil rather as the sea monster Glaucus – evidently modelled on a species of spider crab – is weighed down by the shells and seaweeds that encrust it.

Fragments of Aristotle's youthful works tell of similar beliefs. A dear friend, Eudemus, had died on a Sicilian battlefield. In a monograph devoted to his memory, Aristotle has Eudemus' soul returning to its home. Another early work, the *Protrepticus*, compares the relationship between body and soul to the unpleasant Etruscan habit of binding their captives face to face with a corpse – the soul being the living partner in this macabre *pas de deux*. Of this passage one scholar said, 'Surely we are here in the presence of a sick, if strong and beautiful, mind.'

Later in life, Aristotle wrote a whole book about the soul, *de Anima*, or *The Soul*. Devoid of Platonic moralizing, it is resolutely scientific in tone:

> Some types of knowledge may be especially fine and worthwhile for their precision or because their objects have greater value and elicit greater wonder. It is for both these reasons that we should treat the study of the soul as one of extreme importance. However its investigation seems to be of special importance for truth as a whole and the study of nature in particular. For souls are the principle of animal life.

This, to us, is a very strange, even suspect, claim. 'Soul' is a word burdened with many meanings but none in modern science. Perhaps we would do better to abandon translation, but mere transliteration hardly helps matters at all. For us, 'psyche' refers to mental states – in particular, consciousness. To be sure, Aristotle does treat mental states in his book, but he treats them as physiology: the Cartesian problem of consciousness hardly arises. Indeed, *The Soul* is not a psychological treatise at all, but his most general statement about the systems of command and control that enable living things to do what they do.

Aristotle asserts, with fairly little argument, two propositions: that all living things – plants, animals and humans – have souls; and that, when a living thing dies, its soul ceases to exist. These were probably commonplace among fourth-century Greek intellectuals. Plato clearly believes the first and has to argue against the second. But what, exactly, is the soul? Aristotle begins by surveying his predecessors' views.

Everyone, he says, agrees that souls are associated with movement: the ability of living things to breathe, grow, wriggle, swim, walk and fly. A good account of the soul should be able to explain how. He considers the popular idea that souls are made of some physical matter. The usual candidates for soul-stuff are the elements: air, water or fire – only earth seems to be missing as a candidate soul-stuff. He rejects them all. Quite reasonably, he cannot see how *any* element is capable of making an animal move. He considers Democritus' argument that movement is due to the restless motion of the spherical atoms that comprise creature's souls. That, says Aristotle, is about as sensible as Daedalus' scheme for animating a wooden statue of Aphrodite by pouring molten silver into it. Elements are the stuff that souls operate on; they are the substrate of life – not life itself.

He also considers some less mundane ideas. One, proposed by a renegade Pythagorean, is that the soul is a harmony. Aristotle interprets this to mean that it is a particular ratio of elements. This idea also strikes him as simplistic. Yet he sees some merit in it. It has something in common with his own theory insofar as it depends not on the properties of matter *per se*, but rather on the way in which matter is arranged. For Aristotle, when he comes to give his own theory, argues that the soul of a living thing is its form – its *eidos* – *in its body*.

I have argued that, when Aristotle speaks of the 'form' or 'formal nature' of a creature, he often means the information required to order matter into a creature of a given kind. This interpretation is based not only on the various analogies he gives (imprints in wax; letters and syllables), but also on the fact that forms are present even when they are invisible. They are somehow present in an animal's seed and are responsible for the development of the embryo and the appearance and functions of the adult. So an animal's soul is its form, albeit under particular circumstances:

> If we must say something general about all types of soul, it would be the first actuality of a natural body with organs.

The key word here is 'actuality' – *entelekheia*. It is this word, a bit of Aristotelian jargon, that is most distinctive about his theory of the soul. He often uses it in opposition to 'potentiality' – *dynamis*. The opposition runs deep into his physical theory. Any change, in Aristotle's view, is the actualization of a potential. Thus when he says that the soul is an actuality, he's stressing the fact that it's something that previously existed only potentially; that it's something that comes into being from something else. When combined with the claim that the soul of a living thing is 'its form in its body', it becomes clear that he means that the forms of unfertilized seed are mere potentials; and that those forms when realized in growing embryos and functioning adults are souls.

This is still irritatingly abstract. But Aristotle tells us much about the properties of souls and they, in turn, tell us what he's getting at. Some properties are quite general and apply to all that souls do in all creatures; others are more specific and apply just to humans. Four of them are particularly revealing.

First, an Aristotelian soul is not made of matter. That's clear from his objection to Democritus' crude materialism, but it also follows from his definition of the soul as the '*form* in a body'. Second, the soul is associated with the presence of organs, which means that it is a functional property of living things. Third, the soul is responsible for change in living things. By this he means that it regulates the body's processes: growth, maintenance, ageing, locomotion, sensation, emotion and thought itself. Finally, the soul is responsible for a creature's goals, ultimately its survival and reproduction.

Aristotle's use of *entelekheia* to describe the soul tells us how important he thought this was, for the word is partly derived from *telos*, an end or goal. This conception, too, runs deep – into his metaphysics. The soul, he says, is 'an entity [*ousia*] in the sense of a definition [*logos*]'. By this he means that a living thing's soul is the sum of its functional features. If an eye were a living creature, he says, then its soul would be vision. He is so committed to the idea that *functional* features define a creature (or an organ), rather than the stuff it's made of, that he even says that if an eye can't see (because it's damaged) then it really isn't an eye at all. It's an eye 'in name only', like the 'eyes' that Greek sailors paint on the prows of their ships.* He insists that a corpse isn't a man at all since it does not have a soul. From this point of view, a male cuttlefish who copulates with a dead female is not only wasting his time but making a serious philosophical mistake.

---

* In practice he's not always so puritanically functionalist: he'll speak of a mole's eye without qualification even as he tells us that moles are blinded by a layer of skin covering the eyes.

LV

SOULS, THEN, BEAR a heavy burden. They embrace no fewer than three of Aristotle's four explanatory causes – the formal, moving and final – leaving only the material cause for the stuff of which it is made. But for all their evident importance souls remain mysterious. What, after all, can move the stuff of which living things are made, contain its goals, yet be immaterial itself?

Confronted with these demanding criteria, scholars have sometimes concluded that, when Aristotle speaks of the soul, he is invoking some sort of spiritual force. This 'spiritual soul' interpretation comes in different flavours that draw on two disparate intellectual traditions, biology and the philosophy of mind, though the end result is much the same – an unnecessary mystification of what Aristotle means.

In the 'philosophy of mind' version Aristotle is a Cartesian mind–body dualist who believes that mental states are independent of the body; when he speaks of soul he is, in Gilbert Ryle's phrase, invoking a 'ghost in the machine'. Now, when Aristotle discusses the 'active' intellect, there are some passages that do, indeed, lend themselves to this interpretation, but they are the despair of scholars for they are so very inconsistent with everything else that he writes about the relationship of souls to mental states.

For one thing, Aristotle denies that souls are agents. He's particularly clear about this when talking about emotions. He points out that any emotion that we might attribute to our souls (joy, despair) is evident in our bodies as a physiological response (laughter, tears). But then he goes further and argues that our tendency to see these responses as a *consequence* of the soul's condition is wrong; rather, they *are* the soul:

> To say that the soul is angry is as though one were to say that the soul weaves or builds. For it is, perhaps, better to say not that the soul pities, learns or thinks, but that humans do these things.

And:

> Thinking, loving or hating affect not the mind but what *has* the mind, to the extent it has it. Actually, it is when this decays that memory and love

stop existing since they belonged not to the mind but to that composite thing which has perished.

Aristotle is trying to root out the 'homunculus'. He is attacking the notion that there is within us all a small, disembodied person – an *I* – who is thinking our thoughts, hating our hates, loving our loves and controlling, in some mysterious fashion, our bodily machines. He does not have Descartes' problem of explaining how an immaterial soul moves the body.

In the biological version of the 'spiritual soul', Aristotle is a vitalist. To be a vitalist is to suppose that living things have some property that cannot be found in, or derived from, inanimate matter; to deny that living things are really just very complicated machines; to believe in the *autonomy* of life. In the eighteenth and nineteenth centuries a battle raged – particularly in Germany – between biologists and philosophers who thought that living things are just machines and those who did not. The latter were invariably impressed by the very thing that so impressed Aristotle: the goal-directness of living things. Teleology was an invitation to fill the explanatory vacuum with resounding, if empty, phrases: *nisus formatus, Bildungsreib, Lebenskraft, vis vitalis, vis essentialis* and the like. The last scientist of any repute to wear the vitalist badge proudly was Hans Driesch (1867–1941). A brilliant experimentalist in his youth, one of the founders of *Entwicklungsmechanik* – experimental embryology – in middle age he abandoned mechanistic theory and became a committed vitalist, arguing that no machine could *even in principle* construct a living thing. 'But this may mean no more than that the living machine is more complex than any that Driesch has in mind,' was Edward Conklin's sardonic quip in 1914, and now Driesch is known, if at all, only as an object lesson on the perils of abandoning the lab bench for the airy realms of philosophy. In an unfortunate homage to Aristotle, Driesch called his vital force *entelechy*.

Mind–body dualists may still lurk in the darker recesses of philosophy departments, but in biology vitalists are extinct. The goal-directness of living things has been explained by natural selection, which tells us why living things have goals and what those goals are, by physiology and biochemistry which tells us how they achieve those goals and by genetics which tells us where those goals are stored and how they are transmitted from parent to child. Aristotle's final, moving and formal causes – all the work that he attributed to the soul – have been absorbed by, and divided among, the branches of biology. The question, then, is this: did Aristotle, ignorant of this seamless hierarchy of explanation, succumb to Kantian despair and press a traditional word, 'soul', into service as a placeholder for the gap between inanimate material and all the things that living things do? If so, then he is a vitalist. Or did he use 'soul' as a term to embrace the processes by which living things develop and function; processes that he thought – rightly or wrongly – were perfectly explicable in terms of the physical science of his day? If so, then he is a materialist – albeit of a very sophisticated kind.

It is sometimes said that all modern biologists are 'materialists' insofar as all our explanations take account of the brute properties of matter – chemistry and physics. But no biologist is a naive, Democritean, materialist, for all agree that the distinctive properties of living things depend on the arrangement of matter. The elements, though necessary, are not sufficient for life. An ordering principle – information – is needed as well. We are, to coin a term, 'informed materialists'.

That was Aristotle's view as well. It is why he identifies *psychē* as the actualization of *eidos*. It is only a beginning. Soul appears in his book on developmental biology and heredity – *The Generation of Animals*; in his book on animal locomotion – *The Movement of Animals*; in his functional anatomy – *The Parts of Animals*; and, most of all, in his physiology – *The Length and Shortness of Life* and *Youth & Old Age, Life & Death*. Soul, in short, pervades Aristotle's biology. When we survey all that he says about its workings it becomes apparent that he has given a detailed and coherent account, embracing many levels of biological organization, of how animals abstract matter from the environment, transform it, distribute the transformed matter throughout their bodies and use it to grow, maintain themselves, reproduce, perceive the world and respond to it – and that the form, structure or organization of *all* this activity is the soul.

The result is a vision that is at once disturbingly alien and surprisingly familiar. It is alien when we consider that we have been talking about 'soul', a word that by religious and philosophical tradition is usually applied to entities only tangentially connected to the physical world, if at all; but it is very familiar when we ignore the word itself and attend to what Aristotle is trying to understand – the moving principle of life.

LVI

ARISTOTLE DIVIDES THE soul among its functions. All living things have a 'nutritive' soul responsible for nourishment – *trophē* – and all that flows from it, but only animals (and humans) have a 'sensitive' soul that controls perception, appetite and locomotion. (He thinks that plants can neither sense their environment nor respond to it.) Humans also have a 'rational' soul. These sub-souls are components of a larger whole; sub-systems of the soul *tout court*.

The nutritive soul is the first soul to appear in an animal's development. Its powers are wide. It reigns over the acquisition, transformation and distribution of nutrition and hence the growth and maintenance of living things, their destruction by ageing and the perpetuation of their forms by reproduction. It holds them together and stops them from crumbling into dust. To put it more succinctly, when Aristotle talks about a creature's nutritive soul, he's talking about the structure and control of its metabolism.

Metabolism: from the Greek *metabolē* – literally 'transformation'. It's a very Aristotelian word. A metabolism is the system by which an organism acquires matter from the world, transforms it into the stuff that it needs, and then redistributes to the places where it needs it. It is an open chemical system – which is what Schrödinger meant when he spoke of life feeding on negative entropy. Aristotle also perceived that living things were open systems:

> We must understand this [a growing uniform part] in terms of a constant flow of water. Different parts, one after another, are coming into being. This is how the matter, of which flesh consists, grows: some is eroded in the flow and some arrives in addition. Additions to every part do not take place, but every part that belongs to the shape and form.

Aristotle often compares the making of an animal to the making of human artefacts: axes, statues, beds and houses. But here he's saying that an animal is not like a house at all. When an animal self-assembles it doesn't just add bits of flesh to flesh like so many bricks until it's complete; the material dynamics of animals are much more complicated for, even as they grow, they also maintain themselves. There is, in the jargon of biologists, a continual 'turnover' of materials, and growth is due to the accession of matter over and above this turnover. This idea is central to Aristotle's physiology and the starting point for any modern physiological model of growth as well.

The belief that living things *transform* food into uniform parts hardly seems like a stunning insight, yet it seems to have been original to him. He says that his predecessors had two views about how the uniform parts grow. Some held that creatures make more of x (flesh, bone, whatever) simply by eating x. Call it the 'additive' model of nutrition. Others were more subtle: they held that creatures make more of x by eating its opposite. This 'anti-matter' theory is hard to understand and Aristotle's illustration of it, that 'water may be said to feed fire', isn't very helpful, but it does contain – albeit in peculiar form – the idea of transformation. Aristotle acknowledges, then, that the 'antimatter' theory contains a germ of truth. Even so, he's clear that a much more general theory of transformation is needed.

Some chemical notation shows what an advance his theory was over his predecessors'. If the 'additive' theory is written $x \rightarrow x$, and the 'anti-matter' theory is $anti\text{-}x \rightarrow x$, then Aristotle's general theory is $x \rightarrow y$. Or, to use his words, one kind of matter, x, 'passes away' even as another, y, 'comes to be'. But actually that's too simple, for Aristotle thinks that uniform parts arise from a system of serial and parallel transformations: $x \rightarrow y \rightarrow z$ etc. or $x \rightarrow y + x \rightarrow z$ etc. That is, he thinks that metabolic transformations are ordered as a chain or even a network. From this simple, yet pregnant, idea Aristotle constructs an entire system.

In blooded animals food is masticated by the teeth, broken down by the alimentary system, transported to the mesentery, liver and spleen where it is transformed into purer nutrition and then transported further, via the veins, to the heart where it is transformed again. The product of this last transformation is blood, the key intermediate from which all animal uniform parts are derived. Blood is distributed throughout the body via the vascular system and transformed locally into flesh, fat, bone, semen and so on. Aristotle tells us how each uniform part is derived from, or related to, the others. Flesh is derived first from the 'purest' nutrient, leaving other uniform parts to be derived from the residue; what's left over is excreted. Nowhere does Aristotle present his model *in toto*, but if you ferret through his texts, you can draw a diagram that resembles a modern metabolic network and that accounts for the origin of all the uniform parts, fluids and waste residues that he mentions.* It is Aristotle's vision of the body's economy in full.

LVII

Thistle a chemistry, but it is a poor one.

T HAT MATTER FLOWS through living things, is transformed by them into the various uniform parts that they need to live and is distributed among those various uniform parts in a way constrained by economic laws – these are the elements of any metabolic theory. But any such theory must be underwritten by chemistry. Aristotle has a chemistry, but it is a poor one.

It begins with the traditional four elements. Food and all its derivatives – the uniform parts – are compounds composed of these elements in particular proportions. Aristotle credits Empedocles with this idea, and Empedocles actually gives a formula for bone: $E_2W_2A_0F_4$, where E is earth, W is water, A is air and F is fire. Aristotle, by contrast, is very vague about the formulas of the uniform parts and generally speaks of them as being composed of just earth and water: hard uniform parts (bone, nails, hooves, horns, etc.) have lots of earth but little water; soft uniform parts (fat, semen, menstrual fluid) have little earth but lots of water; flesh is something in between. Aristotle suggests that such a formula is a step towards the definition of any uniform part. That makes sense since the recipe dictates its functional properties.

All this is intuitive enough. But probe deeper, and difficulties appear. Aristotle berates Empedocles for merely thinking of the uniform parts as mixtures – agglomerated heaps of elements. They're not, he says; they're compounds – genuinely new substances. Very well, but how are such compounds formed? Our chemistry is based on a molecular theory of material substance. It is precisely the truth of this theory that makes it so rich, for it permits a multitude of possible transformations – 'reactions' – and countless molecular species each with its own distinctive physical properties. But Aristotle has rejected Democritean atomism, so his compounds are made of completely continuous matter down to the finest microscopic scale.

---

* For a diagram of Aristotle's metabolic network see Appendix II.

How can different kinds of *continuous* matter combine to form a new kind? Aristotle gives us no model or metaphor to explain it and I can't think of one that will. He says that when elements form a *mixis* they are transformed into something entirely new, yet he insists that those elements still exist, or do so 'potentially'.* In fact, the elements *must* exist within a mixture for his chemistry relies on their re-emergence during transformation. The root of the problem is plain. When Aristotle rejected atomism he also rejected any molecular theory of chemical combination. In doing so, he rejected any theory that allows elements to be the building blocks of new substances and yet remain themselves unaltered and available for recycling by living things as they please.

Heat transforms food into the various uniform parts. But Aristotle struggles to understand the nature of heat. He notes that 'hot' and 'cold' can be used in many senses, which is certainly true. It's unfortunate, then, that he uses them so indiscriminately. All living things, he believes, have an internal source of 'vital heat' (except for embryos which obtain their heat from their parents), which is why they feel warm to the touch. This internal fire, which is not the same as conventional fire, is sustained by nutrition – 'Fire', he says, 'is always coming into being and flowing like a river' – and, like all fires, it needs to be fed.† This internal fire drives 'concoction', a process analogous to 'cooking', 'broiling' or 'boiling', all of which – he thinks – drive off a mixture's internal heat and moisture leaving varying proportions of earthy material behind. Concoction seems, and is, a rather crude device, but Aristotle argues that the subtle, iterative application of heat to raw nutrition, blood and then to more derived compounds can yield all the different kinds of matter of which creatures are made.

It may seem that in describing Aristotle's metabolic model and the chemistry that underlies it I have forgotten about the soul. But in fact I have been talking about the soul all along. The system that I have described – the structure of the metabolic network – *is* the nutritive

---

* Here lies a trap for the unwary. For Aristotle, *synthesis* is the formation of a mixture (an agglomeration of parts) and *mixis* the formation of a compound (a new substance). Confusingly, the modern English cognates of these terms – synthesis and mixing – mean exactly the opposite. Translations do not always make this clear.
† The idea of an internal fire resembles, of course, our own concept of cellular respiration which is, literally, a slow combustion. But where for Aristotle the important product of the internal fire is heat itself, for us high energy bonds such as those found in ATP drive 'concoction' – macromolecular catabolism – and heat is just a by-product.

soul, or at least a part of it. The ends to which a creature puts its nutrients, how much of each kind of uniform part it will make and when and where it will do so, its growth, its reproduction and its death – all of these depend on the organization of metabolism and all depend, Aristotle tells us, on the nutritive soul. Yet there is more to the nutritive soul than this:

> Some think the fire itself is the main cause of nutrition and growth. It's not – the soul is; though it may be a contributory factor. Fires will always grow so long as there's fuel but the size and growth of all naturally composed (i.e. living) things is limited and defined: this is the job of the soul not the fire, of defining characteristics not matter.

We must imagine Aristotle sitting in front of a hearth (as Heraclitus was said to do), staring into the fire, occasionally poking it, thinking about the fire that rages inside him, that keeps him alive, that permits his thoughts to flow apparently without cease, devouring the world. 'Fire is always coming into being and flowing like a river' – how very true. But no fire can rage unchecked for ever lest it consume itself. All fires must be fuelled, stoked, damped – *regulated* – if the tenuous flame of existence is to be maintained. That, too, is the work of the soul.

LVIII

ARISTOTLE SAYS THAT tortoises hiss, copulate and have shells. They also have large lungs, small spleens, simple stomachs and a bladder; male tortoises have internal testes and seminal ducts that converge in an 'organ'. So he evidently dissected one. At least one tortoise came under his knife while still alive for he also says that if you cut out a tortoise's heart and then put its shell back on it will keep wiggling its legs. Aristotle doesn't have pets: he has specimens.

He vivisected with an enthusiasm that is no longer fashionable. 'After being cut open along its entire length, it' – the chameleon – 'continues to breathe for a long time.' Insects, too, seem to be able to survive being cut in half for a surprisingly long time. (He appears to assume, no doubt correctly, that his readers know that chickens, goats, dogs and men do not live for long without their hearts.) It all seems rather brutal, but these observations are carefully considered, for when Aristotle vivisects he's after the seat of the soul.

Where is the soul located? The Aristotelian answer is 'everywhere' and 'nowhere'. A creature's soul is, after all, not a physical object but the sum of its functional features. That truism, however, does not preclude the possibility that some organ or other is particularly important in its regulatory functions. In blooded animals – vertebrates – Aristotle supposes that organ is the heart.

This may strike us as an odd choice: why not the brain? But that's easily answered: the soul's first job is *nutrition*, and that's obviously no job for the brain. Very well – but what has the heart to do with nutrition? Everything, replies Aristotle. This is where his physiology becomes strange. To be sure, insofar as nutrition is carried in the blood, the cardiovascular system must *somehow* be involved, but Aristotle puts the heart front and centre in his nutritional physiology. He thinks that it is the main site of concoction; in fact the 'boiling' action of concoction is what keeps the heart in motion. The heart is the main site of concoction because it is also the site of the 'internal fire'. We think that it's a pump; he thinks it's a chemical reactor. He calls it the 'citadel of the body' and says it has 'supreme control'.

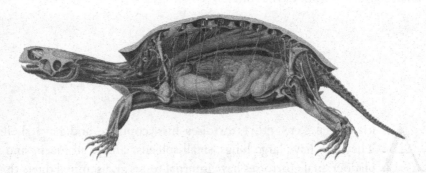

KHELŌNĒ – TORTOISE – *TESTUDO* SP.
LONGITUDINAL SECTION

Of course, not all animals have blood. But at least some bloodless animals have something like blood and something like a heart too. That is why he's so quick to misidentify the cuttlefish's *mytis* as a heart-analogue. He does not, however, make the mistake of applying his cardiocentric model of the soul to all animals. Since certain insects continue moving 'when divided into parts' it follows that each body part must have 'all the parts of the soul'. All the parts? That seems like an exaggeration, though the ability of, say, a male praying mantis to continue coitus even as his mate chews off his head makes the point. In fact, he is probably thinking of centipedes and millipedes since he says that they're like 'concretions of many animals'. All this vivisection leads Aristotle to conclude that plants, insects, reptiles and mammals have successively more centralized souls. He thinks that centralized souls are 'better' than distributed ones. It was during these investigations that he vivisected a tortoise. I haven't repeated Aristotle's observation, but one unusually empirically minded philosopher says he has. He claims to have obtained Aristotle's result too even though he didn't really follow his protocol, having first compassionately decapitated his terrapin.

LIX

ESCHYLUS WAS VISITING Sicily when an eagle, mistaking his bald head for a rock, dropped a tortoise on it and killed him. The tortoise probably died too since the only part of the story that's certainly true is that golden eagles do seize tortoises, carry them aloft and release them in order to crack them open like nuts. Neither the playwright nor the tortoise matters here; one is only an accidental anvil and the other is food; the interest of the story is how the eagle accomplished this feat.

A neurophysiologist, sketching the mechanisms involved, would describe a causal chain that begins with a goal (bodily maintenance), that requires an 'appetitive motivational drive' (hunger), perception (of the tortoise and Aeschylus' head), a variety of computations (how and when to seize, carry and drop the tortoise) and motor-responses to execute them by. He would say that the physiology underlying some of these processes is well understood, that some of them are obscure, and that how it all works together is quite unknown. He would point out that we struggle to give a computational model of a worm wriggling across a petri dish never mind an eagle on the hunt.

Aristotle also attempts a mechanistic explanation of animals in motion. He sketches how the senses work, how they transmit information to the sensorium, how this information is integrated with respect to the animal's goals and how it is transformed into mechanical action in its limbs. This system has even been given (by two classical philosophers) a very scientific-sounding acronym, the CIOM model, which stands for Centralized Incoming Outgoing Motions. Aristotle simply calls it the sensitive soul. By any name its anatomy is hazy, its physiology wrong, but its structure percipient.

Perception obviously requires the transmission of information about the world from the world to within an animal. As Aristotle puts it, perception is the transmission of an object's *form* without its matter. This process begins with the five senses: sight, smell, taste, hearing, touch and their respective organs. He assumes that perception involves a qualitative change in a sense organ. That implies that the perceived object must make contact with it.

It's easy to see how contact-dependent change works in touch, taste, hearing and, perhaps, smell. Vision is trickier. Empedocles and Plato argued that the eyes contain a fire and that light rays from this fire travel to the object of sight. Aristotle trenchantly points out that if this torch-beam theory were true, we'd be able to see in the dark. We might reasonably assume that his own theory of light is merely the reverse: that light rays emanate from some source and enter our eyes where they effect some change. That, however, is Newton, not Aristotle.

Aristotle supposes that some media – air and water – have the property of being either opaque or transparent. When such a medium is exposed to the sun or a fire it becomes transparent. Light, then, is not a ray, a wave or a particle, but a quality, an actualization of a potential. When we look at an object through a transparent medium, its shape and colours initiate movements in the medium that travel to our eyeballs where they effect a change.

Each sense organ perceives certain kinds of changes in the world; this specificity depends on the elemental composition of their uniform parts. To perceive colour and shape, eyeballs must be transparent and so are made of water; to perceive touch, flesh must be made of something solid and so is made of earthy stuff. His account of what actually happens in the eyeball when we see some object is quite opaque. He probably believes that the eyeball undergoes a physical transformation of some sort. Certainly he supposes that contact with sense organs initiates a chain of physical consequences that reach into the body.

The target of this chain is the central sensorium, the locus of perception itself. By ancient anatomical tradition, Plato thought the central sensorium was the brain and, for once, Plato was right. Aristotle, of course, supposes that it is the heart. His main argument for this is that there should be a single, central principle of all the soul's functions. To make his theory work, he obviously needs a physical connection between the heart and the peripheral sense organs. You might expect him to appeal to nerves, but he doesn't know about them. (He uses the term *neuron* but applies them to sinew; Herophilus would identify nerves as a distinct tissue in the following century.) Aristotle therefore supposes that sensory transmission operates via the network of blood vessels as well as various 'channels'. Most of these 'channels' do not obviously correspond to anything in modern anatomy, but one of them is probably, to use its modern name, the optic nerve. His argument rests on the fact that if it is severed by a blow to the head, blindness follows as if a lantern had been snuffed out. It's unclear whether he thinks that sensory information is transmitted by the vessels/channels themselves, the blood or something else; in any event, physical continuity between the peripheral sense organ and the central sensorium is essential.

The core functions of the sensitive soul take place in the heart. It's where raw perceptions are translated into mental representations which, when added to desires, become actions. Aristotle assumes that the sensitive soul's function is to maintain the animal's wellbeing by ensuring, *inter alia*, that it gets enough to eat and that it doesn't get eaten itself. Animals therefore experience the world in terms of pleasure or pain, as defined by the goal of self-maintenance. The eagle perceives the tortoise with pleasure; the tortoise perceives the eagle with pain. However, a given perception may be either pleasant or painful depending on the animal's internal state: an eagle sated on tortoises may disdain another one.

The term that Aristotle uses for the mental representation of some object is *phantasia*. To explain, he personifies: ' "I have to drink," says desire. "Here's drink," says sense-perception or *phantasia* or thought.' He does not, of course, give a mechanistic account of *phantasia* or any other higher cognitive process, but he recognizes the difficulty. After explaining the physiology of smell, he says, 'but smelling is more than such an affection by what is smelly − *what more?* Is not the answer that, while the air owing to the momentary duration of the action upon it of the smelly thing does itself become perceptible to the sense of smell, smelling is an *observing* of the result produced?' Indeed.

Phantasia and desire may be black boxes, but when he explains how they effect action, he becomes very physiological again. Heating and cooling of the heart accompany both of these mental events. These thermal changes initiate motion that is then transmitted to the limbs. To explain how he invokes devices that he calls 'automatic puppets':

> The movement of animals is like that of automatic puppets, which are set moving when a small motion occurs: the cables are released and the pegs strike against one another . . . for they [animals] have functioning parts that are of the same kind: the sinews and bones. The latter are like the pegs and the iron in our example, the sinews like the cables. When these are released and slackened the animal moves. Now in the puppets . . . no qualitative change takes place . . . But in animals the same part has the capacity to become both larger and smaller and to change its shape; the parts change qualitatively when they expand because of heat and contract again because of cold.

The important point about these puppets is that they are automatic (*automata*). They seem to have been mechanical dolls of some kind. He's careful to note that their motions aren't exactly like those of animals since animal motions also involve qualitative change such as expansion and contraction of the kind found in the heart. This gives him another anatomical problem. He has to translate qualitative changes in the heart into mechanical changes and then distribute those mechanical impulses to the limbs, and he has to do this not only without nerves but also without muscles.

It's not that the Greeks were oblivious to muscles. Classical statues of athletes and heroes display them in enviable relief. Hippocratic texts refer to muscles as *myes* – 'mice' – but are vague about what they do. Aristotle avoids the term altogether and calls them *sarx* – 'flesh' – which he supposes has mostly a sensory function. His effectors of local motion are, then, sinews and a substance called *symphyton pneuma*.

Variously rendered 'connate pneuma', 'hot breath', 'spirit' or simply ΣP, *pneuma* is one of the most mysterious, yet powerful, substances in Aristotle's chemistry. It is something like hot air, but its heat is of a special kind, not the heat of conventional fire. It is analogous to the divine 'first element' (*aithēr*) of which stars are made. More mundanely, it gives organic materials special properties: olive oil is shiny, floats on water and does not freeze thanks to its high *pneuma* content.

Pneuma is also a proximate instrument of the soul. *Pneuma*, heated or cooled by the heart, expands and contracts and so moves the heart's minute sinews. These mechanical movements are, in turn, transmitted to the rest of the body. How they do so isn't very clear since he knows that the network of sinews, unlike that of bones and blood vessels, is discontinuous. It's another connectivity problem, not unlike that of getting sensory information from the sense organs to the heart. However it works, he does see that a small change in the heart's motions can become amplified to move the entire animal. This is his motivation for *automaton*-causality. Switching similes, he also compares the way in which animals move to the way in which a great change in a ship's course is brought about by the smallest movement in its rudder.

FARNESE HERCULES. AFTER LYSIPPOS, C. 330 BC

At the end of *The Movements of Animals*, Aristotle summarizes his account of animal motion in a simple geometrical diagram:

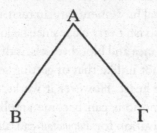

It is reasonable that motions run from the parts to the 'origin' [*archē*]
and from the 'origin' to the parts and to each other. Then the motions
from each letter in the diagram we have drawn arrive at the origin and
from the origin, as it moves and changes, being potentially many, the
motions of B goes to B, that of Γ to Γ, that of both to both. But from B
to Γ it goes by going first from B to A as to a principle, then from A to
Γ as from a principle.

This is a long-winded way of explaining that movements are initiated
and effected at peripheral organs (B and Γ), but that, whatever happens,
they are always mediated by A – the origin; the heart; the seat of the soul.
It is the essence of the CIOM model; all we have been doing is putting
flesh on it.* Aristotle allows that animals can have action without *phantasma* (involuntary movements such as the heart's beating); and *phantasmata*
that initiate action without actual perception (dreams, hallucinations);
and that humans have a whole separate level of cognition, *nous* – reason
– that regulates their actions. Our eagle, however, is a much less complex
creature. Soaring above Sicily's stony hills, it perceives the glint of
Aeschylus' head, constructs an (erroneous) *phantasma* of it as a rock,
responds to its ravenous appetites, fires its *pneuma*, loosens its joints,
slacks its sinews, opens its talons, drops the tortoise, stoops to kill and
simply satisfies its desires.

* For a diagram of the full CIOM model see Appendix III.

LX

WHEN ARISTOTLE SAYS that the heart has 'supreme control' he does not *just* mean that it's in the middle of the sensory and metabolic network – he means 'control' in a very literal sense. He compares the organization of animals to that of a well-governed city. A central organizing principle, the soul, sets things in motion and the rest just follows.

This is most obvious for the workings of the sensitive soul. But it's true for the nutritive soul as well. Deeply impressed by the fragility of life, Aristotle worries that the heart's internal fire will rage unchecked, consume all its fuel and so precipitate a metabolic crisis. He therefore argues that animals must have a variety of devices that keep their fires under control. The most important of these involves air.

Fires, says Aristotle, are regulated by altering the flow of air around them. In the same way air from the lungs regulates the heart's fire. This is how: (i) the lungs expand and draw in cool air as a smithy's bellows do; (ii) the cool air flows to the heart and damps the internal fire; (iii) the heart contracts; (iv) the lungs contract; (v) the newly heated air is expelled; (vi) the heart heats up once again; (vii) the heart expands; (viii) the lungs expand; (ix) the cycle repeats.

It's an ingenious mechanism. Of course, it's all wrong.* And it only works for mammals, birds and reptiles; other animals, he says, must cool their internal fires in some other way. Bees, cockchafers, wasps and cicadas breathe through their skins;† fish do not breathe at all. They don't gulp air,

---

* It is anatomically wrong – the vessels that connect the lungs to the heart are the pulmonary arteries and veins, but in living animals they are filled with blood and not, as Aristotle supposes, air. It is chemically wrong – since Aristotle hasn't read Lavoisier he does not know that combustion is a reaction that combines a component of air, oxygen, with a fuel. (In fact, he explicitly considers, and rejects, the possibility that the internal fire is nourished by air.) This leads him to the notion that the effects of air on fires (or life) must be due to cooling. It is physically wrong – the model depends on the idea that the intensity of a fire is affected by ambient temperature but, of course, this is not so. Besides the heart–lung cycle he also thinks that in blooded animals the internal fire is cooled by the brain and damped down by nutrition; that, too, is wrong.
† In *Youth & Old Age, Life & Death* Aristotle says that this breathing accounts for the buzzing sound that insects make, but in *Historia animalium* he's clear that it's caused by the motion of wings.

indeed die when exposed to it, so they are cooled by the water that they take in over their gills. But many small divisibles (insects etc.) and soft-shells (lobsters, crabs and the like) don't really need much cooling at all since their internal fires are simply not very intense.

When Aristotle explains how the internal fire is controlled, he lays the workings of the nutritive soul bare. This is another dimension of his teleology. He claims that the soul is responsible for the formal, motive and final causes and then shows all three at work. He sets a goal for the body and then shows how it is achieved. Many scholars, struggling to convey what Aristotle means by the soul, have described it as a 'cybernetic system'. The metaphor is consciously anachronistic, but plausible.

In the 1860s Claude Bernard showed that mammals regulate their body temperatures by altering the circulation of their blood in response to signals from the nervous system. Bernard's slogan that 'It is the fixity of the *milieu intérieur* that is the condition of a free and independent life' inspired Walter Cannon to popularize, in 1932, the term 'homeostasis'. In the 1940s Norbert Wiener formalized homeostasis as the product of regulatory systems that contain negative feedback circuits. Coining the term 'cybernetics' for the science of such self-regulating systems,* Wiener argued that they solved the problem of teleology: how torpedoes (the weapon, not the fish) can have goal-seeking behaviours. If machines can have goal-seeking behaviours then so can living creatures. Vitalism was expunged from its last redoubt: 'Many of the characteristics of organismic systems, often considered vitalistic or mystical, can be derived from the system concept and the characteristics of certain, rather general, systems equations' – so von Bertalanffy in 1968.

The Aristotelian soul certainly has many properties of a system. It is a set of interacting units (organs) that form an integrated whole (a body). It has modules (the nutritive, sensitive and rational souls); and these modules have specialized functions and are hierarchically arranged. In some cases (humans) it is centralized; in others (centipedes) it is distributed. It has a purpose: to regulate the functions of life.

---

* Searching for a name for his new science, Wiener began with 'governor', the name given to the device that regulates a steam engine. This led him to its Latin ancestor, *gubernator*, and thence, via an etymological trail, to the word's ultimate Greek ancestor *kybernētēs* or 'pilot' which he considered particularly apt since the steering devices of ships were especially good examples of negative feedback control systems. From *kybernētēs* came 'cybernetics'. It was a felicitous choice since the steersman is an ancient metaphor for control used by both Plato and Aristotle in the context of political hierarchies.

But is the soul a *cybernetic* system? If the metaphor has any power, then it should illuminate Aristotle at his most concrete – that is, when he describes the heart–lung thermoregulatory cycle. Aristotle claims that he's described how the heart beats and the lungs pump. Has he? The answer is not obvious from his verbal account. If, however, we grant his physics, chemistry and anatomy, and diagram his model using the block-and-arrow formalism of cybernetics, the structure of the mechanism becomes clear.* The diagram, which is isomorphic with his text, shows that his model works, but not as he thinks it does. He thinks he's described an oscillator that will make the lungs expand and contract rhythmically; in fact he's described a thermostat. He's worked out how to keep the heart at a steady boil.

But that is no mean accomplishment. For his system contains the essence of any homeostatic device, a negative feedback circuit. It truly *is* a cybernetic system. Credit for first inventing, or at least applying, negative feedback control usually goes to the Alexandrian scientist Ctesibius (*fl.* 250 BC), who incorporated it into the design of a water clock. Perhaps credit should also be given to Aristotle who, two centuries earlier, saw the need for such a device in living things and sketched, however fancifully, how it might work.

This is, of course, an Aristotle for our times. Cybernetics and von Bertalanffy's General Systems Theory became, in turn, the progenitors of modern systems biology, that quintessentially twenty-first-century science concerned with networks that depict the flow of matter and information among the parts of which living things are composed. The systems biologist B. Ø. Palsson puts it like this: 'components come and go, therefore a key feature of living systems is how their components are connected together. The inter-connections between cells and cellular components define the essence of a living process.' Remove the reference to cells and that's pure Aristotle.

Of course, the point is not to make Aristotle seem terribly modern. Rather, it is to better understand *his* answers to some of biology's deepest questions. What gives living things their goal-directedness? Souls do – by which he meant control systems of a complexity sufficient to show goal-directed behaviour. What holds living things together? Souls do – by which he meant the functional interconnections of their parts. How should we study living things? We have to take them apart, reduce them down to their individual bits and pieces. But, having done so, we also have to put them back together again for it is only then that we really understand how they work.

---

* For a control diagram of the heart–lung cycle see Appendix IV.

FOAM

FORMATION OF A HUMAN EMBRYO
BASED ON AN ARISTOTELIAN MODEL

LXI

WHEN ARISTOTLE WANTS, as he so often does, to convince us that living things have an end – a goal – and that they cannot therefore be purely explained by the workings of matter alone, he appeals not simply to the beauty of animal design, the devices by which they keep themselves alive in the face of the world's vicissitudes, but rather to the fact that they develop in a regular way. In the *Physics* Aristotle tackles the claim that he attributes – rightly or wrongly it's hard to say – to Empedocles that order can just 'spontaneously' emerge in the womb. Hence his argument that a child's teeth require some goal-oriented process, underpinned by a formal nature, if they are to appear in the right place at the right time. In *The Parts of Animals* he returns to the attack. Now it's the backbone that worries him. Empedocles apparently also claimed that vertebrae are distinct because the backbone just happens to twist and break during development. That, says Aristotle, can't be right. The seed from which the embryo developed must already have had the *potential* to produce the vertebrae. That is why 'a human being gives rise to a human being' and not a horse.

It's one of his favourite sayings. It is a very deep truth. It is also not a compelling argument for it merely restates the obvious. How, exactly, does a human being give rise to a human being? It is one thing to say that Empedocles is wrong, quite another to show it. In the recesses of the womb, where no one can see, theories are free to breed.

Aristotle's solution is to mount a research programme to find out what's going on in there. He studies a forty-day-old human embryo:

> Place a male embryo, detached at forty days, in anything but cold water, and it dissolves and disappears. In cold water, it coheres somewhat inside a membrane. If that is pulled apart, the embryo is revealed, as big as one of the large ants, its parts clearly visible, including the penis and the eyes. These, as in other animals, are very large.

He doesn't say where he got it from. He seems to have studied more than one. The passage, which is just a description, appears in *Historia animalium*. Its explanation appears elsewhere. This second work contains a mechanistic account of how animals develop, one deeply integrated with his physiological theory; an explanation for why creatures have two sexes when they do, and why, sometimes, they don't; a mechanistic account of the transmission of form from parent to embryo, and a theory of inherited variation, that is to say, a genetics. It also has an analysis of life-history variation and a discussion of environmental influences. *The Generation of Animals* is, in short, a general account of how a human being gives rise to another human being or a fish a fish. *Historia animalium* aside, it is his longest book. It is also his most luminous.

LXII

THE REPRODUCTIVE BIOLOGY is orgiastic in content but clinical in tone. During the mating season, Aristotle says, 'all animals are excited by desire and the pleasure derived from copulation'. Male frogs, rams and boars call to female frogs, ewes and sows; pigeons kiss. Some females show desire by coming into heat. Mares are wanton and female cats wheedle toms for sex, but hinds are reluctant for they find the stags too stiff.

Males fight, of course. Stallions, stags, boars, bulls, camels, bears, wolves, lions, elephants, quail and partridges all have a go at each other in a display of sexual mayhem. Stags round up females, dig holes in the ground and bellow at rivals. Gregarious creatures tend to be combative, solitary ones less so. And if partridge cocks are 'lecherous' and break the eggs of hens, pigeons are much gentler and pair for life – though occasionally females will go off with another male.

All this is but a preamble to the act itself. Aristotle defines a male as an animal that 'reproduces inside of another animal'.* To get inside another animal, most males mount her from behind. Sharks and rays, however, mate belly to belly, dolphins copulate side by side, lions, lynxes and hares copulate back to back and snakes intertwine. He also says that during sex

* This definition seems to exclude egg-laying fishes which, we know, have external fertilization. Aristotle certainly knows that many male fishes sprinkle milt over newly laid eggs, but exactly what he thinks is going on during fish mating is unclear – he admits it's an obscure business.

hedgehogs stand on their hind legs and face each other so that their spines
don't get in the way, that bears adopt the missionary position and that
camels take all day about it.*

It is not, however, copulatory topology but reproductive physiology that
gives the essential difference between the sexes. Aristotle argues that males
and females both produce a reproductive residue, *sperma* – seed. The male
seed is *gonē* – semen – which he supposes is hyper-refined blood and, like
all metabolic residues, completely uniform in composition. The female
seed is *katamēnia* – menstrual fluid. This latter claim will strike the modern
reader as peculiar, but it's all of a piece with his physiology.

Since embryos require nutrition, and blood is the purest form of nutrition, it's obvious that the menses, which are clearly pretty blood-like,
should be the stuff from which an embryo is formed. Furthermore, the
monthly discharge can be explained as unused seed which, in turn, neatly
explains why girls become fertile only once they have begun to menstruate
and why they cease to do so when pregnant. The menses are quite similar
to semen, but rather less well refined or concocted, which makes sense
since females are, according to Aristotle, colder than males. Women may
have souls, but they also have cold hearts.

As always Aristotle wants a theory that covers all animals (or at least all
blooded animals), but the idea that embryos are made from menses has the
obvious weakness that most animals don't menstruate. Undeterred,
Aristotle identifies the blood-like fluid that cows and bitches discharge
when they're in heat as their menses.† Hens, obviously, never discharge
anything resembling blood so he points to 'wind eggs', the dwarfish, yolkless eggs that they sometimes produce as being a kind of avian menses.‡

---

* Hedgehogs, bears, camels, lions, lynxes and hares do not copulate like this.
† Aristotle devotes pages to sorting out female vaginal discharges including: urine, vaginal
lubrication, pathological discharges, post-partum bleeding, menstrual bleeding in humans
and oestral bleeding in animals that come into heat. He argues, correctly, that the first
three have little to do directly with reproduction but, incorrectly, thinks that menstruation (in humans) and oestral bleeding (in dogs and cows) are the same thing – *katamēnia*
– that is, the *sperma* (seed) which the mother contributes to the embryo. These two secretions are, in fact, quite different. Only primates truly menstruate and Aristotle knew only
one primate, humans, at first hand.
‡ Aristotle's claim that virgin chickens lay *hypēnemia* (literally 'wind eggs') is not true now.
All supermarket eggs, which are large and have perfectly formed yolks, are produced by
virgin chickens. However, the first eggs produced by a pullet – a young chicken – are
often small and yolkless; substitute 'virgin' for 'young' and you have Aristotle's claim. He
may also be strictly correct. Modern breeder hens are very odd birds. They have been
selected for egg production for thousands of years; perhaps ancient breeds only laid the

And, although he thinks that most fish eggs are embryos, he acknowledges that some fish are packed with unfertilized roe that is, as it were, *their* menses. But that's Aristotle: an answer for everything.

LXIII

MALES PRODUCE RATHER little seed; females produce a lot. From this it follows that they have very different genitals. Aristotle, accordingly, has much to say about penises. The seal's is large; the camel's is sinewy; the weasel has a bone in his. Two copulatory append-ages hang from the cloacae of male, but not female, sharks and rays. He's uncertain about birds. In *The Generation of Animals* he says that no bird has a penis, but in *Historia animalium* he states that the goose does.* He says that snakes don't have penises; in fact they have two that emerge during sex. He's vague on the tortoise's penis, which is very large and stiff.

He turns to testes. Most live-bearing tetrapods (mammals) have testes suspended from their bellies, but dolphins, hedgehogs and elephants keep theirs inside near their kidneys. The internal testes of birds and egg-laying tetrapods (frogs, lizards, tortoises) are located near their loins. In all of these animals the testes are connected to seminal ducts (urogenital duct/ vas deferens) that unite in a common duct. The egg-laying animals (birds, reptiles, amphibians) have a common passage for the faeces, semen and urine (the cloaca), but mammals do not.

It's detailed. It's mostly accurate. It's very easy to be lulled into compla-cent familiarity. But then he says something unexpected and you realize, if you did not already, that his notions of how things work are different from ours. Aristotle sees that the testicles have something to do with semen, but not that they make it. Instead, he argues that they store it and regulate its flow. His reasons are characteristically complex.

Testes, he says, can't be needed for generation since snakes and fishes don't have them. They do, however, have semen-filled 'passages' which are,

odd wind egg when virgin. Certainly, at least some bird species only start laying eggs once mated.
* Many anatids (ducks, geese and swans) have a taste for violent, coercive sex and elabo-rate intromittent organs to do it with. A recent report gives the Argentinian lake duck a 20cm-long, corkscrew-shaped, spine-bearing penis.

then, the equivalent of the seminal ducts of birds and tetrapods, the main semen-receiving organ.* (Since semen is a hyper-refined blood, the product of successive bouts of concoction, if it's produced anywhere, it's in the heart, though on this he's quite vague.) It also follows from this that testes are an optional refinement, structures that some, but not all, animals have for the 'better' rather than the 'necessary'.

The testes store semen. This is shown by the fact that the testes of some birds (partridges and pigeons) are filled with semen during the breeding season but depleted afterwards. But in tetrapods their function is to regulate its flow. Observing that the seminal ducts (vas deferens) in tetrapods loop up and over the ureters *en route* to the penis, he argues that this arrangement steadies, or even limits, the seminal flow.† The testes are counterweights that maintain the loop by opposing the natural tendency of the seminal ducts to coil. That's why, in humans, the testicles descend at puberty and why castrated animals are sterile: lop off the testicles and the seminal ducts spring up into the body cavity and so inhibit the flow.

The model is strikingly mechanical. He even compares the testes to the stones that weaver-women use to keep the warps of their looms in place. His account of what the penis does is equally odd. He thinks that it gives the semen a final concoction by the heat generated from friction during copulation. Putting it all together, he supposes that semen is concocted in the vascular system, gathered in the seminal ducts and stored in the testes, which also ensure an ejaculate of the right amount; the penis gives it an extra charge and emits into the female genital tract.

Aristotle's model of the male reproductive system is a based on a live-bearing tetrapod, probably a bull or ram. He refers to an anatomical diagram of it. His model of the female's is also based on some ruminant. He

---

* Aristotle expects testes to be round, but fish and snake testes are elongate so he takes them to be the equivalents of the tetrapod seminal ducts, in fact, the vas deferens. The mistake is a surprising one, since he knows that fish seminal ducts fill seasonally with semen just as bird testes do, which should suggest a similar function. This is comparable to his uncertainty about the identity of fish and bird kidneys since they lack the classic tetrapod shape.

† Since the testes aren't, in fact, counterweights, Aristotle's ingenious explanation for the looping vas deferens must be wrong. So what's the true function of the loop? The anwer, curiously, is that it doesn't have one. It's a contingent, non-adaptive product of a mammalian evolution history in which internal testicles descending from the abdominal cavity, their ancestral position, to between the loins, happened to take an inefficient route. Here Aristotle's teleology over-reaches itself. At least it does so if the standard evolutionary account is correct.

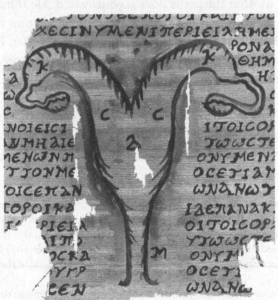

REPRODUCTIVE ORGANS OF A BLOODED LIVE-BEARING TETRAPOD
AFTER *Historia animalium*, Book III
Above: male. *below:* FEMALE

calls the whole structure a *hystera* – 'uterus' – and insists that it is always 'double' which tells us that his description is based on a ruminant since their uteri are, indeed, mostly made of two large uterine horns which humans lack. The uterine horns – *keratia* – then unite to form the *delphys*, which leads to a fleshy, cartilaginous tube with an opening, the *metra*. These are probably the uterine body and the cervix. Stretching unity to the limit, Aristotle tries to bring the reproductive systems of female mammals, reptiles, fishes, cephalopods and insects under a common scheme. He finds this hard to do, which is not very surprising, for they are, in fact, very different.

LXIV

B
UT ENOUGH OF anatomy. What are Aristotle's views on the female orgasm?

He thinks that women want sex – lots of it. Sexual intercourse is *ta aphrodisia*. He describes highly sexed women as *aphrodisiazomenai*. Adolescent girls have to be watched since they have a natural impulse to use their developing sexual faculties. They may even contract bad habits (a veiled warning against masturbation?), but usually settle down after having had a few children. Some women are, however, as wanton as mares. Nymphomaniacs are literally 'stallion-mad' (*hippomanousi*).

The Greeks lacked a technical term for orgasm so Aristotle simply speaks of the 'pleasure' or 'intense pleasure' of sex. But he certainly thinks that women typically have them; his models of male and female sexuality are very similar. In women 'the pleasure of intercourse is caused by touch in the same region of the female as the male' – which suggests that by 'pleasure' he means orgasm and by 'the same region' he means the glans and the clitoris. True, he has a name for the former, *balanos*, but not for the latter, but, to give credit where credit is due, he seems to have found it.

Some women, he says, when experiencing pleasure 'in a way comparable to a man', produce a saliva-like fluid which is different from the menstrual fluid. That must be vaginal lubrication. Sometimes there's a lot of it, more than a man's emission – apparently a reference to female ejaculation. When women have pleasure, and secrete this sexual fluid, it's a sign that the uterus is open and conception likely. He says that blondes are especially wet.

In fact, the question is not whether women experience pleasure during sex, for Aristotle thinks they should and do; rather, it is do they *need* to climax to conceive?* Aristotle disagrees with himself. In *The Generation of Animals* he argues that, although a woman usually experiences pleasure during sex, she can conceive without it and, conversely, can fail to conceive even when she 'keeps the same pace' as her male partner. The female orgasm is nice but not necessary. In Book X of *Historia animalium*, however, the orgasm seems much more important, for there he argues that during sex the menstrual fluid is secreted into an area 'in front of the uterus' (presumably the cervix or vagina), where it mingles with the semen. This secretion apparently happens at orgasm, for both partners have to 'keep the same pace' if conception is to be successful. In fact, infertility is usually due to men 'completing quickly' while their partners have hardly begun to do so ('for in most things women are slower'). To determine whether premature ejaculation is indeed the cause of infertility he suggests that the man in question should have intercourse with other women to see if he can generate children – which shows an admirably empirical spirit. To solve the problem of unequal timing he also suggests that the woman should excite herself with 'appropriate thoughts' even as her lover dwells on his troubles to cool his ardour.

It is hard to say whether the 'orgasm nice' or 'orgasm necessary' theory represents Aristotle's final thoughts. Book X clearly doesn't belong with the rest of *Historia animalium* since its content is mostly clinical; some scholars even doubt that he wrote it. Yet these various theories do share the idea that sex is a collaboration. Both partners are prompted to it for the intense pleasure it gives and, ideally, take their pleasures together; at least they do if they want to conceive a child, and that, for Aristotle, is certainly the point.†

---

* In his *Essays* Montaigne quotes Aristotle to the effect that a 'man ... should touch his wife prudently and soberly, lest if he caresses her too lasciviously the pleasure should transport her outside the bounds of reason'. I don't know where he got this dismal bit of advice from, but it certainly wasn't Aristotle.
† Evolutionary biologists have also puzzled over the function of the female orgasm. Male orgasms are an obvious adaptation, a direct inducement to reproduce, but women, however much they may like one, don't need an orgasm to conceive. If, then, the female orgasm has a function, it must be a rather subtle one, and there are many ingenious accounts of what that might be. Some biologists have even argued that it has no adaptive function at all, but is merely a developmental by-product of selection for male pleasure – the genital equivalent of male nipples. This will strike most of us as implausible.

LXV

I N THE OPENING LINES of *The Generation of Animals* Aristotle says that now he wants to investigate the moving cause of life and that 'To inquire into this and to inquire into the generation of animals is, in a way, the same thing.' It's a rather elliptical way of putting it, but he is stating the problem in its most general terms. He believes that the matter from which parents form their progeny – the seed – is only potentially alive. Somehow this matter must be animated. For us this is the problem of fertilization; for Aristotle it is the acquisition of a soul.

To say that an embryo 'acquires a soul' sounds deeply mysterious, but Aristotle just means by it the acquisition of a set of functioning organs. Or, to put it another way, how the embryo gets its form. Plato parked his Forms in the realm beyond the senses; Aristotle places his forms in the seed. An animal gets its soul from its parents. That, however, leaves much to be explained. From which parent does the soul come? Are the sub-souls – nutritive, sensitive, rational – transmitted as a unit? When does the soul actually appear in ontogeny? When does life actually begin?

Aristotle's approach to answering these questions is empirical. He observes that, compared to the flow of the menses, ejaculate volume is paltry. So he's first inclined to think that fathers supply the embryo's form, and hence its soul, while mothers supply its matter. That's tantamount to saying that the mother just supplies the building material that the father sculpts into a functioning creature. Indeed, Aristotle often speaks as if this is exactly what he believes. Throughout *The Generation of Animals* he returns to a set of parallel dichotomies with which he tries to capture the difference between males and females: hot/cold; semen/menses; form/matter; soul/matter; moving cause/material cause; active/passive – the terms vary, but the contrast is always clear.

Or is it? For all Aristotle's repeated insistence that males and females make utterly distinct contributions to the embryo, when he turns to the details of embryogenesis and inheritance the roles of males and females begin to blur and merge until finally it is hard to tell them apart. Some scholars argue that *The Generation of Animals* contains very different, and incompatible, theories, but perhaps we should read these sexual dichotomies as slogans that further analysis will elucidate and refine. For example, having told us that fathers contribute the embryo's soul, he presents some evidence to show that this is not in fact quite true, and that mothers do, after all, give their offspring life.

Aristotle claims that hen partridges can 'conceive' just by smelling males on the breeze. It sounds absurd, but he really does say it – he says it twice. Partridges aren't the only birds that produce 'wind eggs'; all birds do, but they're particularly common in prolific ones. The business about the breeze isn't important; the fact that virgin birds produce wind eggs is. Does Aristotle really believe that birds can conceive without impregnation? He does – but the trick is to know what he means by it. For us conception occurs when a sperm fuses with an ovum to make a zygote; for Aristotle conception occurs when semen meets menstrual fluid and makes an egg, but since wind eggs can be produced by virgin hens, menstrual fluid can clearly sometimes congeal spontaneously into a conceptus.* The menses are, then, in some sense alive; they have, in his jargon, the potential for a nutritive soul. Aristotle is clear that wind eggs are duds; full conception, leading to a chick, requires a cock, copulation and semen, but he wonders if even this is true for all animals since he speculates that some fishes may dispense with males. The puzzling thing about the khannos is that you only ever catch females.† Perhaps males don't exist. Aristotle is, however, reluctant to discard the need for males without more data ('there haven't been enough observations'), so sticks with his theory that both seeds contain the potential for a nutritive soul, and that only the semen contains the potential for a sensitive soul and specific form – the features that make, say, a sparrow a sparrow rather than a chicken or a crane.

---

* When Aristotle claims that the smell of a cock partridge can cause a hen to 'conceive' he means only that it can induce her to produce wind eggs that never complete development. An experienced pheasant breeder has told me that this isn't so – young hens produce wind eggs regardless of the presence of a male, but I wonder whether the influence of male pheromones on partridge oogenesis deserves further study.

† Aristotle mentions three fishes that might reproduce without males: the khannos, erythrinos and psētta. The khannos is the comber Serranus cabrilla; the erythrinos is probably the anthias, Anthias anthias; the identity of the psētta is a mystery, but is thought to be a textual error for the perkē, the painted comber, Serranus scriba. If these identifications are correct, then all three of Aristotle's 'female-only' kinds are members of the Serranidae. In 1787, Cavolini showed that S. cabrilla and S. scriba are simultaneous hermaphrodites; Anthias is a protogynous (female-first) hermaphrodite in which males are rare. In the simultaneous hermaphrodites the testes are small and hard to see. Aristotle, having failed to find the testes, or any males, therefore moots the possibility that these fish kinds reproduce without sex.

Describing how development works, Aristotle leans heavily on his potential/actuality dichotomy: 'Thus the seed of the hand or face of the whole animal really *is* the hand or face of a whole animal though in an undifferentiated way; in other words what each of those is in actuality, such the seed is potentially . . .' This is at once wonderfully insightful and frustratingly opaque. It is insightful for it captures the idea that the seed contains something – the form – that is not the animal itself but that has, nevertheless, the power both to shape and to become it, and that ontogeny is the process by which this potential is translated into an actual living, breathing, copulating creature. But potential-talk can also seem like a feeble substitute for a physical model of development. What, exactly, are these potentials? Point to one. Or, if you can't do that, at least give us a hint of how they work.

Aristotle evidently feels this tension too and so does attempt a physical model. He begins by asking whether or not these embryo-forming potentials can be transmitted independently of the physical matter of the semen itself. He invokes one of his favourite analogies: human craft. Consider a carpenter making a bed out of wood. In doing so, he's not actually contributing matter to the bed, rather the knowledge of his craft (a potential), manifest as a functional movement, *shapes* matter. Analogously, to contribute a potential, semen need not actually contribute matter to the embryo.

Besides this analogy Aristotle also provides three lines of zoological evidence. (i) Some insects, he thinks, copulate in a peculiar way: instead of males inserting an organ into females, it's the other way round.* In such cases, he suggests, the males does not actually transfer any semen, but only a potential. (ii) When a chicken copulates with more than one cock, the chicks may resemble either father – usually the second in line – but never have 'every part twice'. The thought seems to be that monstrous animals (conjoined twins) might be caused by an excess of seminal material. If so, then you'd expect multiple matings to produce deformed chicks, but they don't, so it's not the quantity of seminal material that's critical, but merely a qualitative 'potential'.† (iii) When male fish spread their milt on eggs,

---

* Perhaps Aristotle is describing the female grasshopper, or some other orthopteran, who, during copulation, bends her long, pointed ovipositor to meet the more modest genitals of the smaller male perched on her back. Even so, it's unclear why such an arrangement forbids true insemination.

† Aristotle thinks that an excess of female seed causes conjoined twinning. But he also wants to argue here that an excess of male seed doesn't. The argument is based on an ingenious, but incorrect, theory of the causes of conjoined twinning. Aristotle is, however, right to suppose that the second-in-line male often succeeds in fertilizing the eggs. The

only the eggs touched by the milt become fertilized. None of these arguments is convincing. Yet Aristotle's aim is clear: he's trying to show that the power of semen to direct development rests not on the transmission of seminal matter itself, but on something else.

What? *Something* in semen must get to the embryo, and if it isn't seminal matter then what is it? To solve this problem Aristotle once again invokes that mysterious stuff, *pneuma*. It's not only an instrument of the sensitive soul, but also a component of the inheritance system. Aristotle searches semen for the signs of activity. He finds it in the fact that semen resembles foam – or does so immediately after ejaculation. The foam is due to a charge of *pneuma* introduced by concoction of the semen during sex. *Pneuma* does not, however, need to be carried in semen since in those insects with strange sex it gets injected directly into the female. The upshot is a theory for how an animal's soul is reproduced in the embryo. The structure of the father's soul is, in effect, encoded in his semen by *pneuma*-tic action.*

We must not think of *pneuma* as the carrier of genetic information itself: it's not Aristotelian DNA. Rather, Aristotle's units of inheritance are much more abstract; they're the movements that *pneuma* induces in the semen. When he describes the motive principle in semen, the movement that is the future soul incarnate, he chose a word both apt and elegant: *aphros* or foam; and he meant by this both literally the foam visible in semen, and the foam visible in the wash of waves receding from a shore. Yet, as the passage makes clear, in choosing this word he was also thinking of something else. It is also for *aphros*, he says, that the Goddess of Love was named.

phenomenon is known as 'last male precedence', is found in many bird species and is due to sperm competition.

* This model of fertilization persisted for centuries through the history of developmental biology. It did so even after 1677, which is when Leeuwenhoek reported the spermatic 'animalcules' that he had seen under his microscope. Fabricius supposed that semen did its mysterious work via an 'irradiant or spiritous faculty'; Harvey, scorning his teacher's terminology, said it works by 'contagion'. Both could just as easily have said it works by *pneuma* or bubbles. Even von Baer's model of fertilization was still very Aristotelian. (It was von Baer who called sperm *spermatozoa* – itself a name redolent of their ambiguity.) It was only in 1875 that Oskar Hertwig definitively showed that the embryo begins with the fusion of a sperm and egg nucleus – and did so by looking at a very Aristotelian creature, the edible sea urchin, *Paracentrotus lividus*. That, however, required a microscope. Proof that chromosomes are the carriers of inherited information had to await Thomas Hunt Morgan's 1910 fruitfly experiments – and even then there were sceptics. In 1928 William Bateson, one of Mendel's earliest champions, and the man who coined the word 'genetics', was still arguing that inheritance operates by a system of intra-nuclear 'vibrations' – that is to say, movements.

APHRODITE

LXVI

ARISTOTLE IS GENERALLY credited with being the first scientist to investigate embryogenesis or, to use his words, 'coming to be'. Was he? The origins of his methods are generally obscure, but a Hippocratic treatise that dates, perhaps, to fifty years before he lived and that was written by, perhaps, Polybus suggests that the foetus of a human being resembles that of a chick. To prove this, says perhaps-Polybus, take twenty eggs, put them under some hens, and open them at daily intervals until they hatch: 'you will find everything as I say insofar as a bird can resemble a man'. Aristotle doesn't reference perhaps-Polybus, oddly so, for he had a

famously good library and often cites predecessors, admittedly mostly
when he thinks they're wrong.

Whether or not he was the first to study the embryology of the chicken,
his description was surely better than any that had come before:

> In hens, it is three days before the first visible sign [of life]: it takes longer
> in bigger and shorter in smaller birds. It is during this period that move-
> ment of the yolk actually occurs, upwards to where the egg's origin is, the
> sharp end where the egg hatches. The heart is in the white, the size of a
> spot of blood. This speck beats and moves as if it were alive. From it, as
> it continues to develop towards each end of the covering which envelops
> it, lead two interleaved blood-vessel tubes containing blood. At this
> stage a membrane with bloody fibrous material actually envelops the
> white, leading off from the blood-vessel tubes. A little later the body too
> can actually be distinguished, extremely small to start with and white.
> The head is visible and in it the eyes, extremely prominent . . .*

Aristotle studied the development of the chick because he could. Fish
embryos are tiny; mammal embryos lie concealed in the womb; but to see
the chick all you have to do is crack open an egg. He also describes the
development of many other creatures, albeit in much less detail. A fish
embryo, as far as he can make out, is very much like a bird's except that it
has only one kind of yolk and no allantois. Even the embryos of live-bearers
(mammals, the smooth dogfish) are quite similar to those of egg-layers
(birds, most fishes and reptiles): both are protected from the outside world
(by the eggshell or the uterus), both are surrounded by an amniotic sac (the
khórion) and both get nutrition via umbilical cords from either the yolk or
the mother's blood. Aristotle knows that cows, sheep and goats have uteri

---

* The whole passage, paraphrased and using modern anatomical terms, runs like this: the
three-day-old embryo has a heart that beats and from which two blood vessels, the left and
right vitelline arteries, ramify into the capillaries of the yolk sac. The body, head and eyes
can be seen. In the ten-day-old embryo, the head is still bigger than the body, the eyes are
large enough to dissect out and several membranes – the chorion, allantois, amnion and
yolk sac – can be seen. These membranes are separated from each other by fluid-filled
spaces; the amniotic sac is vascularized; the yolk has become more liquid and there is less
albumin. The stomach and other viscera can be seen. The twenty-day-old chick has down
all over its body and it is bent so that its head is next to its leg and covered by a wing. The
allantois now contains excreted material and its connection to the chick has been severed;
the yolk sac has been almost entirely absorbed within its stomach. The chick sleeps, wakes
up, moves, looks up and chirps; it is about to hatch. So much for the development of birds.

CHICKEN EMBRYO

studded with *kotyledones* (cotyledons or caruncles) but that most other animals do not.* Still, sometimes, when he's in generalizing mood, it's hard to know whether Aristotle is talking about a chicken or a man.

When doing so he isn't being careless. Rather he's saying something quite important:

> It is not the case that a human being, horse, or any other particular sort of animal is formed at the moment when an animal is formed. The final stage of each animal's development is its goal and what is distinctive about it . . .

Which is to say that when an embryo initially forms, you can see only the general features that make it an animal – the fact that it's got a heart and the basic outlines of its organs. The *specific* features – the features that make a man a man rather than a horse – appear last in development.

That's a beautiful observation. It was made again, in far greater detail, by Karl von Baer who, in his great *Über Entwicklungsgeschichte der Thiere*, 1828,

---

* Aristotle does better than Leonardo, who infamously sketched a human foetus attached to a cow's cotyledonary placenta. On the other hand, he does not have a technical term for the placenta.

called it his 'First Law' of comparative embryology. It would become one of the great generalizations of evolutionary developmental biology.*

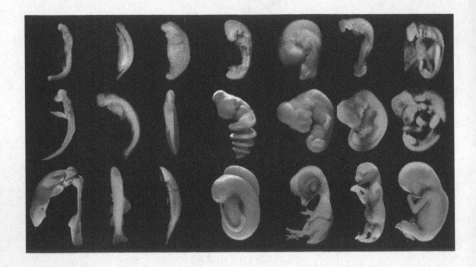

VON BAER'S FIRST LAW OF EMBRYOLOGY IN VERTEBRATES, ILLUSTRATED
LEFT TO RIGHT: DOGFISH, SALMON, AXOLOTL, SNAKE, CHICKEN, CAT, HUMAN EMBRYOS
TOP TO BOTTOM: EARLY, INTERMEDIATE, LATE STAGES

Aristotle's anatomy, although certainly the product of his own dissections, also has much in it that he learnt from fishmongers and butchers, hunters and travellers, physicians and soothsayers. His embryology, however, is obviously about things that he saw for himself. Who but a biologist, eager to uncover the secrets of generation, would spend so much time peering at tiny embryos? If, with a nod to perhaps-Polybus, we acknowledge that Aristotle was not the first to study the chicken embryo, he was certainly the first to see in it the solution to the problem of development.

* One that stood until very recently. In the past few years, transcriptomic data have shown that the very earliest stages of embryos are *also* quite variable. It is now thought that the embryos of different, related species are most conserved at some intermediate stage. In vertebrates, this is around the time of somite formation and neurogenesis. After that, the pattern is as Aristotle and von Baer said it is.

LXVII

WHEN THE MENSES contact semen they congeal into an embryo or an egg. Aristotle uses homespun analogies to explain how this works: 'The case resembles that of fig-juice which curdles milk, for this too changes without becoming any part of the curdling masses.' Or, elsewhere, 'this acts in the same way as rennet acts upon milk'. This is all about making cheese. When rennet, a substance derived from the stomachs of unweaned calves, is mixed with milk it causes them to separate into solid and liquid parts: curds and whey. Aristotle supposes that seminal *pneuma* does the same thing to menstrual fluid, coagulating the earthy stuff out of it, leaving a fluid behind. He probably thought the analogy particularly apt. The active ingredients (semen, rennet, fig-juice) all draw their power from being charged with vital heat; their substrates (menstrual fluid, milk) are both very closely related derivatives of blood.*

The result of all this cheese-making is an embryo enveloped by a membrane floating in a fluid. And now the real business of *pneuma* begins: the manufacture of the embryo's parts. Aristotle claims, and repeats the claim with the vigour of a man who thinks he's made a major discovery, that the heart is the first organ to appear in the embryo. It's a reasonable one if we allow that he means the first visible functioning organ and so exclude the somites and notochord that form well before the heart does. It isn't just a fact for Aristotle: it fits with theory too. The heart *must* be the first organ to develop because nutrition and so the growth of all the other organs depends on it.

Nutrition supplied by the mother, and concocted by her heat, flows into the embryo through the vitelline vessels and is redistributed by the heart and its system of ramifying vessels. He compares the vessels to the roots of a seedling or to irrigation runnels in a field, and the way nutrition seeps

---

* Looking around his kitchen for a chemical analogy to explain early embryogenesis, Aristotle, strikingly, reaches for one based on enzymes (fig-juice and rennet both contain proteases). He hints at the idea of catalysis for he thinks that the active ingredients do not become part of the product; he fails to grasp it for he also thinks that they become consumed in the reaction.

through the walls of the vessels to how water seeps through unbaked pottery. In the final stage the judicious application of heat transforms the nutrition into the flesh, sinews, bones and all the other tissues of which a growing embryo is built.

Aristotle thinks that tissue and organs are made out of raw, unformed material, but first he demolishes the obvious rival idea, that the parts of an embryo – perhaps even the entire embryo – already exist in the parents' seed, but are too small to be seen. His opponents were the Neo-Ionian *physiologoi* who denied that matter of any kind – even tissues – could be either created or destroyed. A late commentator relates Anaxagoras' theory: 'For in the same seed he [Anaxagoras] says that there is hair and nails and veins and arteries and muscles and bones, and they are invisible because of the smallness of their parts but as they grow they are gradually separated out. For how, he says, might hair come to be from what is not hair and flesh from what is not flesh?' Aristotle, however, picks on Empedocles who, he says, believed that organisms self-assemble from pre-formed organs. (He *says* – Aristotle often seems to be an unreliable reporter of the Sicilian's ideas.) In any event, he adduces many arguments against the theory, but doesn't hesitate to make some rather banal points as well: 'Also, if the animal's parts are dispersed within the semen how are they to live? If they coalesced they would form a small animal' – and that, his reader is left to conclude, is obviously absurd.

Aristotle gives his own vision with the aid of two beautiful metaphors. In one, nature paints the embryo:

The parts are all sketched in outline first and only later acquire colour, softness and hardness. It's exactly as if nature were an artist working on a painting. Painters sketch out animals as well – before they apply a bit of colour.

In the other, the embryo is woven like a net:

How, then, are the other parts produced by the semen? They . . . either come into being all together or in succession as Orpheus' poem suggests . . . like the knitting of a net. The former is not the case: some parts are clearly visible already existing in the embryo while others are not . . . the lung is bigger than the heart, and yet appears later than the heart in the original development.

It is the second passage that tells us the real reason that Aristotle doesn't like the idea of pre-formed organs. Any such theory must invoke tiny chickens or bits of chicken in the semen and Aristotle simply does not believe in the existence of things too small to see. This jaundiced view of the invisible world springs directly from his most fundamental theory of matter. Semen is homogeneous: it is composed of neither molecules nor microscopic fowl.

Having established the pattern, Aristotle feels compelled to explain it. *How* do the parts come to be one after the other? He considers the possibility that organs give rise to each other – that the liver actually grows out of the heart – but rejects this on the grounds that each organ has its own form, and the form of one organ cannot exist in another; all organs are made from more basic matter. His own solution depends on a much more subtle kind of causal chain. Semen, he says, initiates movement in the embryo; and, once it does so, this is what happens:

> Imagine A moving B and B moving Γ as in those amazing automatic puppets. Even when their parts are inactive, they retain some sort of potentiality. This means that when some outside force sets one part in motion the next part is immediately activated in actuality. So, in a sense, A does move Γ in the automatic puppets, not because there is any current contact with any part but rather because of its previous contact. The same is true of the semen's origin. The producer of the semen sets things in motion through a past connection as opposed to any current one.

These are the *automata* of *The Movement of Animals*. Using puppets to explain how an animal moves is obvious enough; using them to explain how an embryo develops, isn't. By 'A' and 'B' and 'Γ' Aristotle certainly means the embryo's developing organs. The semen's movements shape the heart, which then shapes other organs which shape yet others, until the picture is painted, the net woven and the embryo complete.

Throughout much of his account, Aristotle seems to be telling us that the making of an embryo is like making a statue: the father is the artist who sculpts and the semen is his hand; the mother is the oven in which her menstrual clay is fired. It now becomes clear that this simile does not capture what he means. He's already conceded that the menses are in a way alive, that they contain the potential for a nutritive soul. *Automaton*-causality gives new meaning to the word 'potential' for it tells us that the menses have a hidden structure and a latent formative power; that it's more like a wound-up clock with a catch that semen merely springs.

Automaton-causality also accounts for the diversity of forms. Embryos start out the same but, as they develop, the causal chains that shape them diverge. He's talking about a creature called the *kordylos*. It's an amphibian: it has gills and swims using a tail that resembles a catfish's but also has legs instead of fins and can live on land.* It is, by nature, intermediate between a land and a water animal. He says that it is so 'warped' because of some event that occurs very early in its ontogeny. He goes on to explain that the environment in which the animal grows up – land or water – influences some 'infinitesimally minute but absolutely essential organ' that, in turn, dictates whether the animal will have the features of a terrestrial or aquatic animal. Much about the *kordylos* is vague, but the general argument isn't: early in ontogeny some small organ is responsible for the many features in which aquatic and terrestrial animals differ from each other: A moves B moves Γ.

LXVIII

W**HEN THE ANATOMISTS** of the Renaissance took to looking inside eggs again, they used *The Generation of Animals* as their guide. They had, of course, nothing else. Aldrovandi (*Ornithologia*, 1600), his student Volcher Coiter the Frisian (*Externarum et internarum principalium humani corporis partium tabulae et exercitationes*, 1573) and Hieronymus Fabricius ab Aquapendente (*de Formatione ovi et pulli*, 1604) scarcely improved on Aristotle's descriptions of the development of the chick – though they made some fine figures.

William Harvey, who revered Aristotle, approached him with a more critical eye. In his *Exercitationes de generatione animalium* (1651) Harvey correctly identified the *cicatricula* (blastoderm), rather than the *punctum saliens* (embryonic heart), as the first manifestation of the embryo. He called it 'The Fountain of All Life', but he also saw, *contra* Aristotle, that blood forms before the heart. It was Harvey, too, who searched for the

---

* The *kordylos* is obviously a newt eft or frog tadpole, but it's unclear that Aristotle knows that it is a larva; he may think that it is a 'dualizing' adult like a seal or a dolphin. When Aristotle speaks of the minute organ that dictates the future development of the animal, zoologists will be irresistibly reminded of the endocrine organs – the hypothalamus, pituitary or thyroid – that control amphibian metamorphosis. Aristotle probably means the heart – he usually does.

coagulum of sperm and menses that Aristotle's theory of fertilization predicted. He dissected newly inseminated does, victims of Charles I's hunting parties in the Royal parks, failed to find the Aristotelian fluids, took the other unifying option and declared (on the frontispiece of his book) *Ex ovo omnia* – 'from the egg everything'.* Acute critic though he was, much of Harvey's embryology remained utterly Aristotelian. 'There is no part of the future foetus', Harvey declared, 'actually in the egg, yet all the parts of it are in it potentially . . .' Note the contrast between *'actually'* and *'potentially'* – Aristotle couldn't have phrased it better himself.

Harvey called this process of actualization 'epigenesis'.† It's here that Aristotle's argument with the Neo-Ionians gets a replay. Many of Harvey's successors, enchanted by the structures revealed by their microscopes, argued that the Aristotelian model was simply wrong. The embryo, they claimed, contained, from its very beginnings, all its parts complete. Some said that they could see miniature embryos in spermatozoa; others saw them in eggs. Historians call this doctrine, in all its varieties, 'preformationism'. Charles Bonnet, a Swiss naturalist, and not a man to shy from logical consequences, proposed that each seed contained within it a fully pre-formed embryo, whose seeds contained within them fully pre-formed embryos, whose seeds . . . and so on to Creation's very start.

The argument between epigenesis and preformationism ran for about two hundred years. For a while the preformationists seemed to have modernity and mechanism on their side. Better microscopes running on optics by Zeiss showed that they didn't. Preformationism was an illusion; embryos really do build themselves.

You can see them do it. All you need is a really good microscope with some fancy filters and a healthy culture of nematode worms. You take a single fertilized egg, mount it on a little pad of agar with a drop of buffer to stop it getting crushed and keep it moist, protect the whole thing with a coverslip and then flip to 1000× magnification. Then you watch. Nothing much happens at first, but then the cytoplasm begins to swirl and deform

* That, however, was really just a lucky guess or, perhaps, a programmatic statement. It certainly wasn't an empirical generalization. Karl von Baer discovered the minuscule, if not actually microscopic, mammalian ovum only in 1827. Less fortunate than Harvey in his patrons, he found it by dissecting a colleague's dog.
† Not to be confused with 'epigenetics' in the modern sense – that is, chemical modifications of DNA, or chromosomal structure, that result in altered patterns of gene expression.

and, quite suddenly, there are two cells where previously there was only one. They divide again and then again and then again – the whole thing happens with remarkable speed and unvarying precision. Cells begin to shuffle about, some duck beneath others; cavities form and bulges extrude; organs – a pharynx, a gut – begin to appear in ghostly outline and then become increasingly defined. The mass of cells contracts, first into something resembling a bean, then a comma, then a pretzel – that is to say, a little worm. Around seven hours after you began to watch, it starts to twitch; by ten it's rolling around in its egg.

There's a lot about Aristotle's developmental biology that seems quite strange. In our biology the parental materials are gametes not fluids; they do not merely come into vague proximity with each other but fuse; the carrier of inherited information is not a pattern of 'movements' but a peculiarly stable macromolecule. And the form of the incipient animal comes not just from the father, of course, but from both parents. Still, you have to admire the sheer audacity of his system. It's all there – a mechanistic account of the most mysterious process in all of biology – how apparently raw matter *comes to be* a living thing, complete with all its parts. And, if you contemplate the invisible gradients of molecular signals, the cascades of transcription factors and the networks of signal transduction proteins driving the cells to their destinations and differentiated forms, it seems that Aristotle's *automaton* logic – A moves B moves Γ – echoing down in ramifying causal chains, captures something very fundamental about how it all works. It *is* 'exactly as if nature were a painter producing a work of art'. If there's a lovelier and truer metaphor for the act of self-creation that made you and me and Aristotle and every other living thing, then I do not know it.

THE VALLEY
OF SHEEP

PROBATON – SYRIAN FAT-TAILED SHEEP – *OVO ARIES*

LXIX

THE POTAMIÁ, WHICH means simply 'river', drops from the Ordimnos massif to the alluvial plain on the Lagoon's north-western shore. One spring day I walked from Anemótia, following its course down. I saw no one else. The hills are barely populated, though they are used, for periodically the path was barred by small dogs that emerged from boxes or barrels to bark at me, straining at the end of ropes. I wondered what they were doing there in this solitude, for there was nothing for them to defend, but later I learnt that their function is to regulate the movements of the sheep that wander through the valley's olive groves. Indeed, turning a corner, I came across a flock that was wandering about apparently untended. The sheep of Lesbos are lean and intelligent. In the olive groves they graze on boughs that farmers have cut down for them, but in the island's arid interior they live on the aromatic plants of the phrygana that grow in the thin volcanic soil. They carry bronze bells around their necks and in the stillness of the hills you can often hear their soft chimes long before they appear.

Aristotle, who has much to say about sheep husbandry, describes how it is one sheep in particular, a castrated ram, who carries the bell and who has been trained to lead the flock and respond to his name. On Lesbos nearly all the sheep have bells, of varying size and timbre, so that as you approach them and they skitter nervously away, a carillon is set in motion that ripples through the flock. One sheep, obviously the leader, planted itself boldly in my path and stared at me with unblinking yellow eyes, and though I was curious to know whether it still had its testicles, I should have had to look under its shaggy coat, and its stance made me doubt that my interest was shared. In Corinth I met a highland shepherd, suitably weathered and taciturn, who confirmed what Aristotle says. At the age of three months, a male lamb, large, disciplined and beautiful, is selected as a future leader. At six months it is castrated, named and apprenticed to a mature ram to learn the command of a platoon of twenty-five sheep. The shepherd added the curious fact that a mature ewe will sometimes usurp control by instinct or force of personality and that, once she does so, she never again bears lambs. He also said that once his commander ram saved him from some mortal danger, but he would not say what it was.

In his round-up of animal variety Aristotle touches on the bio-geography of sheep. He reports that in Pontus (the Black Sea littoral) the rams don't have horns, but in Libya there is a sheep with long horns and both the rams and ewes have them;* that Sauromatic (Bosphorus) sheep have hard wool; that on Naxos they have very large gall bladders but on Euboea they have none; that flat-tailed sheep tolerate winter's cold better than long-tailed sheep and short-fleeced sheep better than shaggy-fleeced, but that crisp-haired sheep suffer most. Syria is home to some singularly outré domesticates:

> In Syria sheep's tails are one and a half foot wide and goats' ears about a foot long, in some cases touching down below at ground level. Cattle also have humps on their shoulders, like camels.

It's not, by itself, a very important observation – just one more piece of natural history lore among thousands. But one wonders: what did Aristotle think these fat-tailed sheep, long-eared goats and hump-backed cattle were? Were they, for him, just local varieties of the same basic sheep, goats and cows grazing on any Greek farm, or were they something quite different? It doesn't seem like a very momentous question, but it is. For upon its answer turns nothing less than one's vision of the order and stability of life:

> Hottentots say great tailed sheep aboriginal at Cape & a thinner tailed kind farther inland . . . Capt Davis in 1598 found cattle in Table Bay with hump on their back & big-tailed sheep.

They have the same data: fat-tailed sheep and hump-backed cattle in exotic locales, very different from anything chewing the cud at home. Yet it's not just the data that matter; it's what you see in them. The second passage is from Darwin's *Transmutation Notebooks*. It's 1837 or 1838, and he's just discovered evolution.

---

* Perhaps this long-horn is another species, the Barbary sheep, *Ammotragus lervia*, since modern North African Berber sheep are notably hornless.

POTAMIÁ VALLEY, LESBOS, JUNE 2011

LXX

THE FIRST CHAPTER of *The Origin of Species* might have been about the glories of the Brazilian rainforest in which the twenty-three-year-old wandered in frank religious ecstasy. Or it might have been about the Kentish countryside where green serenity conceals a vicious struggle for light and life. Or it might have been about the Galapagos, in evolution's Origin Myth the *fons et origo* of the theory itself. Darwin could even have just abstracted the four volumes on barnacles that he'd published only a few years before and told how their cyprid larvae proves their link to shrimps and crabs; he might have described the weirder species with microscopic males ('mere bags of spermatozoa') and gigantic prosociform penises. All this, after all, is the *problem*; it's what he's trying to explain, and you'd think he'd want to grab the reader by showing how wonderful it is. But he doesn't. Prosaically, he begins with pigeons.

He argues that all the pigeon breeds in the world are descended from the common rock pigeon, *Columbia livia*. Thousands of generations of selection by the hand of man have divided and transformed them and that's what happens in nature too. Know the pigeon, *understand* the pigeon, and all the rest follows. Darwin's argument is so familiar that no biologist can look at a pigeon, sheep, goat or goldfish without reading in their feathers or feet or fins an evolutionary tale be it grand or grotesque, or both. Each fat-tailed sheep or a hump-back cow tells a narrative of origin, migration and change that began thousands of years ago in the deserts of the Middle East and the civilizations of the Indus, that spans the mountains of Asia Minor, skirts the scrubby coastal hills of the Levant, continues down the Great African Rift across countless kilometres of veldt, reaches a geographic terminus in Table Bay, and yet – for evolution itself does not stop – continues to this day.

But it's not quite Darwin's story. It is only in the last few decades that molecular genetics and archaeology have truly traversed the meandering genealogies of our farmyard animals to their remote and antique origins in the wild.* Darwin's real point was more profound. He wanted to show that species are variable and that some of this variation can be inherited. Nature *generates* heritable variation – and how:

> The diversity of the breeds is something astonishing. Compare the English carrier and the short-faced tumbler, and see the wonderful difference in their beaks, entailing corresponding differences in their skulls. The carrier, more especially the male bird, is also remarkable from the wonderful development of the carunculated skin about the head, and this is accompanied by greatly elongated eyelids, very large external orifices to the nostrils, and a wide gape of mouth. The short-faced tumbler has a beak in outline almost like that of a finch; and the common tumbler has . . . [etc.]

Darwin needed to understand inheritance. Heritable variation was the fuel on which his evolutionary engine ran, so he needed to know its laws and its limits. He batted the problem about for decades. The tentative jottings of the *Transmutation Notebooks* metamorphosed into the confident claims of the *Origin* which, in 1868, spawned the sprawlingly aporetic *Variation of Animals*

---

* Darwin suspects that hump-back cattle may, in fact, be derived from a distinct species of ancestral bovid; they are now thought to descend from a distinct subspecies, *Bos primigenius indicus*, where European cattle are derived from *B. p. taurus*.

and Plants under Domestication. It was all a failure, the greatest of his scientific life. Yet we now know that Darwin was right to suppose that every species is rich in heritable variation. Indeed, the great lesson of post-Darwinian biology is that diversity goes all the way down. Some of this phenotypic variety is caused by variety in genes; some by variety in environment; much of it is caused by both in ways that are so complex that we can scarcely disentangle them at all.

Darwin grasped some of this. Did Aristotle? Many scholars have thought not. Aristotle, they argue, believed that a scientist's task was to enumerate the 'essential' features of the creatures he studied. 'Essential' features do not vary among individuals, or vary only accidentally (an amputee is still obviously a man, even if he is no longer actually bipedal). In seeking the essence of every form Aristotle ignored the variety that individuals show and placed it beyond science's remit. However different Socrates and Callias – or two sheep – may look, for Aristotle they are 'one in form' and that's the end of it.

It isn't. Yes, he does want to understand the typical, functional – 'essential' – features of his kinds. But he also has a parallel research agenda to understand the useless variety that riddles even the smallest of them, the kinds that can be described as *atoma eidē* – 'indivisible forms'. He doesn't have a term for this kind of variation, so I call it *informal* variation by analogy to the modern biologist's *intra-specific* variation. In Illyria and Paeonia (Balkans), says Aristotle, the pigs have solid hooves like those of a horse rather than the cloven feet that most pigs have. This sounds like an Aristotelian bizarrerie, but it isn't, for Darwin says that such pigs exist in England. Aristotle and Darwin are both clear that they're not talking about two different *kinds* or *species* of pigs but rather variants of the regular pig.*

So the various sheep, pigs, horses and cattle distributed across the limits of the world are just different manifestations of particular forms. Every domesticated animal, says Aristotle, has a wild equivalent; should they be classified into different kinds? No, such a division would be unnatural. Humans are a unity too. Aristotle knows that Ethiopians have black skin and curly hair, yet he takes it for granted that they and Greeks share the same indivisible form.

Darwin claimed that most of the variation visible in domesticated animals is heritable. Aristotle, by contrast, gives most informal variation to

---

* A syndactylous mutant, of the sort known in Louisiana as 'mule-foot hogs'.

the direct effects of the environment. Some places are hot, others are cold; some are wet, others are dry, and such differences make for differences in appearance. The depths of the sea are cold so the sea urchins that live there have long spines. Africa is dry but the Black Sea littoral is damp so the Ethiopians have curls but the Scythians' and Thracians' locks are lank. Egypt's torrid climate ensures that the naturally cold animals – snakes, lizards and the Red Sea's turtles – that live there are very big. On the other hand, scarcity of food means that Egypt's dogs, wolves, foxes and hares are rather small. Bees are more 'uniformly coloured' than hornets and wasps because they have a relatively monotonous diet. All this informal variation is devoid of functional significance. It is not for the sake of anything, but is just the product of material natures, the physical properties of tissues moulded by the vagaries of the world.

Aristotle's environmentalist view of geographic variation is puzzling. Did he not understand that the varied features of domestic animals are inherited? Presumably he'd only read about fat-tailed Syrian sheep and mule-footed Balkan hogs, but surely any farmer could have told him that long-hair and crisp-hair sheep are *breeds*? He informs us that in some districts the sheep are white and in others black, and that this may be due to the water.* That's absurd. In Greece sheep flocks are a chequerboard of white and black and every shepherd must have known, as my Corinthian highlander certainly did, that fleece colour is inherited. It's also not as though the principles of selective breeding were unknown in fourth-century Greece, for in *The Republic* Plato discusses how to breed a better sheepdog. Of course, being Plato, that's only by way of introducing his real interest, how to breed a better human. Granted Aristotle did well to ignore Plato's eugenic fantasies,† but he had at least one other colleague, his closest friend, whose grasp on the causes of intra-specific variety was really rather subtle – and whose data were much better than his own.

---

* In another passage he hints that fleece colour might be heritable, albeit in a strange way – namely, that the colour of the veins under the ram's tongue predicts the colour of its offspring. I doubt that this is true; at least some farmers that I asked about it seemed puzzled.
† In the *Politics* Aristotle suggests that the state should regulate marriage and the production of children with a view to raising healthy children; he even recommends that deformed children be killed. But nowhere does he adduce hereditary arguments for such laws so his argument is not eugenic.

LXXI

HE SEEMS LIKE THE classic epigone. If Aristotle is all pyrotechnics, Theophrastus is candle-powered. His theories aren't as bold; they don't run as deep; they seem mostly borrowed from his friend. When telling us about his plants, Theophrastus never names Aristotle – but he's always there. Yet Theophrastus shouldn't be underrated. That difference in temperament is also a difference in method. Theophrastus is more cautious, less quarrelsome, more empirical, less theory-driven. He carries less metaphysical baggage. This isn't just because his *Metaphysics* is in fragments, where Aristotle's survives intact. When Theophrastus considers alternative explanations and gives you the evidence for each, you don't get the sense – pervasive with Aristotle – that he's already stacked the deck. 'Divine Speech' is not the very model of a modern scientist – but he's closer.

Thracian wheat, Theophrastus says, sprouts late and takes three months to mature; in other places wheat sprouts early and matures in two – why? One obvious explanation is that there's something different about Thracian air, water or soil. He analyses the effects of soils, water and winds on plant growth at length. In Lesbos, a river near Pyrrha is so nutritious that its waters actually kill plants, and people who bathe in it become covered in some scaly stuff. (He must mean the mineral-rich hot springs of Lisvori that emerge just west of the Lagoon.) Although animals, too, are affected by environment, he says that they are less so than plants, since their connection to the soil is less direct.

The allusion is to Aristotle's Egyptian animals. In fact, the whole model is very Aristotelian. But then he points out that if you grow Thracian wheat in other places it *still* sprouts late, and if you grow early-sprouting wheat in Thrace it *still* sprouts early.* Each variety of wheat, he concludes, has its own 'special nature'. He seems to think that the differences between the wheat varieties are fixed, that they are heritable. But, speaking generally, *both* the

---

* Done deliberately, this is known to botanists as a 'common garden' experiment and is used for exactly the purposes that Theophrastus invokes: to unravel the contribution of heredity and environment to phenotypic variation.

environment and hereditary qualities can affect the growth of a plant: 'For when something comes about as a result of two or more things possessing power, the whole necessarily varies with the differences in its sources (and this also happens in animals; for animals get differences due not only to the male and the female parent, but also to the country and air, in short, their food).' Or, to put it in Francis Galton's terms, it's a matter of both nature *and* nurture.

Here, at least, Theophrastus is closer to the phenomena than Aristotle. We sense that the student was a gardener, but that his teacher just peered over the farmyard fence. But the two scientists complement each other. Theophrastus' theories are thin. *How* is variation inherited? He doesn't really say. Aristotle does.

LXXII

HE MAY ATTRIBUTE CURLY hair and straight hair to the effects of climate, but Aristotle knows, of course, that children inherit at least some of their parents' peculiarities. He had at least two of his own, a daughter and a son. Of all the scientific problems that he tackled, the inheritance of subspecific – informal – variation is among the hardest. Its phenomena are elusive: to accurately describe how children resemble their parents requires a grasp of probability; to accurately describe the inside of a cuttlefish does not. And observation, by itself, can't crack a genetic problem: difficult experiments involving the rearing and measuring of many individuals over many generations are required. Darwin, who conducted just such experiments, and even tried his hand at ratios, made no headway at all.

It's no surprise, then, that Aristotle's data on inherited variation are poor. Even so, it's surprising just how poor they are. True, he mentions a few cases of inherited variation, but they're just confused hearsay and he misses much of what he could have seen. No Darwin, he ignores domesticated animals. Of course, he doesn't cross anything (though there are some intriguing passages on hybrids). He devotes pages to variation in human eye and hair colour, but gives no indication that they can be inherited. He's fascinated by teratology – dwarfism, hermaphroditism, conjoined twinning, anomalous genitals, extra appendages – and says that such deformities are often inherited, but sometimes not, which, although certainly true, doesn't get us very far. All in all, Aristotle's grasp of the facts of inheritance is only slightly more sophisticated than the musings of any newly minted father:

> Some children resemble their parents, others do not. Some resemble their fathers, others their mothers, some in the whole body, some in each individual part, some their parents, some their ancestors, some just a general person. Males may resemble the father, females the mother. Some, though, resemble no relative but do resemble a human being. Some do not even resemble a human being in form but, actually, a monster.

Mendelian ratios are not even a distant dream.

Yet, however weak these data may be, they give Aristotle a list of phenomena to explain, namely, why a child: (i) sometimes takes after its parents; (ii) sometimes takes after its ancestors; (iii) sometimes doesn't take after a relation, but just looks human; (iv) sometimes doesn't look human, but just looks monstrous. Also: (v) why boys usually, but not always, take after their fathers and girls after their mothers; and (vi) why the different features of a child may take after different parents or ancestors. Naturally, Aristotle has a theory to explain all this; and, just as naturally, he must first dispose of someone else's.

Aristotle often doesn't name his opponents, but occasionally we know who they were anyway since we have a text containing the very argument that's aroused his ire. A fifth-century tract called On Generation contains a brief account of a theory of inheritance that Aristotle evidently read. It belongs to the Corpus Hippocraticum but Hippocrates certainly didn't write it. The theory is especially interesting since it crops up in the nineteenth century too. Aristotle effectively demolishes it and so, with one blow, takes two scalps separated by more than two millennia, one of which is Darwin's.

The Hippocratic model is simple. A father's seed originates in his body parts: his hands, heart and all his other organs and tissues give off fluids that travel via the blood vessels to the penis where they are churned, heated and ejaculated. Something similar happens in mothers. The parental seeds mix in the uterus and an embryo is formed which has the features of both parents, weighted by their contributions. It's a superficially persuasive idea. The direct physical connection between body parts and seed neatly explains how the characteristics of the parents' bodies are transmitted to the seed and so to their offspring. Democritus seems to have adopted a version of it, but probably had particles rather than fluids as his units of transmission. In 1868 Darwin published the same idea, with a few elaborations, and called it 'pangenesis'.*

Aristotle took pangenesis seriously. 'Hippocrates' sketched several arguments for it; Aristotle repeats this evidence and even adds to it – but only so that he can knock it down. Over a dozen pages of digressive dialectic he offers fifteen separate objections. One of them turns on *the* great question of nineteenth-century genetics: can acquired characteristics be inherited?

'Hippocrates' argued that if some part of a parent is crippled, then the semen that comes from that part will be weak and the child will be crippled in the same way. Aristotle sees that if this were true then 'children [would be] born which resemble their parents in respect not only of congenital characteristics but also of acquired ones'. (In his translation of *The Generation of Animals*, 1942, Peck says: 'It will be seen that this translation, in spite of its sound of modernity, is a close representation of the original.') Aristotle even speaks of a man from Chalcedon who was branded on his arm and whose child had a faint version of the same mark. The middle-aged Darwin proposed his version of pangenesis precisely because he thought that acquired characteristics might matter in evolution after all. Aristotle, however, will have none of it: 'In fact the children of the disabled are not necessarily disabled just as children do not necessarily resemble their parents.' Pangenesis also implies that if you were to prune some part of a plant then its offspring should grow up ready pruned, but they don't.† Mutilations should be inherited, but they aren't: the relationship between the parents' bodies and the genetic content of their seeds must be much less direct.

---

* The Hippocratic theory is so similar to Darwin's that modern scholars use his label with no sense of anachronism even though it was more sophisticated. In *The Variation of Animals and Plants under Domestication*, Darwin says that William Ogle had told him that Aristotle had known, and rejected, a very similar theory to his own. Since Darwin never read much Aristotle, and certainly not *The Generation of Animals*, there's no doubt he hit upon it independently.

† This is his version of the 'Jewish foreskin' argument: if acquired characteristics are inherited, then why, after millennia of circumcision, are Jews still born with foreskins?

LXXIII

ARISTOTLE'S OWN MODEL of inheritance is a triumph of speculative biology. It is probably one of his most mature theories. It contains his clearest and most detailed account of the mother's role in reproduction. Here, as nowhere else, the two parents become *almost* equal in their ability to shape the embryo. He no longer talks of active forms and passive material, but rather of competing forces in the womb.

In his standard account of embryogenesis, an animal's form is transmitted to the embryo by movements in the semen. They're his units of inheritance, his information-carrying vehicles. He's rather fuzzy about how the menses transmit information, but he must think that they do so as well – after all, they give the embryo at least a vegetable sort of life. But now, seeking to explain the phenomena of inheritance, Aristotle expands his vision. He argues that the semen and menses also seethe with movements that encode the individual features of both parents. The result is a dual-inheritance system: a set of paternal movements that encode the form – that make an embryo grow into a sparrow rather than a crane (or a human rather than a horse); and a set of movements provided by both parents that encode their informal features – that make it grow into an adult that resembles one parent rather than another. The relative power of these informal movements determines whom, if anyone, their child resembles. The idea of embryonic conflict probably came from Democritus or 'Hippocrates', but Aristotle's model is subtler than either of those for he argues that the conflict is asymmetric. The menses' movements exist only *potentially*. They're there, but deactivated, and do their stuff only when the semen's fall down on the job. Even now he can't *quite* give mothers an equal hand.

To show what his model can do, Aristotle begins with the obvious fact that children come in two sexes. Sex determination was just the sort of problem to attract the speculative talents of the *physiologoi*, and Aristotle tackles their theories with glee. Anaxagoras argued that semen from the right testicle produces boys, semen from the left, girls. That gave fathers all the credit. One Leophanes, taking this argument to its logical conclusion, proposed gender selection by tying off a testicle before sex. Aristotle thinks this is nonsense, albeit because he doesn't believe that testicles produce semen at all. Empedocles' theory was characteristically complex. As Aristotle tells it, Empedocles had microscopic male and female parts, derived from each parent, fissioning and fusing in the womb. But it's hard to work out what's going here either because Aristotle dislikes this theory so ('Empedocles was rather slipshod in his assumption,' 'the whole cast of this cause seems to be the product of the imagination', 'besides it is fantastic to imagine . . .') and rather mangles it in the telling, or because it never made sense in the first place (Empedocles *would* write in verse). Aristotle does, however, pick out one weakness. It seems that Empedocles held that foetal sex somehow depends on the womb's temperature, and Aristotle has a piece of evidence that decisively falsifies this claim. If you dissect live-bearing animals, he says, you often find male and female twins in the *same* uterus, so uterine temperature *cannot* determine sex. But in truth he doesn't just say that; he positively crows.

In Aristotle's own theory movements in the semen encode maleness, movements in the menses, femaleness. Since female movements appear only when the male's are weak, every little girl represents a failure in her father's semen. Aristotle tries to integrate his model with his embryo-genetic theory. Semen is hot, the menses are cold, and if the embryo is to be concocted properly, the two seeds must be present in just the right amount. The relative power of their movements are, then, somehow influenced by their relative heat. There's a sleight of hand as he shifts between heat and movements, but the gist of the model is clear. When the semen 'conquers' the menses, the result is a boy; should it fail to do so for some reason then movements latent in the menses flourish and the result is a girl. So daughters are sired by feeble, or at least cold, fathers. This suggests a route to environmental sex determination and Aristotle claims that diet, paternal age, ambient temperature and the direction of the wind can all influence the heat of the semen and hence the sex of the child. Following Aristotle's precepts if you want to father a daughter first take a long, cold shower, try your best and reflect that it beats ligating a testicle.

The conflict in the embryo is, however, just the beginning, for Aristotle distinguishes between the initial specification of sex and its consequences. He argues that the embryonic conflict directly determines just one small part of the embryo that influences the rest of the body to give all the other sexual characteristics. This is *automaton*-causality at work again. The distinction is very much like the modern one between 'primary' and 'secondary' sex determination. In 1944 Alfred Jost castrated foetal rabbits, found that they always turned out female, and so demonstrated that the critical organ of secondary sex determination is the gonad (it produces the hormones that determine the other secondary sexual characteristics: external genitalia, breasts, beards, etc.). Aristotle points out that castrated animals and eunuchs are feminized and infers from this that 'some of the parts are principles [of sex determination], and when a principle is moved or affected many of the parts that are associated with it must change with it'. The conclusion should have been *almost* as obvious to Aristotle as it was to Jost: that the testes are critical to sex determination. But Aristotle is sceptical about testes and in love with the heart, so he argues that it is the embryo's heart from which all other differences flow.

LXXIV

HAVING ESTABLISHED THAT a child's sex is the outcome of a conflict between informal movements in the semen and menses, he argues that other informal, inherited variation is encoded in the same way. So he talks about noses – in particular, Socrates' nose that, famously, was everything the ideal Greek nose was not. The ideal can be seen on the Artemision *Poseidon* or any statue of the time: high-bridged, straight and rather large. Socrates' was small and snub. (In his *Symposium*, Xenophon has Socrates defend his snub nose, bulging eyes, wide mouth and flabby lips – they're beautiful because they work better than yours, Socrates says.) Thus the foam visible in Socrates' semen represents countless, minute movements that encode, very precisely, his characteristics, among them his snub nose.

Socrates was married to Xanthippe, and her menses have movements that encode her informal particulars too. But, as with her gender, they're only potentially there – that is, they're not necessarily expressed. Xanthippe was a notorious shrew and so we'll give her a hooked nose. If Socrates' semen totally 'conquers' Xanthippe's menses, then their son, Menexenos, will be a clone of his father, snub nose and all. But if Socrates' semen fails to conquer, then Xanthippe's latent movements will be expressed and he'll have a hooked-nosed daughter.

One of the oddities of Aristotle's theory of inheritance, then, is that he thinks that most features – he is certainly thinking of facial features here

– are sex-associated. Boys generally take after their fathers, girls after their mothers. I don't know why he thinks this. After all, modern parents happily parcel out their children's features between themselves (with a nod to grandparents) regardless of sex.* Actually, Aristotle recognizes that the association can break down. Should Socrates' seminal movements mostly triumph, but his nose-movement fail, then Menexenos will have his mother's hooked nose. In that case, Aristotle says, the nose 'passes to the opposite' bloodline.

GREEK NOSES. *LEFT*: HEROES'. *RIGHT*: SOCRATES'

The differences between the Aristotelian and Hippocratic theories of inheritance run very deep. Aristotle assumes that inherited traits have a discrete distribution: Menexenos may have Socrates' or Xanthippe's nose, but he can't have a nose that's something in between. 'Hippocrates' assumes a continuous distribution: depending on the precise proportions of parental seed, Menexenos may have either of his parents' noses or anything in between. This implies that Aristotle's hereditary movements are stable: they can be passed down more or less unaltered for many generations; Hippocratic mixtures of hereditary fluids are not: each generation produces a novel mix.

---

* I considered the possibility that people are inherently biased to see a resemblance between sons and fathers and daughters and mothers. So, in a small experiment, I asked thirty-five parents to rate their fifty-five children for ancestral resemblance (father, mother, paternal grandfather, etc.) in a variety of features (nose, eye shape, hair colour, etc.). From this I constructed, for each child, a 'paternal' and 'maternal' similarity score. It was not a very powerful test; even so the scores for boys and girls were indistinguishable, so the bias, if present, must be small. People seem to think that a child's features can come from either parent regardless of its sex. Of course, some of my subjects will have had a dim recollection of high-school Mendelian genetics, and so it's possible that perception was different in ancient Greece, but I doubt it.

This distinction is familiar and fundamental since it divides early modern theories of inheritance too. Aristotle assumes 'particulate' inheritance, 'Hippocrates' assumes 'blending' inheritance.* By particulate I do not mean that Aristotle supposes that actual particles are transmitted – that would be too Democritean – but only that the movements are stable and discrete. This has important consequences.

Aristotle sees that a good theory of inheritance must do more than explain why children look like their parents; it must also explain why children sometimes take after their grandparents or even more remote ancestors. Aristotle is rightly convinced that such reversions are common, but the example that he gives seems improbable. There was, he says, a woman who lived in Elis, a district of the Peloponnese, who had an adulterous affair with an Ethiopian. She bore by him a daughter who was white; this daughter, when she grew up, had a son who was black – even though the child's father was presumably a Greek.† In Book I of *The Generation of Animals*, Aristotle says that his opponent's theory can't explain this and he's right. And, although he doesn't return to the case when setting out his own theory, it's exactly the sort of phenomenon that it can explain. At least it can do so if he adds another layer of complexity to it.

Aristotle argues that sometimes the semen's heat and movements are not quite powerful enough to reproduce the father's features, yet not so weak as to default to the mother's. In that case Menexenos will have his grandfather's nose and will, in turn, transmit his grandfather's nose to *his* sons. Menexenos may even have some more remote paternal ancestor's nose, but this is less likely. Since such failures bring about a permanent, heritable change in the paternal movement we can call them, without anachronism, mutations. Aristotle seems to think that many, perhaps most,

---

* The term 'blending inheritance' is usually associated with the theory proposed, in a hostile 1867 review of Darwin's *Origin*, by the Scottish engineer Fleeming Jenkin. But it was Francis Galton who drew a clear distinction between particulate and blending inheritance. Of course, a continuous trait distribution does not *necessarily* imply particulate rather than blending inheritance for, as R. A. Fisher famously showed in 1918, a continuous distribution *is* compatible with particulate inheritance if we assume that many particles contribute to the phenotype. That, indeed, was the basis of the reconciliation of the Biometricians and the Mendelians and the explanation for why many traits (skin colour, height) can be continuous and yet be controlled by particulate genes. However, the Hippocratic author (i) hasn't read Fisher and (ii) clearly talks about fluids rather than particles, so his theory must be a blending one.
† The data are improbable, but not implausible, for the inheritance of human skin pigmentation is complex. In general, however, one would expect the daughter to be coffee coloured and the grandson to be fairer than that.

mutations cause ancestral reversions. His word for a permanent inherited change in movement is *lysis*, or 'relapse'.

But knocking out Socrates' nose movement doesn't, by itself, explain how Menexenos can have his grandfather's nose, for Aristotle must also explain where the information that specifies his grandfather's nose is located. He therefore argues that the movements of Socrates' semen encode not just his snub nose but also his father's nose, his grandfather's nose, his great-grandfather's nose and so on for . . . how long? Aristotle doesn't say. The movements of Xanthippe's menses likewise encode the noses of *her* female ancestors. But none of these ancestral movements are expressed; they are mere potentials awaiting reactivation in the event of a failure of the active, parental movement. Generations of noses encoded in our bodily fluids – it's a dizzying thought.

Were Menexenos to wind up with his grandfather's nose, or even his mother's hook, that wouldn't be too bad, for Aristotle imagines that some mutations have much more drastic effects. People talk, he says, of a monstrous child who has the head of a ram or an ox, or a calf with a child's head and suppose that they are human–animal hybrids. But they are not hybrids: it's just that the movements in their parents' semen and menses have failed to do their work. His examples – a child with an ox's head (or the reverse) – suggest that he's not only deflating popular belief, but also taking another swipe at Empedoclean preformationism. He wants to make sure that no smart-aleck student sticks up his hand and says: 'I have a friend, who knows a woman, who had a cousin, who gave birth to a child with a calf's head. Doesn't that prove that Empedocles was right?'

It doesn't. Aristotle can explain all sorts of monstrosities by appealing to the movements in the semen and menses. If Socrates' nose movement is very weak then Menexenos could just have a general human nose.* And in the event of total failure he'll have a monstrous nose – and by 'monstrous' Aristotle here means animal-like. Strip away all the human nose-specific movements embedded in the semen and all that's left are the movements that make an animal. This view of mutational effects arises naturally from his von Baerian views of embryonic development. If embryos first develop features common to all living things (the nutritive soul) or all animals (the sensitive soul), and only later develop the characteristics of particular species, then it is easy to see how a failure of the semen to concoct the menses could

---

* Devin Henry suggests to me that Aristotle may have in mind here the features of children with Down Syndrome – trisomy 21 – who, though clearly human, don't particularly resemble any ancestor.

cause development to halt part-way and so deprive a human foetus of its human features. It would be, in his vocabulary, very 'imperfect'.

Any theory of inheritance that seeks to explain reversion (or atavisms, throwbacks, ancestral resemblance, skipped generations – they're different labels for similar phenomena)* must assume that the units of inheritance are stable – that is, are particles in the broadest sense of the word – and that these particles can be silenced for generations and then be reactivated. These two ideas themselves recur. Aristotle explained reversion by allowing his movements to be actual or potential; Pierre-Louis Moreau de Maupertuis, the eighteenth-century *philosophe* who reported the first pedigree of an inherited trait, allowed his hereditary *éléments* to have more or less 'tenacious arrangements'; Darwin, who devoted a chapter of his *Domestication* to atavisms, made his version of pangenesis run on gemmules that could be dormant; Mendel made his *elementen* dominant or recessive. Doubtless there are others.

In contrast to modern – post-seventeenth-century – taxonomy, functionalism and embryology, all of which are directly built upon Aristotelian foundations, there is no reason to think that Aristotle's insight into the logic of inheritance has echoed down the ages. It is much more likely that nature has simply pointed, as she so often does, those who have interrogated her in the same direction. (As Aristotle, in another context, said of Democritus: 'he was merely brought to it [the theory of substantial definition], in spite of himself, by the constraint of the facts'.) Of course, every theory gives a different account of how the units of inheritance combine and transmit, and only one of them was right. Aristotle's was wrong too; but survey the admittedly dispiriting history of early genetics and one can conclude that until 1865, when Mendel presented his *Versuche über Pflanzenhybriden* to a sublimely indifferent world, there was no better theory than his.

---

* In modern genetics 'skipping generations' due to segregating recessive alleles is distinguished from rare 'atavisms' due to mutation. But the distinction is not to be imposed on Aristotle – nor on Darwin for that matter.

RECIPE FOR AN OYSTER

LIMNOSTREON — OYSTER — *Ostrea* SP.

LXXV

THE HARD MEN of Kalloni are its divers. Scorning SCUBA, they go down on the end of hoses attached to diesel-fuelled compressors to harvest oysters, scallops and mussels by the ton. Most of the divers are young, but I met one who could have been sixty. He was as slim as a cormorant and seemed to be built of olive wood. I asked him how many underwater hours he had. Five hundred last year. As a lifetime total, that impressed me since I've only done 150. Yes, he continued, *last year* I did 500, and 500 the year before and each year before that since I was a young man. We mostly dive in winter.

There are no numbers on the size or state of the shellfishery, but the bathymetry hints at former riches. South-west out of Skala, on a bearing for the Lagoon's mouth, the flat bottom starts to be interrupted by circular hillocks that rise and fall on the sonar screen. These are the oyster reefs that the fishermen call *kapalíes*. They say that there are a few thousand of them, or at least that's how many there were. Since the 1950s dredging has flattened many of them. That's now illegal, though fishermen will mutter that some people – always other people – still dredge while the Port Police look the other way, although whether out of indolence or venality is unclear. What is certain is that the oysters and scallops are in decline and that vast beds of mussels are spreading in their wake.

The Lagoon's shellfish are Aristotle's *ostrakoderma* – hard-shells. He describes their anatomies and something of their habits, mentioning, among others, the *límnostreon* (oyster), the *kteís* (scallop), the *pínna* (fan mussel), the *lepas* (limpet) and the beautiful *kēryx*, my trumpet shell.* He dwells on the *porphyra*, a snail that was once fished for a purple substance that it secretes in its hypobranchial gland. He says that there are many different kinds of *porphyraí*, that those in the features of the Lagoon are small, and that snails from different places produce different-quality dyes.

---

* Archestratus, who's interested only in the edibility of marine creatures, says that Aenus on the Gulf of Saros produces large mussels, that Abydus, Parium and Ephesus, all on the Troad, respectively produce oysters and small cockles and smooth clams and that Mytilene produces scallops.

He's probably distinguishing the banded murex, *Hexaplex trunculus*, the source of an indigo dye, and the spiny murex, *Haustellum brandaris*, the source of the true Tyrrhean purple, and local varieties of each. The most common species in Kalloni is *H. trunculus* which infests the bottom of the Lagoon and crowds into any baited trap for, as Aristotle says, although it feeds on bivalves it is also a scavenger. These snails are now a nuisance, but once they were the basis of a major industry. Mounds of broken shells, the earliest Minoan, the most recent Byzantine, can be found scattered across the Aegean. In Aristotle's day the dye was worth its weight in silver.

Describing the anatomy of the oyster, Aristotle refers to its 'so-called' eggs. He plainly means its eggs, more precisely its gonads, which appear in the summer months as a milky sack. However, he denies the oyster, or any other shellfish, its gonads. He even denies them to the sea urchin and suggests that the *ricci di mare* is just where it stores its fat, even though the individual ova can just be seen with the naked eye. But if eggs aren't eggs, how do oysters reproduce? Aristotle, rather surprisingly, says that they don't. Instead, he argues that they generate spontaneously from the stuff in which they grow.

PORPHYRA — PURPLE MUREX — *HEXAPLEX TRUNCULUS*

LXXVI

WHEN ARISTOTLE SAYS that some animals generate spontaneously he means it:

Some animals are generated from animals whose shape is appropriate to their kinship. Others are generated spontaneously and not from kin. Some of these are generated from rotting earth and vegetation, as occurs with many insects, while others are generated inside animals themselves out of the residues from various parts.

Cockles, clams, razorfish and scallops generate spontaneously on sandy bottoms; oysters grow in slime; the *pinna* in sand *and* slime; ascidians, limpets, *nereites* (a snail, probably *Monodonta*), sea anemones and sponges on rocks; hermit crabs come from soil. In Cnidos there's a kind of mullet that springs from sand or mud, as do some kind of small fry. Fish lice are generated from the slime of fish. Worms (*helminthes*) generate spontaneously in our guts. Insects and suchlike seemingly generate spontaneously everywhere: fleas are produced in putrefying matter; lice in the flesh of animals; ticks in couch grass; cockchafers and flies in dung;* horseflies from timber; pseudoscorpions in books; clothes moths in clothes. Other insects come from the morning dew on leaves. Fig wasps are generated spontaneously in figs. Every conceivable habitat, it seems, produces its own form of life. When Aristotle begins to talk about spontaneous generation, it becomes apparent that he thinks that the inanimate world is endlessly fecund.

He does not ask his readers to accept this vision at face value, but provides evidence for it. A naval squadron, he says, once anchored off Rhodos and a lot of earthenware was thrown overboard. The pots collected mud and then living oysters. Since oysters can't move on to pots, or indeed anywhere, they must have arisen from the mud. He has another oyster anecdote. Some Chians once transported a lot of oysters from the Lagoon at Pyrrha to their island south of Lesbos, and deposited

* The scarab beetle, however, is said to lay eggs or larvae in dung.

them in a strait 'where the currents meet'. The oysters grew in size, but did not multiply.

The argument has a lovely symmetry: first he shows that oysters can appear without reproduction; then he shows that they don't reproduce; and then he explains away the structures that less scientific thinkers believe are their reproductive organs.

But, as Thomas Kuhn was so fond of reminding us, empirical evidence, no matter how good, is never decisive when big scientific issues are at stake – you also need a new theory. Aristotle apparently feels this tension too. (In *The Heavens* he remarks, 'It is, however, wrong to remove the foundations of a science unless you can replace them with something more convincing.') Here he addresses the problem by giving a recipe for making an oyster. Place water (seawater is especially good) and earthy material in a cavity of some sort, mix, heat with *pneuma* (abundant in seawater) or else place in the sun. The mixture will concoct and then foam. Some putrefied residue will form – a by-product of the concoction. After a while, earthy material will begin to congeal and form the shell; the living material will be inside. Life: it's as simple as that.

Most of Aristotle's spontaneous generators are bloodless animals, invertebrates. There's one spectacular creature, however, whose reproductive organs he does not have to argue away for it does not have any. 'The eel', says Aristotle in Book IV of *Historia animalium*, 'is neither male nor female and produces no offspring.' In Book VI he adds: 'eels do not come into being by copulation nor do they lay eggs'. These two sentences contain four factual claims, all of them wrong. *Contra* Aristotle, eels *are* either male or female, *do* pair, *do* lay eggs and *do* produce offspring. Nor does this exhaust the list of Aristotle's errors about the eel; in fact, he got little right about them. For all that, when zoologists consider Aristotle on eels, they have tended to the charitable. That is because he is the *Ur*-Hero of one of their greatest quests. He showed them that the eel was a problem.

The problem is that eels don't have gonads. Cut an eel open, says Aristotle, and you never find milt or eggs. In this he's perfectly correct – at least in Greek waters. Where the gonads of fish are generally filled with semen or eggs, the eel's are generally not. Aristotle's response is to add the eel to his list of spontaneous generators. He has, of course, to consider rival theories. One turned on the shape of the eel's head. Some eels have broad heads that give them a vaguely frog-like aspect; others have delicately narrow snouts, and some people evidently thought that the dimorphism was sexual. Aristotle disposes of this briskly: 'One difference between male and female eels, so people claim, is that the male's head is larger and longer, the female's smaller and more snub-nosed. But what they are talking about is a difference not of gender but of kind [genos].'* And then there were those who claimed that eels are viviparous. For these theorists he has only scorn: 'People say that eels can occasionally be seen with what looks like hair, worms or sea-weed attached to them. These are ill-considered claims resulting from observational failure.'

His own solution is an account of the eel's ontogeny. They are, he says, generated from the gēs entera, the 'guts of the earth', a sort of worm that grows near the edges of rivers and marshes where there is plenty of sun-baked putrefying matter. The gēs entera are the mothers or hosts of the infant eels – Aristotle is not very clear; at any rate if you open one you will sometimes find small eels inside. It's an ingenious and wholly specious theory. He's probably talking about burrowing arenicolid worms; their casts litter beaches and mudflats and do, indeed, look like small heaps of coiled intestines.† I suppose that the gēs entera are an attempt to make sense of eels. Many of the other marine spontaneous generators (clams, oysters, snails, sponges, etc.) are, in his view, very simple animals; he often compares them to plants. An eel is nothing like a plant: it's a large, fiercely active, blooded predator, and there are lots of them. Aristotle was bold; but even he baulked at manufacturing a major fishery of metre-long animals annually out of mud, and so he gave them a larva.

* Many scientists have measured the heads of many eels, but eel craniometry is an incon-clusive science. Some investigators have agreed with Aristotle that head shape is the mark of different species, or at least races, of eels; others have sided with his opponents and attributed the difference to sex; yet others suggest that the difference lies in a purely plas-tic response to diet.
† Platt identifies the gēs entera as earthworms; Peck as parasitic nematomorph worms of the genus Gordius; neither explains why. D'Arcy Thompson's suggestion that the gēs entera are related to casentula, the name that Sicilian fishermen give to the leptocephalus larvae, seems unlikely since the leptocephalus is rare in inshore waters and does not live in mud.

LXXVII

ARISTOTLE'S THEORY OF spontaneous generation had a baleful effect on early modern science. Descartes, Liceti, even Harvey, were all in its thrall. Van Helmont, no fool, reported spontaneous generation of mice from a mixture of rags and wheat. The theory's fall was, oddly, precipitated by a passage from Homer. There has been a battle and corpses litter the ground. Achilles weeps by the remains of his friend Patroclus, and implores his mother, the nymph Thetis of the silver feet:

> But I am very much afraid that the flies might in the meantime alight on the open wounds of Patroclus and breed worms in them. Then his corpse will be defiled, for there is no life left in him, and his flesh will rot.

Francesco Redi, who had learnt his Aristotle at Pisa, read *Iliad* XIX and wondered whether Homer might have been right after all. The experiments that he carried out in the laboratories of Ferdinand II, Archduke of Tuscany, were simple and convincing. He placed into separate flasks, dead snakes, some river fish, eels from the Arno and slices of veal. Some flasks he sealed with paper or else a kind of fine muslin called *velo di Napoli*; others, the controls, he left open. The open flasks generated swarms of flies; the sealed ones did not. He followed the fly through the course of its life cycle and published his results, in 1668, in a work titled *Esperienze intorno alla generazione degli insetti* – 'Experiments on the Generation of Insects'.

Leeuwenhoek took care of the oyster. In 1695 he bought a bushel of the animals at Zierikzee in the Zeeland deltas. Wielding his microscope, he cracked open their shells, peered into their mantle cavities and described the oyster's sperm and eggs. He also found thousands of veliger larvae complete with embryonic shells. He does not name Aristotle, but reserves his ire for contemporaries: 'I place these observations before the world thus stopping the mouths of those pig-headed fellows who still pretend that Shellfishes are generated spontaneously from mud.' Leeuwenhoek also saw the deeper problem with spontaneous generation. In an unconscious echo of the *Iliad* he wrote: 'For if from exhalations there come forth

Animals, why is it that after a Battle in which fifty thousand men and more are defeated, who are left putrefying in the Field, there do not come forth a great many young children or Adults, or something resembling a Human Being or a Horse, for while in a Battle many Men are defeated, many Horses are also killed.' Once you allow that some animals can arise spontaneously, why not all?

More than a century later, biologists sieving Europe's coastal waters with nets of silk found the larvae of Aristotle's remaining *ostrakoderma* in the plankton. In 1826 John Vaughan Thompson identified the cyprid larva of a barnacle in Cork Harbour; in 1846 Johannes Müller fished the strange *Pluteus paradoxus* from the German Bight and watched it transform into a sea urchin; in 1866 Anton Kowalevsky discovered the tadpole larva of a sea squirt in the Bay of Naples. The larvae themselves are exquisite. At 100 diameters magnification they look like machines made of Venetian glass. Their discovery changed the order of nature. Where Aristotle thought that the sea squirt was the lowliest of animals, Kowalevsky's discovery that its larva had gill slits, a dorsal nerve cord and a notochord showed that it was chordate. Far from arising from rock slime, the sea squirt was our close relation.

By the early nineteenth century most animals had been removed from the roll call of spontaneous generators. Not all, for parasitic worms, with their opaquely complex life cycles, were still on it. Microbes remained suspect until Pasteur's experiments of 1859. In Northern Europe, popular belief in the eel's spontaneous origins persisted until the end of the seventeenth century.

In Greece it still does. We were long-lining for eels at the Vouváris' mouth when Dimitris, who knew that I was interested in that sort of thing, mentioned that scientists do not know where the eel comes from, but that it grows out of mud. This seemed to be based on local lore and personal observation rather than reading Aristotle. Of course, scientists do know where the eel comes from. And it was Redi who described what Aristotle only hints, that the eel is a migratory fish. Adults live in rivers and lakes, sometimes for many years, migrate down to the sea, embark on an obscure voyage, reproduce and die. Their progeny then return as glass eels which, between January and April, invade European estuaries by the millions en route to the rivers that their parents came from.

And yet the eel's gonads remained missing. Some thought, *contra* Aristotle, that the eel was viviparous. Leeuwenhoek claimed that he'd found an eel's womb full of young eels ready to be born. In fact he'd found an eel's bladder full of parasitic nematodes. In 1777 Carlo Mondini, professor at Bologna, finally identified the ovary of the eel. It proved to be a frilled ribbon of tissue that runs the length of the animal and that had previously been mistaken for fat. The testes were more elusive; they were found by Simon Syrski at Trieste only in 1874. It is easy to see why Aristotle missed them: the gonads of both sexes are more or less empty until the eel is well at sea. (Even now, few gravid animals have been caught; one was taken from the stomach of a mid-Atlantic sperm whale.) Aristotle did not know that; but then neither did Sigmund Freud who, as a twenty-year-old research student, dissected 400 eels in a search for their sperm, failed to find them and then turned his mind to more tractable problems. Some years later Grassi and Calandruccio working at Messina showed that the weird pelagic leptocephalus was the eel's true larva. It was only in 1922 that Johannes Schmidt sailing in the *Dana* finally located where the eel mates, dies and comes to be: 22° 30' N, 48° 65' W, the Sargasso Sea.

LXXVIII

ARISTOTLE'S LOVE OF spontaneous generation seems perverse. We can hardly censure him for not knowing where eels spawn, or for not having seen the oyster's larva, but why does he think that flies are spontaneously generated? After all, he knows that flies copulate and engender maggots and *also* knows that maggots develop into flies. The conclusion is – or should have been – obvious. Even Homer understood it. Yet he resists it, and so fails to give the fly the life cycle it manifestly has.

The inconsistencies are not just empirical. Spontaneous generation also runs against the grain of some of his deepest theory. For Aristotle, order does not, *cannot*, depend on the properties of matter alone but also requires a formal cause. A sexually reproducing animal gets its form from its paternal parent; form is the information that shapes the dynamic organization of the soul. But spontaneous generators, by definition, lack parents. So how do they come to be? Does a snail not have a soul?

Aristotle's recipe for a spontaneous generator is evidently an attempt to solve, or at least paper over, some of these problems. It's obviously based on his model of sexual reproduction. There's a substrate (a material cause) analogous to the mother's menses; there's a source of movement (an efficient cause), a source of soul-heat analogous to the *pneuma* in semen; there's concoction, foam, and the emergence of order and life. It *is* an explanation, but a very thin one. In the absence of a father, what ensures that a particular form of animal will be produced, an oyster, say, rather than a clam? Why are there so many different kinds of spontaneous generators?

Aristotle's answer isn't very clear, but specificity somehow depends on the exact mix of ingredients. That's why he is at such pains to tells us the exact kind of habitat in which each of his spontaneous generators can be found: the larva of the dung fly, *myîa*, comes from dung, but the larva of the horse fly, *myōps*, comes from timber. It also has to do with the shape of the cavity in which the cooking takes place. Together, these variables determine how 'honourable' – he means roughly how 'complex' – a creature will emerge from a given reaction. But, since the raw materials are collectively ubiquitous, it seems likely that life will emerge anywhere; indeed, as he blandly assures us, 'in a way, all things are full of soul'.

It's a wonder that he finds this account convincing. It is scarcely different from the materialist theories that he so dislikes and it shares all their flaws. In *Physics* II, 8 Aristotle insists that spontaneous events do not 'normally come about in a given way' – they're unusual, even rare. Yet oysters, clams, flies and fleas are among the most abundant animals that he knows; how, then, can they be the product of spontaneous events? Aristotle also insists that spontaneous events do not have goals, but only appear to. Yet, by his own account, spontaneous generators have – reproductive parts aside – the same sort of organs that sexual animals do. An eel may not participate in eternity, yet it is otherwise as much a teleological construct as a sardine: they both have mouths, stomachs, gills and fins that they use in exactly the same way. Having brought forms down from their Platonic realm, and having made them the centrepiece of his theory of inheritance and ontogeny, he then apparently discards them. And he does so because he cannot work out where eels keep their gonads and how oysters mate.

And so puzzle remains. Aristotle believes in spontaneous generation even though the animals that he knows best all have parents. He believes it even when his own data on particular animals – those pesky flies – point the other way. He believes it even though to make it work he has to distort his own – brilliant – theory of development. He believes it even though it contradicts his metaphysics and gives the game, the hard-fought game, to his materialist opponents. He believes it even though there's a simple alternative explanation ready to hand. So why does he believe it?

The beliefs of any scientist depend at once on the theories he inherits from his predecessors, the theories he formulates for himself and the evidence of his own eyes. Aristotle does not tell us where he got the theory of spontaneous generation from, but it was certainly a commonplace in his day. Theophrastus says that many of the *physiologoi*, Anaxagoras and Diogenes among them, thought it was true. It was likely wrapped up with origin-of-life theories.

A passage in the *Problems*, a pseudo-Aristotelian text, makes the connection explicit. The author, probably one of Aristotle's students, is wondering why some animals generate spontaneously while others need sex to do so. He starts by asserting that *all* animal kinds ultimately originated from the 'compounding of certain elements'. But, he continues, as the *physiologoi* have explained, a full-scale zoogony requires 'powerful changes and movements'. We are, evidently, to imagine large-scale chemical turmoil of the sort present when the cosmos was young. (Prebiotic-soup scenarios of belching volcanoes and flashing lightning bolts spring irresistibly to mind.) These days, however, things are calmer, so the only animals spontaneously generated are small; big ones have to reproduce sexually.

The average Greek certainly didn't need elaborate theories of that sort. Popular belief held that the cicada was spontaneously generated from soil; as a mark of their autochthony Athenian girls wore golden cicadas in their hair. Then, too, bread, meat, wine, wood, cloth – almost any organic substrate at all – will, if left unattended for long enough, spring to life and produce seething swarms of animals. Even a tub of water will develop an ecology. What could be more natural than to suppose that the creatures originated there? Even cautious Theophrastus admits that some plants really do spontaneously generate.

It can't be that Aristotle's love of spontaneous generation is *just* relict popular or Pre-Socratic belief. He's usually so quick to correct his predecessors and so scornful when he does. Still, intellectual inertia can't be completely discounted either. Perhaps what happened was this. Aristotle, beginning, as he always does, with popular belief and expert opinion (but ignoring Homer) holds that some animal kinds are spontaneous generators. It is, for them, his null hypothesis. He then begins to investigate them, accumulating evidence that will tell either for or against the idea. He takes an empirical stance, refusing to believe that an animal has a complete life cycle unless he's seen the whole thing. For example, he says that some writers assert that all grey mullets are spontaneous generators, but that this is empirically wrong – only one is. He does not consider that spontaneous generators are the product of invisible seeds since he's generally sceptical about the existence of microscopic objects such as atoms. He develops an explanation for spontaneous generators that is as consistent as possible with his model of sexual generation and simply elides its difficulties. He retains the Pre-Socratic epithet 'spontaneous' for generation of this sort even though his own definition of spontaneous events, given in the *Physics*, is much more restricted. Here, as so often, he uses a single term in several quite different senses and forgets to tell us which usage he is employing.

It's not a very satisfying solution, but Aristotle does not help us find a better one. He rarely expresses doubts or speaks of struggles, but almost always projects the confidence of a man who has a grip on the phenomena and a good explanation for them. Sometimes, however, we do glimpse an Aristotle in two, or even three, minds. He seems to be unable to decide whether the *porphyrai* – the muricid snails of the Lagoon's muddy bottom – generate spontaneously or not. In spring, he says, the *porphyrai* gather together and secrete a 'honeycomb' upon which the baby snails can be seen crawling. He's obviously talking about their egg cases but, as with the oyster's gonads, he doesn't see that. Instead, in *Historia animalium*, he suggests that the baby snails generate spontaneously from the mud underneath the 'honeycomb'. In *The Generation of Animals*, giving a slightly different account, he suggests that the honeycomb is a seed-like residue which gives rise to baby snails, rather as a plant gives rise to buds. And then, in another passage in *The Generation of Animals*, he moots the possibility that they might be sexual after all: 'The only animals of this kind [the *ostrakoderma*] in which copulation has been seen are the snails; but there has been no visual confirmation of whether copulation results in reproduction' – more research is needed.

Aristotle's empiricism is also apparent when he treats insects. He thinks that most animals are spontaneous generators; and it may be supposed that he does not know about complex life cycles at all, but this is not so for, in a beautiful passage, he relates the cicada's:

> The large and the small cicada copulate in the same way, belly to belly. The male inserts himself into the female, not the female into the male as with other insects, and the female cicada has a cleft pudendum into which the male inserts himself. They lay their eggs in uncultivated land, and bore a hole with the sharp point they carry at the rear, just like locusts . . . The cicadas also lay their eggs in the stakes people use to prop up their vines: they bore a hole in the stakes; and also in the stalks of the squill. Their spawn seeps into the ground becoming numerous in wet

weather. The larva, when it has grown large in the ground, becomes a cicada-mother [mature nymph]...As the solstice approaches, they emerge at night: the covering immediately splits and they become cicadas instead of cicada-mothers. At once they become black, harder and larger and start to sing. In both kinds the males sing but the females do not.

And he adds that, if you flex the tip of your finger near one, it will crawl on to your hand.

TETTIX – CICADA – CICADIDAE

FIGS,
HONEY, FISH

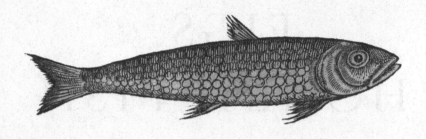

MEMBRAS — PILCHARD — *Sardina pilchardus?*

LXXIX

ONCE, AT ERRESOS, I saw a slain tuna. It had been caught far up the coast and was being dismembered on a taverna table: whetting his knives, arms red to the elbow, the proprietor gestured vaguely towards Troy. Aristotle speaks of the fish at length, but evidently did not dissect one, for he says nothing about its anatomy – its hot blood, its heart as big as a child's, its skeletal struts, precision-milled mandibles and aerofoil fins – the whole astonishing hundred-kilogram, steel-blue, armour-clad, organic killing-machine aspect of the thing. Instead, he speaks of its life.

In the spring, he says, the female *thynnos* 'conceives' or fills with roe. As summer approaches it migrates into Pontos Euxeinos, the Black Sea. Fishermen, spotting the churning shoals from watchtowers, net the glistening bodies at night while the fish sleep. The *thynnos* spawns only in the Black Sea. As it does so, it becomes lean and spent and infected with a parasite that looks like a scorpion and is as big as a spider, clearly a fish louse of some sort. The young fish grow very fast and in autumn depart for the Aegean's depths where they hibernate, wax fat and then return to Pontus again.*

All earthly creatures necessarily die. Seeking a vicarious immortality, they therefore reproduce. As Aristotle puts it: they 'return on themselves' not as individuals, but as forms. Their life cycles are not autonomous, but are ruled by higher cycles – the moon and the sun revolving around the earth. The moon times the menstrual cycles of women; the sun, veering along the ecliptic, gives the seasons to which all creatures adjust their lives. There is, then, an endless amount to say about when and where animals mate, give birth, hibernate and migrate.

Most animals, Aristotle says, pair in spring, but there are many exceptions. Humans pair and give birth at any time of year, but – a further qualification – men are more ardent in winter, women in summer. And the

---

* The Atlantic blue-fin tuna, *Thunnus thynnus*, is now extinct in the Black Sea but migrated and spawned there in historical times, as Aristotle says it does. (But *contra* Aristotle it spawns elsewhere in the Mediterranean too.) A more serious error in his account is that he says that tuna lay 'sack-like' egg cases, but the tuna spawns many small, free-floating eggs. He may be thinking of the pelagic spawn of the frogfish, *Lophius*. He's confused about its spawning behaviour too.

alkyōn, which appears near the setting of the Pleiades (early November), builds an elaborate nest and breeds at the winter solstice (December), when it is often calm and men speak of *halkyonídes hēmerai* – 'halcyon days'. He means the Eurasian kingfisher, a winter migrant in Lesbos, whose blue-green flash can often be seen in the marshes and creeks that surround the Lagoon. It's only a charming coincidence that Linnaeus named the bird for Atthis, the sparkling girl whom, of all her pupils, Sappho loved most.*

THYNNOS – ATLANTIC BLUE-FIN TUNA – *THUNNUS THYUNNUS*
ERISSOS, 2012

An annual procession of fishes spawns in the sea. The first fish to do so, in early spring, are the *atherinai* (sand smelts) which, when spawning, rub their bodies against the sand. Then, Aristotle says, comes the *kestreus* (a grey mullet), the *salpē* (salema) in early summer followed by the *anthias* (swallowtail sea perch?), *chrysophrys* (gilthead), *labrax* (European sea bass) and *mormyros* (striped sea bream). The *triglē* (red mullet) and the *korakinos*, whose identity is unknown, breed towards autumn, as does the *salpē* again. The *maenis* (blotched picarel), *sargos* (white sea bream), *myxinos* and *khelōn* (two more grey mullets) spawn in winter. Some fishes spawn at different times of year in different places.†

---

* But *Alcedo atthis* does not, as Aristotle says, breed in winter, in Greece, in a large nest by the sea; it breeds in spring, in Central Europe, in a burrow in a riverbank. Aristotle does say that there are two kinds of *alkyōn*; one of them may, then, be a tern, but its breeding habits don't match the *alkyōn*'s either. D'Arcy Thompson argues that Aristotle's account of the *alkyōn* is heavily influenced by astrological myth, but Peck disagrees. It's true, however, that one of the Pleiades was known as *alkyōn*.
† It's hard to judge the accuracy of these spawning times. In part that's because, for all the

Shunning extremes, many animals conceal themselves against the blazing Aegean sun or Boreas' winter blasts. At the height of summer all sorts of animals – snakes, lizards, tortoises, a variety of fishes, snails and insects – disappear. At the setting of the Pleiades, bees hide themselves in their hives and fast so that by winter they are almost transparent. The bear, which has mated in the month of Elaphebolion and fattened itself over the summer months, gives birth and goes into hibernation for three months.* Other animals move to more temperate climes. Aristotle records the raucous vernal and autumnal migrations of the cranes between Africa and the Eurasian heartland.

You don't, of course, have to be an Aristotle to notice the progress of the year. None of this, however, is a paean to the seasons in the fashion of an Alcaeus, Simonides or Thoreau. Aristotle's aim is to show how animals adjust their habits according to the seasons to ensure that they can breed, raise their young and get food – the fishes that flood into Pontus all do so because it has richer food and fewer predators than the open sea, and because its sweet spring waters aid the growth of their young. He also wants to explain how each animal kind has a certain comfort zone, how the thermal tolerances of 'weaker' animals are narrower than those of 'stronger' ones – which is why the quail, a weak sort of bird, migrates in front of the stronger crane, and the mackerel in front of the tuna. Above all, he wants to show how living things depend on the structure of the physical world. Their life cycles naturally echo the cycles of celestial rotations: 'Nature's

effort that icthyologists – Rondelet, Cuvier and D'Arcy Thompson among them – have expended on identifying Aristotle's fishes, we still don't know what quite a few of them are. Take the *korakinos*: all we know about it, from Aristotle and other sources, is that it lives over rocks and spawns late in the year. It has been variously identified by Cuvier, Gesner and D'Arcy Thompson with the damselfish (*Chromis chromis*), the shi drum (*Umbrina cirrosa*) and brown meagre (*Sciaena umbra*), all of which, however, spawn in early to mid-summer. Besides the list given here, taken from *Historia animalium* VI, 17, Aristotle also tells us about the spawning seasons of various fish elsewhere in *HA* and they don't always agree. For example, he tells us that the *sargos* spawns in spring and autumn (*HA* 543a7), autumn (*HA* 543b8) and thirty days after Poseidon (*HA* 543b15, *HA* 570a33) – roughly January. In fact, *Diplodus sargus* spawns between January and March (FishBase). Looking at the most securely identified fishes, I estimate he gets it right about half the time. That said, the data in FishBase need not apply to Greek waters.
* As given above, the life history of the *arktos* (Eurasian brown bear, *Ursus arctos arctos*) makes sense. If Elaphebolion is roughly March/April, and it hibernates in December, that would give a gestation period of about nine months, not too far off the 7.5 months recorded in the panTHERIA database. Unfortunately, Aristotle also says – *and says in the same chapter* (*Historia animalium* VI, 30) – that the sow is pregnant for only thirty days, that is, gives birth in May. Many editors have tried many expedients to sort this out.

aim, then, is to measure the generations and endings of things by the meas-
ures of these [celestial] bodies.' But nature, he warns, doesn't always achieve
its aim, for matter can be intransigent.

LXXX

THERE IS, BURIED within the structure of his physical system, a
threat to the very integrity of Aristotle's world. Science, he says, is
the explanation of change; and the world certainly has change
enough to be explained. Storms sweep in from the sea; rains fall, rivers
gush. Landslides obliterate, mountains erode, volcanoes erupt. Living
things – *countless* living things – live. By his own account, none of this can
be taken for granted. That is because the world has an in-built tendency to
stasis. But here I must be more precise for, when I say that 'the world' has
this tendency, I do not mean that the cosmos as a whole does, but only the
part of it that corresponds roughly to what we call Earth. It's the part that
Aristotle, with greater precision, calls the 'world under the moon'.

The root of the problem is elementary. Literally so. The sublunary
world is, in Aristotle's view, composed of four elements. Each element has
a natural home in the cosmos towards which it tends to travel and where,
having arrived, it rests. The natural home of elemental earth is at the centre
of the sublunary sphere; that of water is just above earth; of air just above
water; of fire just above air. We're roughly in the middle of this system,
hence we usually see fire and air move up, water and earth down. These
elementary tendencies are, for Aristotle, as pervasive as gravity is for us.
But where gravity keeps our world together, the elements' movements
threaten to turn Aristotle's world into an onion. Indeed, to a first approxi-
mation, the world does look like an onion – it has a core of earth surrounded
by successive layers of water, air and fire. However, were the elements
perfectly sorted out – were the world at complete equilibrium – then it
would be silent. It would be a world locked in elemental *rigor mortis*. Life
itself would not, could not, exist.

Aristotle sees the problem posed by his theory of elements and proposes
an elaborate solution. Something, he argues, must continually deflect them
from their natural place of rest; something must stir the sublunary pot. To
keep the elements in motion he begins by giving them a cycle. Each element

is a combination of two dichotomous sets of more fundamental properties, opposed 'potentials': hot *v.* cold and dry *v.* wet. Thus elemental earth is cold and dry, water is cold and wet, air is hot and wet and fire is hot and dry. These potentials dictate the possible bi-directional transmutations – fire ↔ earth; earth ↔ water; water ↔ air; air ↔ fire – the cycle is complete.

This elegant scheme cannot, by itself, stop the world from turning into an onion. To cycle through their transmutations the elements must come into contact with each other. So he looks to those celestial mixers, the sun and the moon. As they approach and retreat by day, month and year, they heat and cool the world upon which they shine. The summer sun heats the soil to produce a hot, wet, air-rich vapour that forms clouds that, with the arrival of winter, cool and transform back into the cold, wet, water-rich substance we call rain: 'We must think of this as a river, flowing up and down in a circle, and made partly of air and partly of water.' By a similar cause the winds blow too – 'even the wind has a sort of lifespan'.

Aristotle explains these processes in *Meteorology*. Much of it is about cycles. It's an argument against entropy; a model of how the world can both change and persist; of how it maintains its dynamic equilibrium. The surface of the earth, he says, is always changing, but so slowly that we scarcely notice it. During the Trojan war Argos was marshy and Mycenae productive; now the Argive marshes are cultivated and Mycenae is dry. Egypt is also desiccating which is why the Nile has changed its course. Some people, he says, think that such changes show that the earth has been drying out since it was formed, but that's a narrow point of view. Rather, as one part of the earth dries, another subsides into the sea because its interior parts are continually 'growing and decaying'. Aristotle is trying to penetrate Deep Time, but his evidence comes from Homer.

All this biometeorology raises the question of how Aristotle conceives the sublunary world. Does he think that it is an organism? Are the meteorological cycles *life* cycles? Are they *for the sake* of anything? There is a passage in *Physics* II, 8 – a much discussed one – in which Aristotle seems to say that the winter rain falls *for the sake of* the spring crops, implying that the physical world is set up for the sake of the living things that inhabit it, perhaps even for man who, after all, reaps what he sows. In his *Meteorology*, however, he says nothing of the kind. Its cycles are explained entirely in terms of material and moving causes; final causes appear to be completely absent. The sublunary world does have homeostatic mechanisms that keep it going, but they are much simpler than the cybernetic feedbacks that he invokes when explaining living things. The organic language is metaphorical; the earth does not have a soul.

He is trying to convey his sense that the cosmos, the seasons, the elements, life itself, are all in some way a unity; that they are all linked together in their coming to be and passing away: cycles within cycles within cycles.

LXXXI

A S ARISTOTLE LISTS THE spawning of the fishes by the seasons, so Theophrastus lists the blooms. The first flowers of spring are the stock and the wallflower. Then comes the poet's narcissus, the bunchflower narcissus, the windflower and the tassel hyacinth. These are the flowers that the garland-makers use. The dropwort, the gold-flower, the peacock anemone, the field gladiolus, the alpine squill and all the other mountain flowers come next; the wild rose blooms last, and is the first to end, for its time is brief.

For all the flowers that Theophrastus knows, he does not know what they are for. He sees the stamens and pistils, but does not know that they are a flower's sexual parts, that pollen is the male seed, and that the glory of their scents and colours exists only to seduce their pollinators. The Loves of the Plants (to borrow from Erasmus Darwin) were unknown to him.

The reasons for his ignorance are, at first glance, obvious. Stamens and pistils are rather small; pollen is smaller yet. Then, too, many plants can be grown from cuttings (a true gardener, Theophrastus is quite a bore on this) and there's no sex going on there. Perhaps he was also influenced by Aristotle's definition of a male as 'an animal that reproduces inside another animal', which doesn't even begin to work for plants. When explaining how plants reproduce, Aristotle just asserts that they 'contain both the male and female principle'.*

Except when it comes to figs. In *Historia animalium*, Aristotle tells a curious tale:

---

* He doesn't mean that they are monoecious – he simply hasn't identified the sexual parts of flowers. It is sometimes said that Aristotle thinks that plants, putatively maleless animals such as the *khannos* and bees are parthenogens, but since he typically says they have the 'male and female principle' combined they are more like selfing hermaphrodites. He's just not clear enough on the mechanics of reproduction in these creatures to enable us to draw the modern distinction.

Wild fig fruits contain what people call *psênes*. It starts off as a larva but once the skin [pupal case] has split open, the *psên* flies out leaving the fruit behind. It then enters the cultivated fig fruits via their openings and causes them not to drop off. That is why smallholders plant [wild] fig trees close to cultivated ones and attach their fruits to them.

As Asian as sheep, there have been figs in the Aegean since before Homer's time. These days, as anyone will tell you, the best figs on Lesbos come from Erresos. The groves there are as cool and green and full of life as the surrounding hills are hot and dry and barren. Many fig varietals are grown on the island: *apostolatíka, vasílika, aspra* (white), *maura* (black), *díphora* (double-bearing, spring and autumn), but the most famous is the *smyrna*, named for the city in Asia Minor. It's the one whose fruit you see in the markets, as big as a child's fist with midnight-purple skin and crimson flesh.

Aristotle's cultivated fig could have been any of the ancient cultivars. The *psên* is the fig wasp, *Blastophaga psenes*, which emerges from the fruit, just as he says.* The wild fig, which today is called the *ornos*, flourishes in riverbeds. The business of tying wild to cultivated figs to ensure their fruiting is called 'caprification' since the wild fig is fit to be fed only to goats. Once widespread, the practice is now rare in Greece; on Lesbos farmers simply plant wild and cultivated fig trees in a ratio of 1:25.

All this is clear enough, but still leaves us puzzled. What, exactly, is the relationship between the wild and cultivated figs? And how does a wasp that originates in one keep the fruit of the other from falling? Theophrastus, who came from Erresos and so knew all about figs, discusses these questions at length. He repeats Aristotle's story and adds a few details such as the fact that the fig wasp has an insect predator, the *kentrinês*, probably the parasitic wasp *Philotrypesis caricae*.† He also gives some hypotheses to explain how wasps keep figs on trees. The details don't matter – they're mechanical and quite wrong. More intriguingly, both he and Aristotle consider the possibility that this business of the two kinds of figs has something to do with sex.

In *The Generation of Animals*, when talking about the sexes, Aristotle returns to figs and says: 'There is some small difference like this [among the sexes], since even in plants of one and the same kind we find some trees

---

* Fig wasps are one of Aristotle's spontaneous generators; in fact they have wonderfully complex life cycles.
† The identification lies in the name, which is derived from *kentron* – 'sting'. *P. caricae* lays its egg in *B. psenes* larvae from outside the fig via a spectacularly long ovipositor. Against this identification, Theophrastus suggests that it preys on the adults as they enter the fig.

that bear fruit and others that do not but assist in concocting the fruits of those that do. This is what happens in the case of the [cultivated] fig and wild fig.' Theophrastus gets even closer, for he compares figs directly to date palms. Evidently drawing on an Eastern report by Herodotus or Callisthenes, he says that date palms have 'male' and 'female' flowers and that farmers assist the formation of dates by scattering the 'dust' – pollen, obviously – from the one on to the other. That, he continues, is much like the practice of tying wild figs to cultivated ones, and both are like a fish scattering its milt over eggs.

They adduce the analogy. They come so close to the truth. The two kinds of figs are just different sexes of the same kind. A fig is not so much a fruit as an agglomeration of tiny flowers sealed within a fleshy shell. 'Wild' and 'cultivated' figs are both *Ficus caria*, but the first contains both male and female flowers, the second only female; the wasp carries pollen from one to the other. Figs won't mature without having been pollinated, hence the need for their proximity. Our two Greek scientists toy with the idea. Yet neither just says that plants have sex too.

An object lesson, then, in the dangers of theory. Or is it? Perhaps not. The fig is a protean plant. To fruit, Theophrastus notes, *some* figs need wasps and the business of caprification that goes with them, but most don't. 'Isn't that strange?' Yes it is. Some fig cultivars, it's now clear, require pollination, but others do not for they are asexual mutants, and both types were widespread in the fourth century.* Theophrastus, generalizing from the fact that some figs could dispense with sex, concluded that all figs could do so. No reason, then, to believe that plants need sex at all.

In the seventeenth and eighteenth centuries Tournefort at Paris, Pontedera at Padua, Cavolini at Naples – even Linnaeus at Uppsala – all tackled the mystery of the fig. They made some progress but left it unsolved. In 1864 Guglielmo Gasparrini, Professor of Botany at Naples, reviewing the fig from Aristotle to the results of his own extensive experiments, came to exactly the wrong conclusion. The 'wild' and 'cultivated' figs, he said, were quite different species, indeed belonged to different genera, and the need for the proximity of wild figs to fruiting trees was mere peasant superstition. Unlucky Gasparrini: he, too, studied an asexual strain and, like Theophrastus, generalized too far.†

---

* Some tendentious archaeology dates the origin of the asexual strain at around eleven thousand years ago.
† In 1881 a consortium of California farmers, including Governor Leland Stanford, learnt

But that's the scientist's perpetual dilemma: how far to push his data. Aristotle's generalizations tend to the sweeping; Theophrastus is more cautious. It makes him a duller read. Both occasionally question the quality of their evidence, but neither ever expressed the doubts that Gasparrini did at the end of his fig monograph, and piteous they are:

> Having now reached the term of my labors I cannot conceal a certain anxiety which has secretly grown up in my mind. I fancy I hear from all quarters that the custom of tying wild to cultivated figs being of such ancient date, and having been upheld by so many distinguished men of science, both ancient and modern, cannot but be founded on experience, against which no theories, no subtleties of science, are of any avail. Verily does the rise of such ideas in my breast so agitate me, that many times in the midst of my labors my breath has been stopped by the fear that some fact illy understood has drawn a veil over my mind.

Unfortunate, but that's how it goes. Much of science is about navigating between the general and the particular, between unifying the phenomena and dividing them, and sometimes you just get it wrong.

LXXXII

BEFORE THE RISING of the Pleiades, there is no honey to be had. Honey comes with the morning risings of the stars – when *alkyōn* (Eta 25 Tauri) and *sírios* (Sirius A) first appear in the pre-dawn sky – and when the rainbow descends to the earth. Honey can be harvested when the wild fig fruits. This is all in late May and June.

Honeybees delight Aristotle. More pages in *Historia animalium* are given

about fig pollination the hard way. Dreaming of fig plantations spanning the San Joachin Valley, they imported 14,000 cuttings from Smyrna. The cuttings flourished and grew into trees which bore figs, all of which shrivelled, turned yellow and fell. The Californians accused the Smyrna merchant, an unfortunate Syrian, of sending them the wrong kind of fig, which he denied. US Department of Agriculture and California State Board of Agriculture scientists were commissioned to get to the bottom of the matter. Wild figs teeming with wasps were duly imported and California's fig industry was born.

to them than to any other animal, bar man himself. He describes their foods, predators, diseases, the various products they collect or make, their uncanny industry and the complexity of their social lives. He says they are divine.

You wonder how he knows so much about them. The Arab encyclopaedist al-Damîrîb al Din (d. 1405) claimed that Aristotle had a glass hive made so that he could see the bees going about their work. The bees, resenting his curiosity, smeared the inside of the glass with clay. The last detail alone makes the story unlikely, and the Arab is vague about his source. I am not sure that Aristotle saw the inside of even an ordinary hive. Beekeeping was, however, a major industry in fourth-century Greece and he certainly interrogated beekeepers at length.

Among the bee problems that Aristotle discusses is the origin of honey. You wouldn't think it a problem at all. Every child knows that bees make honey from nectar that they collect from flowers. As expected, Aristotle describes how bees enter calyced flowers and gather up the sweet juices with an organ that resembles a tongue.* He also says that honey from white thyme is better than red; Theophrastus speaks of white and black thyme.† It all seems very clear. But in other passages, some directly adjacent, he contradicts himself. Honey *doesn't* come from flowers since, if beekeepers take honey from the hive in autumn, when there are plenty of flowers about, it is not replaced. Instead it falls from the sky. His tone is that of a man correcting popular error.

That's what the business of honey and the stars is all about. Honey production is tied to the astronomical calendar in some quite direct way; it seems to be a phenomenon rather like dew. That seems absurd. It isn't – or isn't very. His sky-honey *is* 'honeydew', the droplets of sweet fluid that suddenly appear in the spring on branches and leaves in the woods. They are, though he does not know it, the excretory product of aphids and other sap-feeding bugs.‡ These days, about 65 per cent of Greek honey comes from honeydew. (Greeks, who can connoisseur you to death on such matters, disagree as to whether flower or pine honey is the tastier.) As for why

* Aristotle believes that bees collect honey. So he does not know that nectar is processed into honey in the hive by evaporation and enzymatic reactions.
† There are a dozen or more *Thymus* species in Greece, and many hybrids, so I don't know exactly what plants they mean.
‡ I asked a Corinthian beekeeper where honeydew comes from. Corinth is a region famous for its honey since antiquity, and my informant was a man who came from a family that had kept bees for several generations and he had read a good deal about them. Even so, he did not know that honeydew is excreted by hemipterans.

Aristotle disagrees with himself, someone has been meddling with the text. I suspect Theophrastus, who wrote a book, *On Honey*, now lost, which, as far as can be made out from second-hand reports, said that it comes from three sources: honeydew, flowers and 'reeds', the last of which may be Indian sugarcane.

If the origin of honey is problematic, the origin of the bees themselves is even more so. It's not that Aristotle doesn't know about the development of bees, for he describes it in some detail with moderate accuracy. A bee deposits the brood in the cell and then incubates it like a bird as it develops into a larva. While the larva is still small it lies obliquely in its cell; later it sits upright, eats, excretes and clings to the comb. It changes into a pupa, gets sealed up in its cell, grows feet and wings and then breaks out and flies away. No, the problem is: which bee deposits the brood?

As with the 'wild' and 'cultivated' fig, there are too many actors on the stage. If there were just two kinds of bees in the hive, they would be like any other animal and it would be a fairly simple matter to work the sexes out. But there are *three* kinds of bees – workers, drones and leaders – and no one has ever seen any of them copulate.* 'The generation of bees is a great puzzle,' he says, but it's just the sort of puzzle he loves.

In *The Generation of Animals* he sets out the hypotheses in a logical sequence. For the sake of economy, I tabulate them. Bees might be:

1. spontaneously generated;
2. the progeny of some other kind of animal;
3. the progeny of bees.
    If 3, then they might be produced:
    3.1. without copulation;
    3.2. with copulation.
    If 3.2, then the following mating, offspring combinations might be possible:

* Actually, Aristotle distinguishes *six* 'kinds' – *genē* – of bees: (i) a small, round, multi-coloured working bee; (ii) a big, sluggish drone bee; (iii) a red leader bee; (iv) a black, broad-bellied robber bee; (v) a long bee that makes bad combs and resembles a wasp; (vi) a black, multicoloured leader bee. He's clearly dealing here with three sexes or castes (worker, drone, queen) and at least two of the several subspecies of *Apis mellifera* that are found in Greece. He seems to reduce the problem by assuming that (i)–(iii) are related, as are (iv)–(vi), since they come from the same hives. But he never makes it clear that workers, drones and queens are really the same kind – that is, have the same form or *eidos* as regular males and females do, and that the other kinds do not. This seems to raise questions as to what, exactly, the ontological status of a *genos* is for Aristotle.

3.2.1. $w \times w \to w$ (and so for the other kinds);

3.2.2. $q \times q \to w + d + q$ (or some other homotypic mating);

3.2.3. $w \times d \to w + d + q$ (or some other heterotypic mating).

Where *w*, *d* and *q* stand for worker, drone and queen, × stands for a cross and → for the offspring.* Pages of analysis follow that draws on general zoological principles and bits of bee lore to eliminate the possibilities in succession.

Quickly disposing of the idea that bees are spontaneously generated, or produced by some altogether different kind of animal that doesn't live in the hive, he turns to bee genders. He begins with some sexual stereotypes or, to put it more kindly, empirical generalizations. Males have offensive weapons (horns, tusks), females don't; females look after the young, males don't. Neither workers nor drones, however, fit these moulds since workers have stings, but look after the brood, while drones are stingless and do nothing. So workers and drones must be neither male nor female, but a bit of both. They're like plants or those sexually ambivalent fishes that putatively reproduce without copulation. (He knows that drones sometimes swarm out of the hive, but not that they are in pursuit of a virgin queen with the intent of having her at altitude.)

He then investigates which bee generates which. Aristotle reports that drones can appear in hives that have neither drones nor a queen but only workers. (Remarkably, he's right – they can.) So workers must generate drones. Workers never appear in hives without queens. So queens generate workers. Neither workers nor drones generate queens (he asserts), so queens must generate themselves. We have the order of generation: $q \to q + w \to d$. Finally, he considers whether any of them copulate. Workers are sexually ambivalent so can reproduce without copulation. Besides, if they did copulate, someone would certainly have seen it, and no one has. Since workers don't copulate, we can assume that queens don't either. Only one possibility remains: a multi-generational sequence of asexual reproduction in which queens generate queens and workers, and workers generate drones, and drones generate nothing.

---

* Given total ignorance of who copulates with whom, and what the products of those copulations might be, there are many more possible combinations that he does not consider, but that does not matter since he'll prove to his satisfaction that bees do not copulate at all.

It's a strange system, though not nearly as strange as reality.* Nor is it as strange as it is often supposed to be. Should you read Aristotle on bee reproduction, you will find that I have edited out one of its seemingly outré aspects: I have called his 'leader' bee, the bee that generates all the others, the 'queen', since that's what it's called today. He, however, often calls it the *basíleus* – 'king'.

Some scholars have accused Aristotle of gender-bias. (She's *female*, after all, so why not *basíleia*?) But Aristotle is innocent. He cannot possibly believe that the leader bee is male. In his biology males are no more capable of reproducing by themselves than they are in ours. Driven by his data, he take the leader bee to be neither male nor female but a mix of both, and simply calls it by its popular name. Social wasps, he says, have a leader wasp commonly called 'the mother' – and, for them, that's the term he adopts.

HONEYBEE – *APIS MELLIFERA*
LEFT TO RIGHT: MELISSA – WORKER; *KĔPHĔN* – DRONE;
BASILEUS – 'KING', OR *HĔGEMŎN* – 'LEADER' (OUR QUEEN)

Aristotle likes his scheme. There's a three-generation sequence of bees that ends with the sterile drones. It has, he says, a kind of order in that each member of the sequence differs from the others in only one way. (Queens are big and have stings, workers are small and have stings, drones are big but don't have stings.) 'Nature has arranged this so well that the three kinds will exist for ever, even though they do not all generate.' He has a life cycle for bees.

Aristotle's analysis of the mystery of bee reproduction is a model of how

---

* In fact, queens are reproductive females, drones are males and workers are (normally) sterile females. Virgin queens produce drones parthenogenetically. Queens mated to drones produce workers or queens depending on how much royal jelly the larvae are fed and other factors. Readers may be puzzled by Aristotle's claim that workers can produce drones in the absence of a queen. But in any hive a certain fraction of workers do have ovaries and can, if needed, start producing eggs that hatch into drones. Such are the complexities of a haplo-diploid sex-determination system coupled to environment-dependent caste formation.

he often does science. It resembles the procedure described in the *Nicomachean Ethics*. He starts with the 'appearances', collects the best explanations for them and deductively eliminates them according to the evidence. By the time he's done he seems to have demonstrated an essential property of bees.

Has he? Aristotle thinks that demonstrations deliver truth. Yet, as he well knows, any demonstration is only as good as its premises and his – though he does not admit it – are weak. They rely on generalizations that he must know are at best 'for the most part' true. (Do males *never* take care of young? Then what about paternal care in the *glanis*, his catfish? Do females *never* have offensive weapons? Then what about cows?) Then, too, his data come from beekeepers, no more reliable than fishermen. His text is littered with 'they says'.

So a doubt niggles, and his discussion ends on a tentative note. I can't pretend it's typical. He's usually so confident that no one will ever surpass his work, so *final*. But here, for once, he looks to the future and tells us that he hasn't worked everything out; more than that, he tells us how future discoveries will be, or should be, made. And though we may respect Aristotle in his more Olympian moods, the moods to which great scientists are prone, here we have to love him:

> So this, at least as far as theory goes, seems to be the situation on the generation of bees – in conjunction, that is, with what people believe to be the facts about their behaviour. Not that there is, currently, any proper understanding of what those facts are. If in the future they are understood, it will be when the evidence of the senses is relied on more than theories, though theories have a part to play so long as what they indicate agrees with what is seen . . .

– since that is exactly what happened.

LXXXIII

IN MARCH THE SWALLOWS arrive in Lesbos. They come from Africa on the *khelidonias* or swallow wind – Theophrastus uses the phrase, presumably a synonym for the *ornithiai anemoi*. Aristotle, who lists swallows among his migratory birds, also speaks of the intelligent way in which a pair will build a nest from mud and straw, rear their chicks and keep their house neat and clean.* He seems to admire the little birds. But, dispassionate as always, he also says that if you poke out the eye of a nestling swallow it will regenerate. He really believes this. He repeats it three times. And, though it seems like a bizarre thing to say, I'm not sure that he's wrong. He may actually have done it.†

Behind this claim lies a study in comparative embryology. Surveying the ontogenies of the various creatures that he's studied, he rates their progeny on a scale of 'perfection' that tries to captures how much they change between emerging from their mothers and adulthood. Holometabolous insects such as butterflies have very imperfect progeny. (He thinks that the growth of a caterpillar is equivalent to the formation of an egg inside the reproductive tract of a chicken, and the chrysalis is the equivalent of an egg.) Cephalopod, crustacean and fish eggs are soft and 'grow' a little after being laid, so they're also low on the scale of perfection. The progeny of birds, snakes, turtles and lizards are more perfect since they have hard-shelled eggs that don't grow, and the young of cartilaginous fishes are more perfect yet since they start out as hard-shelled eggs in the womb but hatch out internally and are born alive. The young of live-bearing tetrapods (mammals) are the most perfect of all.

Having established this rather crude scale of embryonic perfection among his greatest kinds, he allows variation within them too. Perfection now depends on relative size at birth and readiness for independent life:

---

* He also thinks that some over-winter, naked, in holes. In 1862 Philip Henry Gosse (*Romance of Natural History*, 2nd series) was still wondering whether this is true.
† Chicken hatchlings are used in regeneration research since they can regrow lenses and retinas after experimental ablation. Since swallow hatchlings are more altricial than chicken hatchlings, they may well be even better at this, but it will be a hard-hearted, not to say brave, researcher who tries to find out.

imperfect animals are born blind. Within the live-bearing tetrapods, Aristotle says that solid-hooved (horses and asses) and cloven-hooved (cows, goats, sheep) animals have perfect young; the whelps, cubs and pups of multi-toed tetrapods (bear, lion fox, dog, hare, mouse, etc.)* are, by contrast, quite imperfect. Within the birds, jays, sparrows, woodpigeons, turtledoves and pigeons also have very imperfect hatchlings.† And so do swallows. Thus, when Aristotle speaks of poking swallow chicks in the eye, his point is that regeneration is more likely in embryos than in adults, and that since swallow nestlings can regenerate, they're really very foetal when they hatch.

Aristotle's science is not quantitative. It's not, to be sure, resolutely *qualitative* either, for he often uses terms such as 'large and small', 'the more and the less' and 'for the most part'. He can also discuss quantitative relationships, such as body-size scaling, with subtlety. Yet he rarely gives what modern scientists love and need: numbers. When describing the life histories of various birds and mammals, however, he does.

You can even put them into a table. That shows that his data, although spotty, are quite good.‡ As always, he's interested in the associations. The web that he weaves is wide, but five features – adult body size, longevity, gestation time, embryonic perfection, litter (or clutch) size and neonate size – are central. Some of the associations that he detects among these features are fairly obvious: longer gestation times result in more perfect (precocial) neonates. Others are quite counter-intuitive. You might expect, he says, that large animals would have larger litters (or clutches) than smaller ones, but actually that's not so – they have smaller litters. Horses and elephants bear only one infant at a time. He obviously believes that such associations have predictive power. 'Stories are told', he says, 'of the

---

* Aristotle says that a newly born bear is very small and poorly formed, indeed that its limbs are unarticulated, and he also says that the mother bear (as well as the vixen) aids the 'concoction' of her cubs by licking them. This, via Roman exaggeration (Pliny, Ovid, Virgil), gave rise to the expression used by irate parents and coaches to their charges: 'You need to be licked into shape.'
† Zoologists call this the altricial–precocial spectrum. Altricial (= 'imperfect') and precocial (= 'perfect') were coined in 1835 by the Swedish zoologist Carl Jakob Sundevall, who used them as features by which to divide the birds into two taxa, *Aves Altrices* and *Aves Praecoces*. Sundevall does not credit Aristotle for the idea, yet in 1863 would write *Die Thierarten des Aristoteles*. Did Sundevall get it from Aristotle? Or was he later attracted to Aristotle by finding the idea there? Sundevall does credit Lorenz Oken for the idea; but where he got it from I do not know. Aristotle's assessment of birds and mammal neonates along the imperfect–perfect/altricial–precocial spectrum is correct.
‡ Appendix V.

[deer's] longevity, but none of them has been established as true: besides, the period of gestation and the swift growth of the fawns do not suggest that it is a long-lived creature.' Given a positive association between longevity and gestation time, if deer truly were very long lived (something that's hard to observe) then they should also have a very long gestation period, but they don't.*

Aristotle doesn't want just *any* associations: he wants *causal* associations. Any association may be, in his terms, 'accidental' rather than 'essential' and so not require explanation at all. He notices that the negative association between adult body size and litter size is confounded with foot morphology. Solid-hoofed animals tend to be large and have one offspring at a time; cloven-hoofed animals tend to be medium sized and have a few; multi-toed animals tend to be small and have many. Perhaps, then, the litter–body-size association is really all about feet. But no: it's body size that matters. 'The evidence is that the elephant is the largest animal but is multi-toed; the camel, the largest of the rest, is cloven-hoofed' – the foot-type–body-size association is, in fact, poor. Moreover, 'It's not only on land that large animals produce few offspring and small ones many, but also in animals that fly and swim, and the reason is the same. Similarly the biggest plants do *not* bear the most fruit.' Thus, not only is he aware of the possibility of confounding variables, but he also has a solution: to search for the same association in quite different groups of creatures.† In the same way, he asserts that the positive association between gestation time and longevity in viviparous tetrapods (so informative about

* Applying simple linear regression to modern data gives a moderately strong relationship in placental mammals of the form log(lifespan) = 0.77 log(gestation period) + 1.53, r^2 = 0.6. Given that the red deer, *Cervus elaphus*, has a gestation period of 235 days, that predicts a maximum life span of around twenty-five years; the actual recorded maximum life span is twenty-seven years. I examined six of the associations among life-history features that Aristotle reports; they were all correct (Appendix VI). Actually, that's not surprising since all these features are strongly associated with adult body size and hence with each other. However, the important point is that he's looked to see what the *sign* of the relationship is – he didn't assume it from some theory.
† The problem of distinguishing causal from non-causal associations still plagues comparative biology. Evolutionary biologists will recognize the soundness of Aristotle's attempt to solve it. Aristotle fails to see, however, that within oviparous fishes there is a positive association between fecundity and body size, but his claim is strictly correct for he refers to animals that swim (i.e. all fishes) and his comparison is between the selacheans, which are large and have relatively few offspring, and oviparous fishes, which are mostly small and have many.

deer) is *not* causal.* Here, at least, he doesn't jump from association to demonstration; here he considers causation *v*. correlation.

To explain his web of life-history features Aristotle wields all his familiar devices. Considering the association between litter size and adult body size, he reaches for his bodily economics. It's particularly effective since fecundity depends on seed production and seed is the most refined, and hence expensive, nutritional product of all. Expending seed drains the body. That's why (he says) men are so exhausted after sex, fat people are infertile, castrated animals and mules are so big and fierce, big animals tend to have few offspring, and highly fecund animals tend to be small. (The Adrianic fowl† is said to be a super-fecund dwarf.) Sex may be fun, and reproduction may be necessary, but growth and vitality drain away through our genitals.

So animals must choose between bearing offspring and doing something else. Aristotle the ornithologist knows that, each year, some birds (partridges) lay a single clutch with many eggs, some (pigeons) lay many clutches with a few eggs, while others (raptors) lay only a single clutch with one or two. To explain these differences he postulates a resource-allocation network that links wings, legs, body size and fecundity and a few other features besides. A bird may invest in some, but not all, of these features for each has a cost in terms of the others.

Such an analysis is, however, incomplete. He may weave his webs of conditional necessity as wide or as deep as he pleases, but he must also give a *final* explanation for why a given animal has one set of features rather than another. I've suggested that Aristotle often doesn't do so. He often doesn't bother to relate an animal's specific features to an ultimate cause, or if he does, only sketches it in a cursory way. When explaining life-history variation, however, these two kinds of explanation – conditional and final – come together brilliantly, for he argues that whether a bird invests in its parts, or in reproduction, depends on its *bíos* or lifestyle. Raptors, he argues,

---

* So what *is* its cause? It's likely that Aristotle thinks that both features – longevity and gestation time – are causally associated with body size, so: large animals live longer than small ones since they're less vulnerable to fluctuations in the environment; they also have larger offspring; larger offspring require longer gestation times – ergo there is a positive association between gestation time and longevity. He certainly claims these associations, but he doesn't spell the entire argument out very clearly; indeed, he tends to dwell on exceptions to the pattern (e.g. that horses are shorter lived but have longer gestation times than humans).

† An extinct bantam breed from Adria, in the Veneto. In his *Ornithologica* (1600) Aldrovandi discusses them at length but doesn't know what they are.

need powerful wings, large feathers and massive claws in order to catch their prey; partridges and pigeons, which feed on grain and fruit, don't. So raptors invest in wings and claws, have little nutrition left over for reproduction and so lay few eggs; partridges and pigeons do not invest in wings and claws, have lots of nutrition left over, and lay many.

In his analysis of animal life-history variation, Aristotle uses quantitative data to descry the great patterns, parses out causal from accidental associations, and then explains the causal associations as the best possible compromise between physiological necessity and teleological need – that is, between the demands of their bodies and the demands of their world. It is, I think, his most complete and successful analysis of the function of any complex of animal parts; of how the nature of each animal finds the best of the possibilities that are available to it. Birds and tetrapods, however, embrace only a small fraction of possible life histories. His best analysis of why creatures reproduce as they do comes when he talks about fish.

LXXXIV

IN THEOPHRASTUS' YEAR, SUMMER brings the rose campion, carnations, lilies, spike lavender, sweet marjoram and a delphinium called 'regret', of which there are two kinds, one with a flower like the larkspur, the other with white flowers that is used at funerals. The iris also blooms then and so does the soapwort, which has, he says, a beautiful, but scentless, flower.

That is how summer starts. But by late July the Aegean archipelagos are scorched earth. In the olive groves cicadas cling to branches and sing for mates; in the pine forests firefighters slump in trucks and watch for incendiaries. (Theophrastus says that Pyrrha's forest burnt and then regrew; it has doubtless done so many times since.) The rivers that feed the Lagoon run dry. The Vouváris always flows but even its spring is a stagnant pool and the waterfall at Pessa is a trickle. Its mullets are targets for rapier-billed herons while its eels have found sanctuary in the estuary's mud. Its terrapins imitate stones. The phrygana covering the volcanic western shore, once soft, multi-hued and scented, is now just a threadbare cloak of brittle thorns.

Even as the land bakes, the sea waxes. The summer wind that the locals call the *boukadora* (the 'wind that goes inside') blows into the Lagoon from the open sea and turns its surface into foam. In Kalloni's depths the glutted spontaneous generators are spawning. But no one eats spermy oysters and mussels (sea-urchin gonads are another thing); it's fish now, especially the polyvalent sardella. Kalloni's pilchards are usually eaten salted as *sardeles pastes*, but now is the time to eat them fresh. Aristotle says that in summer the fish come into the Lagoon to spawn. If he means the pilchards, then he is only partly right for, although they have migrated in from the open sea, they did so as larvae. Kalloni isn't their breeding ground, it's their nursery, and by August they're grown and fat and are heading out to their spawning grounds in the Aegean itself. Traversing the Lagoon's narrow mouth, they're intercepted by a wall of nets and landed by the ton.

Aristotle's analysis of bird and tetrapod life history, insightful though it is, doesn't take into account the one factor that evolutionary biologists suppose is most important in shaping animal life histories: the pattern of age-specific mortality – that is, whether the risk of death weighs most heavily on juveniles or adults. Considering fishes he makes good the omission. He says that it is their function to be very prolific. Of course, it is the function of every living thing (spontaneous generators aside) to reproduce, but for fishes the goal is especially exigent because of their high rate of infant mortality: 'The majority of the externally laid embryos are destroyed, and that is the reason why fish, as a kind, produce many offspring. For nature uses number to combat destruction.'

To produce so many progeny, egg-laying fishes have a plethora of special features. Females are bigger than males so that they can hold all their embryos. As evidence of this need he points to some small fishes whose uteri seem to be just one mass of eggs. He also cites the *belonē* that literally bursts under pressure from them, though nevertheless survives the experience. He means the pipefish that lives in the eelgrass beds near the head of the Lagoon and broods its embryos in a pouch.* This is also why most fish eggs are only 'perfected' – fertilized – once they've been laid. Were they to become 'perfected' in the uterus, there would not be space for them all. Fish eggs are generally small, but once they're fertilized the embryos and larvae also grow very quickly to 'prevent the

---

* Aristotle does not know that it is the male pipefish that broods its young in its pouch; neither did the Kalloni fishermen that I asked.

destruction of their kind which would occur as a result of their spending a long time over the period of their formation'. Finally, some fishes, such as the *glanis* (Aristotle's catfish) care for their young to prevent them being eaten.

To compensate for the high mortality of their embryos and larvae, egg-laying fish, then, have a whole suite of interlocking adaptations: high fecundity, small eggs, reverse-size sexual dimorphism, altricial development, rapid growth and parental care. Those selachians that, by contrast, give birth to live young have no need to be so fecund because their young are large and relatively perfect at birth and so 'have a better chance of escaping destruction'. When I began this book, I said that scientists make poor historians for they tend to read their own theories into the past, but I also ventured that a scientist might see things that the classicists have missed, precisely because he does. Aristotle's analysis of why tetrapods, birds and fishes have such varied numbers of offspring is as good example as any of this tension. For all that has been written about Aristotle's biology, these passages have been, as far as I can tell, ignored. Yet any evolutionary biologist reading *The Generation of Animals* would seize on them for they are, or seem to be, about life-history theory, the part of adaptationist biology that considers the ultimate currency in which reproductive success is counted. Moreover, the structure of Aristotle's analysis is instantly recognizable. Like ours, it is an analysis of the solutions that animals have found to the varied contingencies of their environments and how bodily economics shapes the form that those solutions take. Our theory is, of course, couched in equations, but that's just a matter of expression; there are other, deeper differences (on which more later). The real question is whether Aristotle's analysis is as important to him, as fundamental, as its analogue is to us. I think the answer must be that it is. For, at the bottom of Aristotle's explanation of organismic diversity, is the claim that the ultimate purpose and desire of each and every creature is to reproduce: the cycle must turn again. So it was for him; so it is for us.

In Skala, to celebrate the catch, there's a panagyri where, for two nights, you can feast for free on piles of grilled pilchards and drink and dance, silver scales flashing like sequins on your feet. The music is self-consciously traditional; most of the songs are the songs of Asia Minor and dwell on Constantinople, the lost heavenly city. (When Smyrna burned in 1922, it was to Lesbos that the Greek refugees first came and where many stayed.) But one song was all about fish:

I got into my new boat and set off from Agios Giorgos. I found some young boys, sailors, fishing: 'You fishermen, do you have fish, lobsters and squid?' 'We have salted sardellas, like beautiful girls. Come aboard, pick them up, weigh them, Take a rope, string them up, and pay as much as you like!'*

LXXXV

I N THE VILLAGES SURROUNDING Kalloni, autumn is when old men reclaim their rights. Scandinavian girls and Dutch families no longer occupy their favourite chairs in the kafeneons; even the English nature-walkers, invariably in couples, have gone home. Old men can then sip ouzo, play backgammon and vociferously debate the issues of the day without having the proprietor, who rather liked the tourists, telling them to keep the noise down and not wave their sticks about.

Life expectancy in the islands is often said to be high – Ikaria, due south of Lesbos, has even been touted as a kind of Aegean Shangri-La where nonagenarians bound about like goats. That's an exaggeration, though there's some evidence that elderly Ikarian women have high survival rates. What is true is that Greeks (and Italians and Spaniards) have, for all their love of cigarettes, higher life expectancies than the citizens of most Northern European countries, and that this appears to be due, at least in part, to a 'Mediterranean diet' rich in vegetables, fruits and nuts, olive oil and legumes and low in meat.

That would have interested Aristotle. 'We must investigate the reasons for some animals being long lived and others short lived and the length and shortness of life in general.' So begins the treatise known as *The Length and Shortness of Life*. From *Historia animalium* the relevant data include, among much else: a report on the *ephēmeron* or mayfly which emerges from little sacks in the River Hypanis near the summer solstice by the Cimmerian Bosphorus and lives only for a day,† the (suspect) claim that the elephant

---

* Cf. E. Pound, 'The Study in Aesthetics', 1916.
† The Southern Bug River, Kerch Strait, Ukraine–Russian frontier, in June. Curiously Aristotle insists that the mayfly is a tetrapod. This is probably because it rests on only four legs, holding the anterior pair in front of it, as if in prayer.

lives for centuries and the observation that most winged insects die in autumn. Perhaps he is thinking of the cicadas, whose desiccated corpses litter the silent olive groves by summer's end.

Summarizing, Aristotle notes that plants tend to live longer than animals; large animals tend to live longer than small ones, blooded longer than bloodless, terrestrial longer than marine. No one feature predicts longevity very well, but together they say something about the relative fragility of kinds. There are plenty of exceptions: some plants (annuals) are very short lived; some bloodless animals (bees) are long lived; some large animals (horses) do not live as long as smaller ones (men). As so often, he descries the great patterns and notes the exceptions.

He wonders whether it is possible to give a general explanation for death in all its forms, and whether the fact that some individuals and kinds live for so much longer than others has one explanation or many. These are, he admits, difficult questions. He therefore designs his theory of ageing to cope with both the general patterns and the exceptions. Indeed, he discusses the exceptions precisely to get simplistic explanations ('big animals live long because they're big') out of the way and so open the door to a much more sophisticated account.

To explain lifespan diversity Aristotle begins with the observation that living things are warm and moist. They are particularly so when young; the old are cold and dry and so are the dead. 'This', says Aristotle with conviction, 'is an observed fact.' He then argues that animals differ in the quantity and quality or heat of their moisture. Using these variables he cooks his explanations. Large animals and plants have (relatively) more hot-moist matter, and so live longer, than small ones. The exceptional lifespans (for their size) of humans and bees are due to the same cause. Marine bloodless animals (invertebrates) may be always wet (they live in the sea), but even so they're short lived because of the low-heat-content stuff of which they're made. Talk of heat content seems vague, but it's all about fat. Of the various uniform parts, fat has a very high heat content and resistance to decay. (Of all foods, olive oil keeps in the kitchen.) Fat, in Aristotle's view, is life-promoting stuff.

None of this, however, explains why animals should age at all. Aristotelian animals are continually sustained by nutrition and have complex regulatory devices for keeping their metabolisms in check. As we age, something must deprive animals of the warmth and moisture that they need to stay alive. Aristotle thinks it is reproduction that deprives the body not only of the material it needs to grow some parts but of life itself. This gives him another way of explaining lifespan variation. It's a matter not just of how much hot-moist matter animals start with but of how fast they spend it. Salacious animals, he says, age more quickly than continent animals. Mules, which are sterile, live longer than horses or donkeys; the cock-sparrow, an unusually lecherous bird, doesn't live as long as his hen. Plants pay the same cost of reproduction. Annual plants die each autumn because they expend all their nutrition on seed. Aristotle seems to see a body as a bank account that is continually filled by the income of nutrition, but that is drained even more swiftly by the expense of maintenance and reproduction, and that, once overdrawn, dies. This is biological economics with a bite.*

Turning to plants, Aristotle argues that the reason that they generally live longer than animals is partly due to their oiliness. But he often gives competing explanations for natural phenomena if he thinks the facts merit them, so he argues that plants are also long lived because they are capable of regeneration: 'For plants are always being reborn; that is why they live so long.' Roots, trunks and branches may die, but new parts grow up beside them. Moreover, as demonstrated by cuttings, 'the plant possesses potential root and stalk in every part of it' – indeed, a cutting is 'in a sense a part of the [parent] plant'. And although he knows, or thinks he knows, that some animals can regenerate organs – snakes and lizards can regenerate their tails and nestling swallows their eyes – only plants can be continually reborn; only they have the 'vital principle in every part'.† By 'vital principle' Aristotle means the soul.‡

Aristotle, then, believes that lifespan can be influenced by a variety of mechanisms. In another treatise, on *Youth & Old Age, Life & Death*, he gives a

---

* This idea is identical to the modern notion of a 'senescence cost of reproduction'. The evidence for it is the same that Aristotle gave: various experimental manipulations that reduce reproductive effort increase longevity. The standard explanation – a diversion of resources to somatic maintenance, and hence longevity, rather than reproduction – is also Aristotle's rephrased in terms of energy rather than 'hot, moist' matter or fat. The truth of this explanation remains unclear. That we are still discussing ageing in Aristotle's terms is, perhaps, a mark less of his sophistication than of our physiological naïvité.

† Snakes can't regenerate their tails. Aristotle is probably thinking of the legless European glass lizard, *Pseudopus apodus*, which can. Common on Lesbos, they are easily mistaken for snakes.

‡ Here he anticipates that quintessentially twenty-first-century concern of biomedical science: the search for totipotent stem cells from which new organs can be built at will. He would delight to learn of *Hydra*, that tiny sea-anemone-like creature, rich in stem cells, able to regenerate all parts, and one of the few animals that does not apparently age at all. It is an animal that has its 'vital principle in every part' complete.

theory more tailored to vertebrates. Here he asserts that death is always due to the exhaustion of vital heat. Blooded animals have particularly active metabolisms and so are especially susceptible to the vagaries of the chemical conflict within them – all that concoction. That is why they have such elaborate homeostatic devices. The reason *they* die, then, is because these devices, in particular their thermoregulatory systems, fail. He even defines the life cycle in thermoregulatory terms: 'Youth is when the primary organ for cooling grows, old age is when it is destroyed. The middle period is the prime of life.' The destruction of the 'cooling organs' – lungs and gills – occurs because, as the animal ages, they become more 'earthy', less flexible, and finally simply seize up. Should that happen, metabolic meltdown ensues followed quickly by heat-death. Or, as Aristotle puts it, the animal 'suffocates'.

Aristotle notes that the words for old age (*gēras*) and earth (*geēron*) are similar.* The etymology is false and, in any case, does not explain why lungs and gills become more 'earthy' with age. Perhaps he thinks that they accrete earth rather as smokers' lungs accrete tar. Or perhaps he thinks that, losing warmth and moisture, they just become relatively earthier. The latter idea appeals since it links his two material explanations for ageing – and in fact he explain wrinkling skin in exactly this way.

Yet there is an interesting difference between the two theories. Where the cost-of-reproduction theory is deterministic – there is a simple cause-and-effect relationship between the depletion of fat reserves and risk of death – the homeostasis-failure theory has a stochastic element. This is evident in passages in which Aristotle claims that old creatures are more susceptible to *variation* in their external environments, their health or the state of the internal fire. Old animals die from even trivial ailments as 'a brief and tiny flame is extinguished by a slight movement'. Small animals are especially vulnerable because they have 'but little margin in either direction'. It's a picture of creatures beset with metabolic challenges that cause the vital heat to wax and wane in ways that if large or young they might well survive, but when old or small tip them over the edge.†

It's unclear why Aristotle thinks that ageing must be explained in different ways in different taxa. However, his various explanations all depend, in

---

* Aristotle's theory of the ageing of land parallels his theory of organismal ageing: land is born moist and as it grows old it dries.

† Old mammals are poor at thermoregulation, but that is surely less a cause than a consequence of deeper processes. But Aristotle's belief that ageing is caused by the decay of regulatory networks, and death caused by stochastic environmentally induced crises, may yet prove prescient.

one way or another, on a creature's metabolism and the devices that regulate it – that is, the workings of the nutritive soul. And that, in turn, implies that a creature's lifespan is not, for the most part, a matter of chance: it's written in its form; it is part of what makes it one kind of animal rather than another.

The peculiar fascination of Aristotle's theory – or theories – of ageing is that they are answers to still-unanswered questions. The proximate cause – or is it causes? – of senescence are, for us, hardly less mysterious than they were for him. There is, of course, no shortage of scientists prepared to assert with confidence no less sublime than his that they know ageing's secret, though if they do then they have failed to convince their colleagues of it. But, then, many of their explanations have scarcely more empirical content than his, and some have rather less.

There is, however, one question to which our answer is better than his. Aristotelian and modern science both demand teleological or, if you prefer, adaptive explanations for most visible and ubiquitous biological phenomena. Hearts, feathers, teeth and genitals are adaptations; they exist for the sake of survival and reproduction. But what can the purpose of ageing be? Death has no obvious utility.

Aristotle sidesteps the question. He says that it is just the 'nature' of living things on earth to age and die. All that remains to be discussed is how and when. Darwin sidestepped it too. He said even less. The omission was glaring. August Weissmann, Darwin's German disciple, tried to fill the gap – and it's as if he's rebutting Aristotle. 'I believe', he wrote, 'that life is endowed with a limited duration, *not because it is contrary to its nature to be unlimited* [italics mine], but because the unlimited existence of individuals would be a luxury without corresponding advantage.' He then argued that old animals, worn and torn, are useless, even harmful, to the species and so evolution has devised ageing just to get them out of the way.

Modern evolutionary biologists demur. They point out that 'good of the species' arguments are weak and at best a last resort. They argue instead that ageing is the result of the absence of natural selection. Most animals and plants have a constant risk of death from external causes such as accident and disease and since the dead cannot reproduce old age is invisible to natural selection. This invisibility means that bodies are designed to work when young but fall apart when old. When, therefore, we ask what ageing is for we must give the peculiar answer that it isn't *for* anything; it is, instead, the evolved consequence of there being no reason to stay alive.

There is, however, a bit more to be said about Aristotle's account of the ultimate reason we fall apart. For it is not merely living things that are subject to the forces of destruction: *every* natural object under the light of the moon is. Animals, plants, tissues, rivers, rocks, the very elements themselves are forever decaying. But this isn't the Second Law of Thermodynamics *avant la lettre* for, in Aristotle's world, everything that is destroyed is regenerated again, even if only as another individual of the same form. More than that: the finite span of any creature's life is *a consequence* of the perpetual elemental turmoil within. The ultimate reason, then, that we must be born, live, age and die is that we, too, are held in the swirling embrace of the cycles of the physical world.

LXXXVI

THERE IS A story of Lesbos that depicts the island as it was nearly two millennia ago.

Two children, a goatherd and a shepherdess, are herding their animals in the hills beyond Mytilene. They started life as foundlings, were raised in humble homes, but now are grown to uncanny adolescent beauty. The flowers are in bloom; the bees buzz in the meadows; songbirds fill the woods. Infected by the season, the youths gambol like lambs, catch crickets and weave flowers into garlands. Together they steal into the woods until they reach a vast hollow boulder from which a rivulet emerges on to a mossy sward. It is a sanctuary: statues of Nymphs (bare arms, loose tresses, belted waists) stand poised in a frozen circular dance amid flutes and panpipes left by generations of shepherds. Daphnis bathes, innocent as a fish; Chloe is stung by love; the smiling Nymphs watch, flowers draped around their stone necks.

Some scholars dismiss the pastoral landscape described by Longus in his novel as an idyllic invention. Others, however, say that the geography is true to life. One even places Daphnis and Chloe's grotto as the source of the Vouváris in the hills to the south-east of the Lagoon. For my part, I favour the waterfalls at Pessa, fed by the adjacent Mákri, where there are deep rock pools shaded by pines and inhabited by tiny freshwater crabs. It's where the local youths now go to bathe. But the exact location of the grotto doesn't really matter. It is, after all, just a story. What matters is that, once again, it is spring.

THE STONE
FOREST

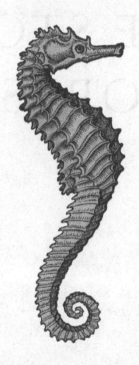

HIPPOKAMPOS — SEA HORSE
— *HIPPOCAMPUS* SP.

LXXXVII

SKALA KALLONI'S RESIDENT pelican was called Odysseus and lived in a kennel by the beach. He belonged to the peripheral economy of pets and strays that depends on the unsaleable fish that are occasionally thrown their way. I have seen the pelican, six cats and a collie staring at the rigging of a single boat – apparently owned by a fisherman of famous generosity – with the blank optimism of commuters waiting for the tube. Odysseus would gape and try to field the fish thrown his way; but pelicans, or perhaps just this pelican, have poor bill–eye co-ordination and so he had learnt to pick them up from the wharf, a manoeuvre that involved neck-twisting contortions for which he was obviously not designed.

Odysseus was an arrogant beauty with pink-flushed plumage and a lemon-yellow bill. But he also had an air of pathos for he had only one leg. If a fish intended for him fell into the water, he would merely stand on the harbour wall and morosely watch it sink. On summer days I have sometimes stretched out on the warm stone wharf to look into the water's depths. Odysseus would then hop over and, out of boredom or misanthropy, gnaw on my shoes until told sternly to desist, at which point he would ruffle his feathers and glare at me with his tiny blood-shot eyes.

Skala's harbour teems with larval fish, too small to identify, that are chased about by flickering schools of small silvery sea bream that are, in turn, preyed upon by the resident squadron of implacable black cormorants. Just beneath the water's surface the walls of the wharf are covered with a brown alga amid which dozens of hermit crabs clamber about, their tomato-red claws vivid against the chalk grey of their worm-eaten shells. They are clumsy and the sea anemones that they carry about on their shells don't help. The anemones are called *Calliactis parasitica*, but they are mutualists that protect the crab with their stinging nematocysts and receive, in return, a movable feast.

Further down the wall the community becomes richer. Black mussels, crystalline sea squirts, hydrozoan colonies spun of gold and yellow threads, green and brown masses of sponges compete for *Lebensraum*.

Small spider crabs amble past herds of *Holothuria forskali*. I once grabbed one of the flaccid creatures and asked a fisherman, mending his nets, the local name. '*Gialopsolos* – know what that means?' Yes – sea cucumbers are also sea pricks everywhere. Unusually, Kalloni also has a *gialopmoya* or sea cunt, a large, beautiful but noxious scyphozoan jellyfish. It has a deep orange bell that pulses gently and tentacles which stream behind it a metre long.

There are also seahorses down there. You don't often see them from the wharf, but they're caught in the nets and, since not even the cats can eat them, just discarded. Often I found them expiring in the sun, flexing their armoured tails in a vain search for something to coil about. I always threw them back but, I suspect, to no good end for they never righted themselves and whirred away, but just spiralled limply into the gloom.

Skala has no archaeology, so let us place Aristotle at Pyrrha, the small *polis* on the Lagoon's south-eastern shore. It is a sweet early-summer morning and the sea is flat. *Sirios*, the Dog Star, which rose just before dawn, has disappeared in the sharpening sun. Aristotle has, perhaps, breakfasted on figs and honey and milky cheese, but now he's sprawled, face down, on Pyrrha's stone wharf, an irate pelican molesting his feet, pulling up sponges, sea anemones and sea squirts. Ripped from the water, dumped on the stone, they form a gelatinous mess, a bit repulsive to the touch.

Aristotle has an ontological problem with sponges. It's not that they were unfamiliar since they were found in every household. In the *Odyssey* sponges are used to cleanse the suitors' stains from the furniture. In *Agamemnon* Aeschylus compares death to a sponge that wipes all our mortal traces away. No, Aristotle is perfectly well acquainted with sponges; his problem is that he's unsure whether they are animals or plants.

His world seems so neatly structured. Sharp lines divide the living and the dead, the animal and the vegetable. In his official ontology living things have souls, dead things don't; animals have sensitive souls, plants do not. No one could mistake a stone for an olive tree or an olive tree for a goat. It all seems very clear. Until we get to sponges. On the one hand sponges are like plants since they're rooted to the rocks from which they grow, and from which they presumably get their nutrition. On the other they do one very unplant-like thing: they can sense, and respond to, touch. People say – Aristotle's reporting a diver's tale – that if a sponge becomes aware that it's about to be plucked from a rock, it will contract and resist. He adds that the people of Torone deny this, but that everyone agrees that the *aplysia* (*Sarcotragus muscarum?*) can sense touch.*

It's not just sponges that seem to bridge the plant/animal divide. Look into the harbour and it's all ambiguity. The *tethya* (sea squirts), *knidai* and *akalephai* (sea anemones), *holothourion* and *pneumon* (either or neither of which might be, from his meagre descriptions, a sea cucumber or a jelly-fish) and the *pinna* (giant fan mussel) are all dualizers, but of a much more radical sort than a dolphin, ostrich or bat. Other ambiguous things grow in more distant seas. Theophrastus tells of a stony, scarlet, deep-sea growth. He calls it the *korallion* and means the precious red coral, *Corallium rubrum*. He speaks about it in his book on stones, between pearls, lapis lazuli and red jasper. Is coral, then, a mineral? Probably not – for it appears again in his *Enquiries on Plants* as a kind of deep-sea plant that grows near the Straits of Gibraltar and that resembles a sow thistle. There are also tree-like growths in the Gulf of Heroes that are about three cubits (135 centimetres) high, resemble stone when they emerge from the sea and display vivid flowers when submerged. Theophrastus has heard of the great fringing coral reefs that run for two thousand kilometres from Aqaba to the mouth of the Red Sea.

Animals have three faculties that plants don't: sensation, appetite and locomotion. They are all faculties of the sensitive soul. All of Aristotle's plant–animal dualizers lack at least one. Sea squirts are sessile but respond to touch; sea anemones are also sessile but can, sometimes, detach them-selves and grab their food; the *holothourion* and the *pneumon* are free living and can move, or at least flop, about but don't have perception; the fan mussel, one of the *ostrakoderma* and therefore similar to snails and oysters, is 'rooted' in the ground like a plant (he means the anchoring byssal threads). Since all of these creatures have at least one faculty of the sensi-tive soul, Aristotle probably supposes that they are, on balance, animals. But he never really says so. That's because he's less interested in the solu-tion to the taxonomic problem than in the reason it's a problem at all. The really interesting point is this:

_____

* The claim that sponges can sense touch and contract has long attracted derision. Even D'Arcy Thompson dismissed it as a fable. But *Suberites* and *Tethya*, two genera found in the Aegean, do visibly contract upon being touched; *Chondrosia* and *Spongia* may do so as well. How they do this in the absence of a true neuromuscular system is obscure. It would be interesting to test Aristotle's report on sponge resistance by an experiment.

Nature proceeds from the inanimate to the animals by such small steps that, because of the continuity, we fail to see to which side the boundaries and the middle between them belongs. For, first after the inanimate kind of things is the plant kind, and among these one differs from another in seeming to have more share of life; but the whole kind in comparison with the other bodies appears more or less as animate, while in comparison with the animal kind it appears inanimate.

The living and the dead, the plant and the animal, form a finely graded continuum. At one end are inanimate, almost formless entities such as stones; at the other, animals running on bi- or even tripartite souls. As one moves along this continuum, from dead things to plants to animals, the characteristic features of each class appear in a step-wise fashion. But the fact remains: it's hard to draw lines in the sea.

LXXXVIII

'**N**ATURE PROCEEDS . . . by such small steps.' Or, to express the same thought obversely and in Latin, *Natura non facit saltum* – nature does not jump. The tag is familiar. It was one of Darwin's favourite slogans; in the *Origin* alone it appears seven times. Huxley, famously, thought it a needless weakness of the theory.* It is a recurring motif in Aristotle too: explicitly when he speaks of plant-like sponges or how, in some animals, bone seems to blend into fishbone; implicitly when he says that snakes are elongate lizards (are, indeed, their 'kindred'), that seals are 'deformed' tetrapods and that apes seem so nearly human.

It's not just a matter of a single slogan. When you read Aristotle you can't help but be reminded of Darwin. Aristotle constructs hierarchical classifications and uses the word *genos* – family – for his taxonomic category. That seems to imply similarity by descent, for what is a family if not a

---

* It wasn't – but it has been the source of endless controversies over the tempo and mode of evolutionary change, notably in the 1970s when Eldredge and Gould proposed their theory of punctuated equilibrium to explain patterns observed in the fossil record.

group of genealogically related things? He distinguishes between analogous parts and those that 'are the same without qualification' – that is, parts that are homologous. What meaning can that possibly have if not an evolutionary one? So, too, his account of how the embryos of different animals are remarkably similar when they first form and only later diverge. As von Baer's 'First Law' of embryology it was one of Darwin's most telling bits of evidence for evolution.

And then, in the writings of both men, there are the explanations by the dozen of how this or that animal's organs are designed to work with each other or in the particular environment in which it lives. Many philosophers and scientists have tried to draw a line between Aristotelian teleology and Darwinian adaptationism. ('Teleonomy', an ephemerally popular weasel-word, was coined to invoke teleology without being too blatantly Aristotelian.) Such semantic quibbles obscure the similarities. Aristotle's functionalism is as resolute as Darwin's – and that of most evolutionary biologists.

Indeed, reading Aristotle, it's easy to suppose that he is struggling towards, or even has, a theory of evolution. He isn't and hasn't. Nowhere in his works does he claim, as Darwin did, that all animals are descended from some remote common ancestor. Nowhere does he suggest that one kind of animal can transform into another. Nowhere does he lament some kind that has gone extinct. *Genos*, he says, is a word that can be used in several different senses – but there's no hint that, in the biology, he's using the genealogical one. When he says that 'nature makes small steps' he means it in a static sense – that one can observe fine gradations between forms. Darwin means it dynamically – that species can transform, but do so gradually. Nowhere does Aristotle appeal to anything resembling natural selection as a force for either stasis or change.

Yet he had all the ingredients. Natural selection is an explanation – the only rational one going – for adaptation. Aristotle understands adaptations and grasps that they must be explained. As scientific explanations go, natural selection is simplicity itself, requiring only a grasp of three concepts: that creatures are variable, that at least some of this variation is inherited and that some of these variants survive and reproduce by virtue of their phenotypes while others do not. Aristotle's own theory of quasi-stable inheritance gives him the first two; Empedocles' selectionism gives him the third. Aristotle, it seems, lacks nothing but the insight, or perhaps the will, to put them together.

It is a diverting though perhaps futile game, to speculate why this is so.

After all, a prepared mind may be necessary for the formulation of a new idea, but it is clearly not sufficient. Did not Huxley declare, upon having natural selection explained to him, 'How stupid of me not to have thought of that'? Hindsight is so easy.

And powerful. It is not impossible for a biologist to read Aristotle and put all thoughts of evolution out of mind, but it is very hard. Evolution underpins all our theories and explains all our observations. We see its handiwork everywhere. We are bred to do so as greyhounds are to run. And there is another difficulty. Darwin looms so vast against his predecessors that we tend to credit him for everything. Historians write of German *Naturphilosophen* and French transcendental anatomists but, as far as biologists are concerned, they write in vain – 1859 remains year zero. 'Ever since Darwin . . .' – it is our story, our origin myth. It is not one that I would, or could, destroy. But I do ask this. Should you come across an apparently Darwinian thought in Aristotle, pause and reflect that you may be recollecting an Aristotelian thought in Darwin.

LXXXIX

I F THIS IS SO, it is not so because Darwin had read much Aristotle. *Transmutation Notebook C* contains a promissory note: 'Read Aristotle to see whether any my views is ancient?' It's dated June 1838, about two years after the *Beagle* had reached Falmouth. After that there is little more until the fourth edition of the *Origin*, 1866, in which, discussing some possible proto-evolutionists, Darwin quotes a knotty bit of *Physics* II, 8 – Aristotle on Empedoclean selectionism. But he did so only because a correspondent had sent him the passage and he rather muddles it up. In fact it is certain that Darwin knew little about Aristotle that wasn't fragmentary or second-hand before 1882 which is when William Ogle, physician and classicist, sent him a copy of *The Parts of Animals* that he had just translated along with the following letter:

Dear Mr. Darwin,
    I have given myself the pleasure of sending you a copy of a translation of 'De Partibus' of Aristotle; and I feel some self-importance in

thus being a kind of formal introducer of the father of naturalists to his great modern successor. Could the meeting occur in the flesh, what a curious one it would be!

Ogle's translation is lovely. Truer translations, with deeper commentaries, have since been made; but just as D'Arcy Thompson illuminated *Historia animalium* with a naturalist's insight, so Ogle illuminated *The Parts of Animals*. When Aristotle tells us that 'All female quadrupeds void their urine backwards, because the position of the parts that this implies is useful to them in the act of copulation,' Ogle has a footnote to tell us that this is so.

It was just the book to send to Darwin. A few weeks later, Darwin replied to Ogle thanking him for the book:

> From quotations which I had seen I had a high notion of Aristotle's merits, but I had not the most remote notion what a wonderful man he was. Linnaeus and Cuvier have been my two gods, though in very different ways, but they were mere school-boys to old Aristotle.

This, by his own account, is when they first truly met. And though we would love to know what Darwin thought as he read *The Parts of Animals* and encountered one of the few minds in history that equalled his in scope and power, and then on the same subject, sadly we do not. Darwin's reply to Ogle was one of the last he ever wrote, for by April of that year he was dead. It may seem, then, that my suggestion that Darwin's works are infused with Aristotle is no more than wishful thinking, but that is not so. When Darwin said that his 'two gods' – Linnaeus and Cuvier – were mere schoolboys compared to Aristotle, he was insufficiently precise. He should have said that old Aristotle taught them.

XC

ARISTOTLE'S CLASSIFICATION OF the animals is the starting point of our own. Linnaeus got many of his European species from him, either directly or via the sixteenth-century encyclopaedists. The vaults of the Linnean Society at Burlington House, Piccadilly contain Linnaeus' copy of Aristotle's zoological works (Gaza's translation, printed at Venice in 1476), and his copy of Gesner's *Historia animalium*. You can follow their names through the successive editions of *Systema naturae* until they appear as modern species in the magisterial tenth. Aristotle speaks of the *sēpia*; Gesner of the *Sepia*; Linnaeus (1758) of *Sepia officinalis* – the name by which we know the cuttlefish today.

Aristotle's higher taxa – the *megista genē* – are also the basis of ours. In the first edition of *Systema naturae* (1735) Aristotle's *zōotoka tetrapoda* appear as the Quadrupedia (only renamed Mammalia in the tenth edition). Some other Aristotelian taxa are shuffled around or subordinated but recognizably intact: Aristotle's *ostrakoderma* become Linnaeus' Testacea; his *entoma* + *malakostraka* Linnaeus' Insecta.

Aristotle's influence on Linnaeus is not only apparent in his actual taxa. At least some of his taxonomic terminology, most obviously *species* (*eidos*) and *genus* (*genos*), are ultimately Aristotelian or Platonic. It is also often said that Linnaeus' classification methods were based on the Aristotelian logic of division. Historians disagree about that, and I am inclined to doubt it. It is, however, quite clear that a complex of Platonic and Aristotelian ideas shaped how Linnaeus, and other Pre-Darwinian naturalists, saw the structure of the natural world.

Around 1260 Albert Magnus, the first modern European to study Aristotle's zoology, expressed Aristotle's claim that 'nature proceeds . . . by such small steps' in much the same terms that he did: 'nature does not make [animal] kinds separate without making something intermediate between them; for nature does not pass from extreme to extreme *nisi per medium*'. By the early seventeenth century the obverse version, *Natura non facit saltum* (or *saltus* – plural), was a commonplace. In his *Philosophia botanica*, 1751, Linnaeus elevated it to methodological principle. 'This is first and foremost what is required in Botany – *Nature does not make jumps.*' Perhaps that is what Darwin was remembering when he quoted it.

The idea that nature does not make jumps is closely allied to another: that nature is organized into a linear scale that runs from rocks to God via plants, animals and humans. The *scala naturae* – the Ladder of Nature – as it came to be known, appears in the cosmic structure of *The Timaeus* which is nothing if not hierarchical. It is also one of Aristotle's themes. Every natural thing may be, for him, a form and matter – *eidos* and *hylē* – compound, but the relative importance of the components varies. In rocks *hylē* predominates; in living things *eidos* does. Among living things there's a ladder of increasing complexity too, plants to humans, running successively on uni-, bi- and tripartite souls. In *The Generation of Animals* Aristotle elaborates this ladder of life within the animals and underwrites it with embryology and physiology.

He begins by linking his scale of progeny 'perfection' (how advanced they are at birth) to their parents: 'Nature's rule is that perfect offspring will tend to be produced by a more perfect sort of parent.' Parental perfection depends on intrinsic heat, hot being better than cold. Heat is reflected in the composition of their uniform parts, hot animals being more fluid and less earthy than cold ones. Heat also reflects anatomy since hot animals have lungs and more elaborate thermoregulatory devices than cold ones. Hot animals also tend to be larger, live longer and be more intelligent than cold animals. The result is a ladder of perfection that reaches from the live-bearing tetrapods down through the selachians, egg-laying fishes, crustaceans and cephalopods, larva-bearing insects and beyond to the spontaneous generators such as sponges, sea anemones and sea squirts which are little more than vegetables. Although this ladder accounts for much of the large-scale variety in their features, Aristotle is far too good a zoologist to believe that any animal can be unambiguously perched on a given rung of the ladder of zoological perfection. His attribute associations are always just 'for the most part'.

The Ladder of Nature was adopted by Neoplatonists, Christian theologians and early modern philosophers. It underpinned Leibniz's cosmology. Vastly expanded and much transformed from its Attic origins, it reached the apogee of its influence in the eighteenth century, which is when it appears in *Systema naturae.** Linnaeus' version of the Ladder of Nature is quite

---

* A. O. Lovejoy, the Harvard historian, traced the origin and fate of these ideas, along with the Platonic 'principle of plenitude', in his classic work of intellectual history *The*

Aristotelian. Biologists forget that he classified not just plants and animals, but *all* Earth's natural products – *Per regna tria naturae* runs the subtitle: there's a taxonomy of stones too. The three great Kingdoms of Nature's Empire – *Animale*, *Vegetabile* and *Lapideum* – are explicitly ordered by declining complexity; the book begins with *Homo sapiens* and ends with *Ferrum* – iron.

It all seems clear cut. It wasn't. In the eighteenth century naturalists struggled, as Aristotle had, to classify rock-like plants and plant-like animals. Successive editions of *Systema naturae* record their efforts. In the first edition, 1735, the lowest of the low animals are in the Order Zoophyta, literally 'Plant-Animals'. It contains the sluggish, barely sensate creatures – sea cucumbers, sea stars, medusae and sea anemones – that had worried Aristotle. (It also contains, rather weirdly, the cuttlefish.) Sponges, corals, gorgonians and bryozoans aren't even animals; they're plants and, within them, the lowest of the low. They belong to the Order Lithophyta, literally 'Rock-Plants'. Over the next fifty years they are all upgraded. By the last, posthumous edition of 1788–93, Aristotle's plant–animal dualizers have full-animal status. The Order Zoophyta still exists but now it contains all the creatures – corals, gorgonians, bryozoans and, of course, sponges – that once inhabited the Lithophyta. Rock-Plants have become Animal-Plants. Linnaeus found these ambiguities attractive. He defined the Zoophyta as 'Composite animals efflorescing like vegetables' and said that this was where the boundaries of the three Kingdoms met.

Of the many naturalists – Trembley, Peyssonnel, B. de Jussieu to name just three – who sorted the plant-animals out, one deserves special mention. John Ellis was a London merchant who liked to press sea life into artistic arrangements. Fascinated by the materials of his art he took to studying them. In 1765 he went down to Brighton by the sea, placed a living sponge in a glass bowl and saw that it sucked water in and out through its 'little tubes'. That, he said, in a letter to the Royal Society, is how a sponge receives its nourishment and discharges its excrement, from which it follows that sponges must be animals too.

> If we consult the ancients, we shall find that in the days of Aristotle the persons who made it their business to collect these substances [sponges] perceived a particular sensation, like shrinking, when they pulled them

Great Chain of Being, 1936. He found them in Augustine's and Thomas Aquinas' theology, Leibniz's cosmology and Spinoza's ethics, and in the writings of *inter alia* Addison, Locke, Pope, Diderot, Buffon, Herder, Schiller and Kant.

from the rocks; and, in the time of Pliny, the same opinion continued
that they have a kind of feeling or animal life in them; but after that no
attention was paid to this kind of knowledge . . .

He felt, with some justice, that he had vindicated Aristotle. Few were
convinced. Sponges only really became animals in 1826 when the Edinburgh
zoologist Robert Grant demonstrated their motile larvae.

Thus the Platonic–Aristotelian vision of nature as a ladder of perfection and its influence. Yet there is another vision of nature's order that can
also be found in Aristotle. Throughout much of his biology he does not
speak of a Ladder of Nature, but only of his great, natural groups of
creatures, all of which do much the same kinds of things – eat, sense, move,
reproduce – but do them in very different ways using very different devices.
Just as both these visions appear at different places in Aristotle's texts, both
also appear in post-seventeenth-century zoology. Sometimes they even
coexist, albeit uneasily.

Even as the Ladder of Nature triumphed, naturalists such as Pallas were
protesting that animals, in all their diversity, could not, should not, be
forced into a linear scale. In 1812 Cuvier divided the animals into four great
groups that he called *embranchements*: Vertebrata, Articulata, Mollusca,
Radiata. 'It formed no part of my design to arrange the animated tribes
according to perceived superiority,' he wrote, 'nor do I conceive such a plan
to be practical.' He elaborated his scheme in *Le règne animal*, 1817. Bold, clear,
detailed and synoptic, it made Cuvier famous. It is here that he celebrated
Aristotle as his great precursor who, he said, had left hardly anything for his
successors to do. Cuvier's classification, however, doesn't look very
Aristotelian at all. The great division between the bloodless and blooded
animals (reformulated by Lamarck as *animaux sans vertèbres* and *animaux à
vertèbres*) is abolished entirely; the hierarchy of Classes, Orders, Families
and Genera becomes enormously expanded; few of Aristotle's *megista gene*
survive intact. There is, however, an Aristotelian element to the scheme.
Just as Aristotle delineated each of his *megista gene* as a complex of functioning parts, so Cuvier delineated his *embranchements*. It is precisely this element
that would give rise to one of the most bitterly fought and consequential
battles in the history of zoology.

XCI

I N OCTOBER 1829 TWO tyro anatomists, Meyranx and Laurencet, submitted a manuscript to the French Académie des Sciences purporting to show that if one took a tetrapod and folded it in half backwards so that its tail touches its head (an exercise performed only on paper, I believe), it looked a lot like a cuttlefish. I do not know if their demonstration was inspired by Aristotle's analysis of cuttlefish geometry in *Historia animalium* and *The Parts of Animals*, for their manuscript appears to have vanished. In any event, the cuttlefish – blameless in itself – was the *casus belli* for a clash between two worldviews.

The protagonists were Georges Cuvier and his colleague at the Muséum d'Histoire Naturelle in Paris, Étienne Geoffroy Saint-Hilaire. Geoffroy, the elder of the two, had nurtured Cuvier's career (had indeed got him his job), but by 1830 the younger man had eclipsed his mentor. Cuvier's *Leçons d'anatomie comparée* had revitalized comparative anatomy; his *Le règne animal* was the standard for classification; his *Recherches sur les ossements fossiles de quadrupèdes* established the fact of extinction in the fossil record; his *Histoire naturelle des poissons* dwarfed everything else that had ever been written about fish. Napoleon had put him on the council of the Imperial University; the restored Bourbons made him a baron and then a peer of France – but to list Cuvier's works, titles, jobs and honours would take pages. Geoffroy's major contribution, by contrast, was the two-volume *Philosophie anatomique* (1818–22), an idiosyncratic collection of essays on comparative themes and teratology that espoused a *Naturphilosophie*-influenced 'transcendental morphology'.

The seeds of the dispute lay in Cuvier's 1812 classification. Following his hero Aristotle, Cuvier claimed that, within each of his four great *embranchements*, animals had fundamentally the same structures, shaped into their various forms by the contingencies of function. Animals in different *embranchements*, on the other hand, had merely analogous organs. Each *embranchement* was separated from any other by a gulf, one across which nature did not, could not, jump.

Geoffroy disagreed. One of nature's romantics, he was inclined to see unities where others saw differences. There was, he said, a grand Unity of Plan that spanned all the animals, a unity that transcended the walls of Cuvier's *embranchements*. Considering the exoskeleton of an insect and the vertebrae of a fish, Geoffroy proposed that they were one and the same structure. To be sure, insects have an *exo*skeleton (hard parts surrounding soft) while fish have an *endo*skeleton (soft parts surrounding hard) but where other anatomists saw this as ample reason to keep them distinct, he explained with the simple confidence of the true visionary that 'every animal lives within or without its vertebral column'. Not content with this application of his all-revealing system, he went on to show how the whole anatomy of a lobster was really very similar to that of a vertebrate − if you only flipped it upside down. Where lobsters carry their major nerve cords on their ventral sides (bellies) and their major blood vessels on their backs, the reverse is true for vertebrates (as indeed it is).

All this pained Cuvier for the violence it did to his *embranchements*; no, it outraged him. For years he stewed and sniped. When, in 1829, Meyranx and Laurencet submitted their paper to the Académie, Geoffroy was delighted. The wall between another two of Cuvier's *embranchements*, the Vertebrata and Mollusca, had been breached. He urged immediate publication. It was too much for Cuvier. Leaping to the defence of his much violated *embranchements*, he denounced the cuttlefish paper in session. It was all Geoffroy's fault; he didn't really blame the young men. Geoffroy replied, and for three months in 1830 the two zoologists were embattled at the Académie. The fracas went public; Goethe and Balzac championed Geoffroy, but the consensus was that Cuvier had destroyed him on points.

It is sometimes said that it was a debate about evolution, and it is true that Geoffroy was flirting with the idea. At the time, however, it was more a debate about the power and meaning of Aristotelian science. Geoffroy looked through the weird geometry of the cuttlefish to the unity that lay beneath, and argued that all its organs were the same as a vertebrate's, merely rearranged. For Cuvier, this was wrong in too many different ways. It was anatomically wrong: cephalopods, he showed forensically, have a variety of organs that vertebrates don't; it was conceptually wrong: there could be no identity across nature's great gulfs; it was historically wrong: a perversion of Aristotle's doctrines. Meyranx and Laurencet, a hapless pair, never did get their paper published. Cuvier, however, published his rebuttal.

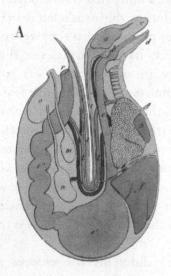 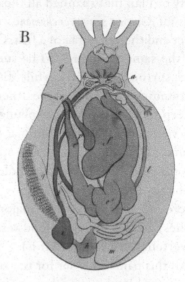

VERTEBRATE AND CEPHALOPOD GEOMETRY, COMPARED

Appealing to the authority of antiquity, Cuvier declared that the study of resemblance among species is 'the object of a special science that is called comparative anatomy, but it is far from being a modern science, for its author is Aristotle'. In reply Geoffroy spoke of how he had broken those ancient bonds: 'I did not content myself with Aristotle's account. At first I had never failed to cite Aristotle in my works . . . but I wanted to receive more advanced instruction from the facts themselves.' Cuvier sneered that, where Aristotle had built a monument of facts, Geoffroy was merely doing philosophy. It was not Cuvier's best-judged line.

A semantic fog enveloped the field. Both claimed that the cuttlefish's organs and a tetrapod's were 'analogous', but they clearly meant very different things by it. Cuvier's usage was closer to Aristotle's; Geoffroy, in a bold move, appropriated the term to mean precisely the opposite, what Aristotle called 'the same without qualification' and Owen, in 1834, would call 'homologous'. By March 1830, however, such terminological matters were no longer at issue. Nor, for that matter, were cuttlefish or classification. The protagonists were divided on something far more fundamental – how form should be explained.

Cuvier was the greatest functional anatomist of his age. It was his proud boast that he could classify an animal from only a single bone. Animal parts are correlated so that 'the form of the tooth implies the form of the condyle;

that of the shoulder blade, that of the claws, just as the equation of a curve implies all its properties'. This was the apotheosis of Aristotle's method. Cuvier's great explanatory principle, the Conditions of Existence that he expounded endlessly, was Aristotle's conditional necessity elevated to a law:

> Natural history has a rational principle which is particular to it, and which is usefully employed on many occasions: that of the conditions of existence, commonly known as final causes. Nothing can exist unless it unites the conditions which make its existence possible; therefore the different parts of each being must be coordinated in such a way as to make possible the whole being, not only in itself but also in its relations with those around it. The analysis of these conditions often gives rise to new general laws, as rigorously demonstrated as those of calculation or experiment.

In an age of scientific laws, Geoffroy had one of his own. Function, he declared, does not determine form; rather form determines function. Calling the vertebrate breastbone as witness, he explained the varying proportions of its parts in purely physiological terms. The hypertrophied sternal keel of a bird, to which the flight muscles attach, stunt other bones by 'diverting to its own profit the nutritive fluid' that might have fed them. No Cuvierian functional harmony there, just economics. He called his discovery the *loi de balancement* – Law of Compensation – and proclaimed it a great discovery. Goethe had already anticipated him. But Geoffroy probably got it from *The Parts of Animals*, for the *loi de balancement* is Aristotle's

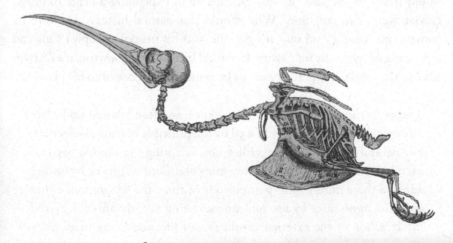

SKELETON OF A HUMMINGBIRD

'what nature takes from one part it gives to another' elevated to law. The Great Cuttlefish Debate of 1830 was, then, about many things: the unity of animal life, the identities of organs, the terminology by which those identities should be described and, above all, the causal explanation of organic diversity. It is testament to the scope of his thought and its protean quality that much of it was Aristotle *contra* Aristotle.

XCII

I T WAS THE LAST great scientific debate that Aristotle attended. Its protagonists lived but two centuries ago yet conceptually they are closer to him than they are to us, for they all wrote on the far side of 1859. *The Origin of Species* transformed the very terms of Aristotle's science or else rendered them obsolete: *genē* (and *embranchements*) became true families that descend from a common ancestor; dualizers ceased to dualize and became convergent solutions to adaptive problems; parts were no longer 'analogous' or 'the same without qualification', but analogous or homologous in a way, a new way, that depended on their origins in the tree of life. Geoffroy's Unity of Plan was explained by descent by modification; Cuvier's Conditions of Existence by natural selection.

It is sometimes said that Cuvier got his teleology from Kant, but for Kant teleology was just a 'heuristic fiction' and an invitation to despair. 'There would never be', he said, 'a Newton capable of explaining a blade of grass.' Cuvier was more sanguine. 'Why should not natural history also have its Newton one day?' ('And now it's got one' was his unspoken reply.) One can feel a pang of sympathy for Cuvier. If natural history has a Newton, it's Darwin who, in the *Origin*, is generous even as he puts his predecessor in his place:

> The expression of conditions of existence, so often insisted on by the illustrious Cuvier, is fully embraced by the principle of natural selection. For natural selection acts by either now adapting the varying parts of each being to its organic and inorganic conditions of life; or by having adapted them during long-past periods of time: the adaptations being aided in some cases by use and disuse, being slightly affected by the direct action of the external conditions of life, and being in all cases subjected to the several laws of growth.

Notice how subtly Darwin changes Cuvier's meaning. When Cuvier invokes the conditions of existence he is generally trying to explain the fit of an animal's parts to each other; when Darwin does, it is to explain the fit of an animal's parts to its environment. The difference is only one of emphasis. Anyone who seeks to understand the design of living things necessarily studies them as wholes in themselves and in their worlds; the three great students of animal design, Aristotle, Cuvier and Darwin, all kept at least an eye to both.

In the *Origin* Geoffroy's Law of Compensation appears under the heading 'correlation of growth'. 'I mean by this expression', says Darwin, 'that the whole organisation is so tied together during its growth and development, that when slight variations in any one part occur, and are accumulated through natural selection, other parts become modified.' Darwin's idea is more general than Geoffroy's, for he allows that the connections need not be economic. But he credits him (and Goethe) for the insight.

These concepts continue their scientific run. But they have transmuted again, for if Aristotle, Cuvier and Geoffroy all wrote on the far side of 1859, Darwin joins them in writing on the far side of 1900 or, if you prefer, 1953.* Aristotle's principle of conditional necessity is now just as often applied, if not by that name, to molecules or even genes. The platy, *Xiphophorus maculatus*, and the swordtail, *Xiphophorus helleri*, are small Mexican live-bearing fish. If forced they will cross-mate and the hybrids, a bit weirdly, can be hybridized again. Some of these second-generation hybrids develop melanomas that spread like mould upon a grape. Natural selection has adjusted the platy's 20,000-odd genes to work harmoniously together in the task of building a platy; the swordtail's genes harmoniously build a swordtail. But platy genes are not designed to work with swordtail genes and so the misbegotten hybrids, whose genomes are a grab-bag of their parents', die riddled with tumours.

They are true Empedoclean monsters. Geneticists call the genetic interactions that give rise to such effects 'fitness epistasis', but it's just a translation of Cuvier's 'conditions of existence' or Paley's 'relations' between parts, or Aristotle's 'conditional necessity'. In this guise the concept wends its way through Muller and Sturtevant's account of speciation mechanisms, Wright's shifting-balance theory, Kondrashov's explanation for the maintenance of sex, Kauffman's N-K landscapes and much more. Wherever it appears, the idea is always the same: you can't mix different animals up.

---

* The rediscovery of Mendelian genetics and the elucidation of the structure of DNA.

Aristotle's 'what nature takes from one part it gives to another' can also be couched in genetic terms. Genes that influence apparently different parts of an animal's body are said to have 'pleiotropic effects'. The term applies whether they do so by virtue of sharing flows of information, matter or energy. There is a mutant strain of nematode that has a life expectancy about half as long again as a normal worm's, but that lays far fewer eggs – the cost, apparently, of a long life. The mutation is said to have an 'antagonistic pleiotropic effect' for it increases one feature while decreasing another. The geneticist's 'pleiotropies', Darwin's 'correlations of growth', Geoffroy's 'Law of Compensation' and Aristotle's 'what nature takes from one part it gives to another' are all, then, related ideas. In its modern guise it underpins the evolutionary theory of life history and ageing as, in its ancient guise, it underpins Aristotle's. Wherever it appears, it expresses the same idea: that the parts of animals are irreducibly bound one to another.

Perhaps Aristotle's most important legacy is one that I have not touched on at all, but that also runs throughout the history of zoology. It is his insistence that the organic world is structured into natural classes that our classifications should not tear apart. For the moderns – Linnaeus and almost all systematists since – this idea became the search for a Natural System of classification. Darwin told us what such a system means and why it exists. 'I believe', he wrote, 'that propinquity of descent, – the only known cause of the similarity of organic beings, – is the bond, hidden as it is by various degrees of modification, which is partially revealed to us by our classifications.' Now the problem is to recover the shape of that hidden bond, the topology of the great Tree of Life. It's being revealed by scientists using very fast search algorithms that feed on terabytes of DNA sequences. Now the animals are divided into three great Super-Phyla (plus a few basal groups such as sponges) which, in turn, are divided into thirty-odd Phyla which, in turn, are divided into ever smaller groups unto species, which although not exactly innumerable can scarcely be said to have been numbered for there may be anywhere between 3 and 100 million of them on earth. The leaves on Darwin's great tree are almost uncountable.

The great tree, a metaphor for the history of life, serves as a metaphor for the history of ideas too. That nature does not make jumps; that there is a Ladder of Nature; that natural groups of animals exist; that those groups should be defined by the homology and analogy of organs; that organs are shaped by their functional *and* economic relations – all these ideas, I claim, can be found in Aristotle. They have also structured modern zoology for much of its history; they still do. We may wonder, however, whether they are the *same* ideas.

It all depends, of course, what you mean by 'the same'. Ideas are the organs of our thought, and like the organs of a cuttlefish and a tetrapod, they may be 'the same' by virtue of common descent or 'the same' by virtue of being independent solutions to similar needs. Aristotle himself was fond of remarking that the same ideas have occurred to many men at many times. (If that seems trite, no doubt that is because it is self-reflexively true.) For the cluster of ideas that I have discussed here, however, I believe that a good case can be made for identity by descent, intellectual homology if you will. Linnaeus, Geoffroy and Cuvier and their predecessors read Aristotle; Darwin read them; we've read Darwin. The genealogical thread is clear.

Among historians the tracing of conceptual genealogies across the ages – the 'history of ideas' (ideas pure and simple) rather than 'intellectual history' (their social and cultural context) – is rather unfashionable. They point out that thinkers of every age appropriate their predecessors' terms and concepts and apply them to their own ends; and that they do so even when the underlying structure of their thought has transformed their sense entirely. Philosophers call this process 'conceptual shifting' and delight to spot it as terriers do rats. Scientists – always sloppy with terminology, forever pushing new theories – are notorious for it. The ever-mutable meanings of 'analogy' and 'homology' are a case in point. Aristotle is addicted to it too – his *eidos* and *psychē* are expressly not Plato's.

Historians are right to stress this particularity, though not to the extent of denying the logic of modification by descent, a logic that applies with equal force to the realm of ideas as it does to life itself. In a way, it's just a matter of how you look at it. Focus on the cuttlefish at home in its cuttlefish world and its weird geometry appears as a solution to its own cuttlefish problems. But take a broader view, and it looks more like a small twist on a basic plan that was laid down a long time ago.

That zoologists drew their ideas from, struggled against or simply used Aristotle, and that they did so for centuries, seems incomprehensible to us now. Darwin eclipsed his predecessors; he became to us what Aristotle was to them: an authority to inspire or merely invoke. Yet, though we have forgotten their ultimate source, Aristotle's ideas, transformed and applied in ways that he could not have imagined, remain with us.

XCIII

A
RISTOTLE NEVER MADE the evolutionary leap. Of course he didn't. After all, he did not stand, as Darwin did, on the shoulders of Linnaeus, Buffon, Goethe, Cuvier, Geoffroy, Grant and Lyell. He heard no transformist whisperings from Paris and Edinburgh. He saw neither the mockingbirds of the Galapagos nor the fossilized giants of the Argentinian pampas. That he had the materials for an evolutionary theory at hand is, of course, evident only in hindsight. We may read Aristotle in Darwin, but not Darwin in Aristotle. By the same token Aristotle's system cannot be anti-Darwinian. His opponents were the *physiologoi* and Plato, none of whom were evolutionists in the Darwinian sense. Many of them were, however, evolutionists in a much looser sense for they gave naturalistic accounts of the origin or transformation of species. Aristotle, radically, rejected them all.

Creationism and evolution are rivalrous siblings. Both propose that the past was a very different place. Both propose that the creatures that we see in the world were not always there but have an origin in time. In the Greeks it's not always easy to tell them apart. The *physiologoi* may have rejected the myths but, as I said, the divine can often be found lurking somewhere in their thought. Xenophanes of Colophon (*fl.* 525 BC) is said to have argued that all living creatures originate from earth and water, though we don't know how he got them to do so. We know Empedocles' zoogony in confusing detail. First, there are those separated body parts, then their fusion into various improbable forms, then the selection process and, finally, the survivors sort themselves out by habitat. Democritus, too, evidently gave a naturalistic zoogony, but we know nothing about how it worked except that it ran on atoms.

The Pre-Socratic zoogonies are not, usually, transformist. Empedocles' creatures, having acquired their features, stick with them. But Anaximander of Miletus (*fl.* 525 BC) appears to have believed that humans are related to fish. The sources disagree how. One says that Anaximander held that humans originally resembled fish, another that humans arose from fish; another that they were born from a *galeos*. That may be a reference to the smooth dogfish, Aristotle's *leios galeos*, which nurtures its embryos in the womb via a placenta and umbilical cord and then gives birth to pups.

And then there is *The Timaeus*. Let us, for just a moment, treat Plato's origin myth with the seriousness that it doesn't deserve. Animals are degenerate humans. The gods transformed the silly, if harmless, men who studied the heavens (astronomers) into birds. Men who used their hearts rather than their heads became land animals: their forelimbs were drawn to earth; their heads deformed for lack of use. Truly stupid men acquired earth-bound bodies and many legs (centipedes?); the utterly thick were made legless (snakes or worms). Vicious men plumbed greater depths. Unworthy of breathing air, they were condemned to live as fish and snails in the muddy waters. Or else they became women.

Anaximander derives humans from fish, Plato fish from humans. The two theories have an appealing progressivist/degenerationist symmetry. Aristotle mentions neither. Indeed, he says very little about origin-of-life or species theories. When attacking Empedocles he, rightly or wrongly, treats his zoogony as embryology. But he was aware of them. In *The Generation of Animals*, discussing spontaneous generation, he says that if, 'as some allege', all animals, even men, were originally 'earth-born', they would have spontaneously generated from the earth as larvae – and he is thinking of the eel's *gēs entera*. Who, exactly, alleges? Anaxagoras? Xenophanes? Democritus? Diogenes? It doesn't really matter: he's just toying with the idea, pointing out that *if* there had been a zoogony, his nutritional physiology shows how it would have worked. As far as he's concerned, it never happened. As far as he's concerned, all sexually reproducing animal kinds have always existed and always will.

Our conceptual world is structured on a Manichean conflict between creationism and evolution. The conceptual world of the Greeks, before and after Aristotle, was structured on a conflict between creationist and naturalistic explanations for the origin of its living inhabitants. For Aristotle, there's not much to choose between them. Both fail to grasp one of the most salient features of the biological world: its regularity.

For Aristotle, the origin of any individual of a given sexual kind requires the existence of two others of the same kind. To make a sparrow you first need two other sparrows. His slogan, 'a human being gives rise to a human being', applies, *mutatis mutandis*, to all sexual kinds. Only parents – more precisely, the father – can supply the form, the *eidos*, required to make a new individual. This theory, taken literally, implies an eternal regress of sparrows. Aristotle takes it literally.

Aristotle's theory of sexual reproduction, and its metaphysical basis, is incompatible with any zoogenic or transformist theory. His theory of inheritance is too. I have argued that Aristotle has a dual-inheritance system. The formal system is the father's unique contribution to the embryo and transmits the *logos* – the set of functional features that will enable its offspring to live in its environment and, if it is male, reproduce its form in turn. The informal system, due to both parents, is responsible for variation among individuals of a kind – Socrates' *v.* Callias' nose – and encodes accidental variety. This division of labour between these two inheritance systems has profound consequences. Aristotle is perfectly prepared to allow that an individual can have a new mutation that gives him some novel feature, a snub nose say; but he does not – *pace* Socrates, who thought his snub so useful – seem to allow that it can be adaptive. In his view, all errors of development, inherited or not, are either devoid of functional import (odd-shaped noses) or deleterious (missing organs). The production of females aside, he never even hints that a mutation might benefit an animal. In his world every creature is, within the limits of its physiology, perfectly adapted to its environment; there is no room for improvement. Were he to meet Darwin he would ask – and ask rightly – where are these 'favourable variations' of which you speak? When a father's *sperma* fails to concoct the embryo, all I see is death, deformity or, at best, a girl. Darwin would have been unable to answer. Happily for him, his successors can – though not without difficulty.

This theory of inheritance obviously closes the door to evolution by natural selection. That troubles us but not Aristotle, for he never argued with Darwin. Could Aristotle have developed a theory of evolution? Perhaps. He'd have to throw some of his own theory overboard and the result wouldn't necessarily be Darwinian.

In middle age Linnaeus became convinced that new, stable species of plants could arise, and had arisen, by hybridization. Aristotle may have believed this too. In the *Metaphysics* he says that mules are 'unnatural'. In the zoology he doesn't. He certainly believes that, in general, only animals of the same kind copulate and produce offspring,* but he also says that animals of different kinds can sometimes mate and produce offspring; or at least they can do so when they are not too different in form, size and

---

* He doesn't define kinds this way – he's not applying a Biological Species Definition – it's just an observation.

gestation period.* He gives an elaborate explanation for why mules are sterile, but clearly thinks it's an exception since his hybridization limits are otherwise generous. He thinks that crosses between different kinds of hounds, wolves and dogs, foxes and dogs, horses and asses, and various raptors all yield fertile progeny. He moots the possibility that the 'Indian dog' is the F2 progeny of a male tiger and a dog bitch (if the tiger doesn't eat the dog), and that the weird *rhinobatos* (guitarfish, *Rhinobatos rhinobatos*) is the progeny of the equally weird *rhinē* (angelshark, *Squatina squatina*) and a *batos* (probably a skate, *Rajiformes*) – but here he's on uncertain ground and knows it.

That new animal kinds might arise from hybridization is inconsistent with Aristotle's oft-stated claim that the form of a kind comes from the father. If a hybrid is to have the functional features of both parents, as the *rhinobatos* presumably has, then its *eidos* must come from both. As I read him, Aristotle doesn't believe that, but there are enough difficulties in his texts to make it plausible that, at some time, he did.

But, had Aristotle taken the road to evolution, I think he'd have taken the road that Geoffroy Saint-Hilaire took. In the second volume of his *Philosophie anatomique*, 1822, Geoffroy laid the foundation of teratology, the science of monsters. He noticed how teratological deformities have a certain order and how they often resemble some normal species or other. He named one human deformity *Aspalasoma* because its urogenital anatomy resembled a mole's, *aspalax*. Such observations became transformist musings. 'Nothing is monstrous, and all nature is one' was one of his more gnomic sayings.

This is in the spirit of Book IV of *The Generation of Animals*: 'Even what is unnatural does, in a way, conform with nature.' Monsters *are* unnatural, but mostly because they are rare. Aristotle's impulse is to naturalize them by explaining them in terms of the normal processes of embryogenesis. Indeed, the 'cause of monstrosities is very close and, in a way similar to, the cause of deformed animals . . .' By 'deformed animals' Aristotle means here *naturally* deformed creatures. Moles are deformed because they are

* Zoologists tend to suppose that hybridization is rather rare. But 10 per cent of bird species can hybridize, and there are many cases of hybrids giving rise to apparently stable species. That's even more true of plants. Among Aristotle's hybrids, dog × wolf can hybridize and produce fertile offspring; there are no verified cases of dog × fox hybrids, though Aristotle claims that the Lacossian hound is one. There are reports of chicken (*Gallus domesticus*) × partridge (*Alectoris* sp.) hybridization in captivity, but the phenomenon is clearly so rare that we may doubt whether Aristotle's information is accurate.

blind; seals are deformed because they have flippers instead of proper limbs; lobsters are deformed because they have asymmetrical claws. They violate, in some way, the norms of the wider kinds to which they belong. In drawing this parallel, he means only that the moving causes of unnatural and natural deformity are the same. Unlike Geoffroy, he does not mean that deformity gives rise to new species. Aristotle never took the evolutionary leap.

He might have. Plato showed him how. Moral vice obviously won't transform a man into a fish, but a mutation, a *lysis*, might. Or at least it might transform a human into a tetrapod. Sometimes Aristotle's language suggests as much. There is a passage in *The Parts of Animals* in which he's explaining why tetrapods walk on four feet rather than two. He says that tetrapods have relatively heavy upper torsos compared to man. This excessive top hamper has two consequences. First, it causes their bodies to become unstable and hence lurch towards the ground; second, it inhibits the soul's activity centred on the heart. For these two reasons tetrapods developed – *egeneto* – bent over and then, for the sake of stability, nature gave them forefeet instead of arms.

Developed? In what sense did tetrapods *develop* four feet? Why the dynamic language? Why not just say that this is the way they are? It's not as though tetrapods are born walking upright, or that the world was once filled with cognitively crippled bipedal horses and sheep staggering about on their hooves. Presumably he's speaking metaphorically. Still, you can see where this comes from. He had the recipe down pat; he had used it so many times. Take an idea from *The Timaeus*. Discard the moralizing. Add some common-sense biology. Present as science.

XCIV

I T IS SOMETIMES SAID that Aristotle could not have been an evolutionist for want of evidence. This seems plausible. There is one class of evidence that Darwin had, and had in abundance, that Aristotle apparently did not: fossils.*

Aristotle did not know that in bygone ages the earth pullulated with creatures now extinct. He did not know that Lesbos and the Troad once looked – and, as these things scale, not so long ago – like the Serengeti with a fauna to match.† It is precisely such evidence, the argument goes, that was required before the theory of evolution could take wing. In November 1832, when Darwin arrived in Montevideo, Volume II of Lyell's *The Principles of Geology* – the one about the fossil record, biogeography and the transmutation of species (arguments against) – was waiting for him in the post.

Yet the argument is too simple. For, although Aristotle never mentions a single fossil in his works, or anything that can be construed as one, it is implausible that he knew nothing about them. More precisely, it is implausible that he was never confronted with *prima facie* evidence for the previous existence of life forms that were, in his day, at least locally extinct.

A roll call of Greek travellers and *physiologoi* before, contemporaneous with and immediately after him described stony objects that resemble animal remains. Beds of seashells located in unlikely places were particularly likely to attract attention. Xenophanes reported shells from a mountain in Sicily. He also reported the imprints of fishes and other marine life in stone from Syracuse, Paros and Malta. Xanthus of Lydia (*fl.* 475 BC)

---

* Perhaps one other: biogeography. Aristotle certainly doesn't think that all animal kinds are cosmopolitan, but he clearly does not, and cannot, have Humboldt's and Darwin's sense of the sheer strangeness of biotas in different parts of the world. But that's perhaps more problematic to a creationist rather than an eternalist. The former may well wonder why the Creator made all those different biotas; the eternalist will just accept their presence as given.

† In his *Geography* Strabo, who has a strong sense of tectonic instability, suggests that Lesbos was once connected to Mt Ida on the Asia Minor shore. In the Pleistocene Lesbos was indeed connected to the mainland of Asia Minor.

saw beds of stranded seashells in Anatolia, Armenia and Iran. Herodotus, Eratosthenes of Cyrene (*c.* 285–194 BC) and Strato of Lampsacus (*fl.* 275 BC) all puzzled over seashells in the middle of the Egyptian desert near Karnak. That the sea must have once covered the land was obvious to them; they just disagreed how.

FOSSIL SHELLS FROM CALABRIA

In *On Stones* Theophrastus describes 'dug up' – *oryktos* – ivory.* He does not give its origin, but the megafaunal deposits of Samos, Kos or Tilos to the southeast of Lesbos seem like a good guess. The late Pleistocene through Holocene levels contain the remains of a dwarf elephant species that may have survived until four thousand years ago. The deposits have been known since at least the Archaic Period. In Samos, the bones of giant extinct animals were displayed, *Wunderkammer*-style, at a cult temple to Hera. Local myth had them as the remains of ancient monsters called 'Neades'. A bone dug up near a seventh-century altar belonged to the extinct Miocene giraffe, *Samotherium*.

Lesbos' own megafaunal fossils are more modest. You can see them at the little natural history museum in Vrissa, a village just above the

---

* In the *Meteorology* Aristotle also talks about *oryktos* things. Here there is potential for confusion because in English translations (e.g. H. D. P. Lee's Loeb) *oryktos* is sometimes given as the Latin *fossile* whence our 'fossil'. Since Aristotle's *fossiles* are clearly inorganic stuff such as lumps of sulphur, it's easy to assume that he's confused and supposes that they have an organic origin. But *oryktos* and *fossile* only mean 'dug up'. It is only in relatively modern times that 'fossil' acquired its current meaning as the petrified remains of once-living creatures – the sense in which I use it here.

Lagoon. Kostas Kostakis, the caretaker, is particularly proud of the giant tortoise whose remains were found near Vatera. A life-size reconstruction made of fibreglass has the dimensions of a VW Beetle, but the fossils themselves are a bit disappointing. The whole thing has been extrapolated, no doubt accurately, in a Cuvierian fashion, from some leg bones, claws and scutes.

No surprise, then, that Aristotle does not speak of giant, extinct Lesbian tortoises. But how did he miss the vast petrified forest that litters the island? In the pyroclastic hills west of the Kalloni, the trunks of extinct conifers, complete with root systems, emerge from the phrygana like sawn-off temple columns. In the little port of Sigri massive stone trunks lie on the beach. They have lain there, immovable, since they were felled by a volcanic eruption 20 million years ago. Aristotle says nothing about them; Theophrastus, too, is silent. In his *Enquiries into Plants* the latter mentions 'petrified reeds' from the shores of the Indian Ocean (bamboo? coral?), but of the petrified forest of Lesbos not a word. Yet Sigri is the next port over from Erresos, his home town. As a boy he could have played on those trunks. They, too, now have a museum, a glorious one.

The mystery may have a prosaic solution. It may be that Theophrastus, at least, knew all about the petrified forest and wrote about it. Diogenes Laertius records a Theophrastan work that may have been titled *On Things Turned into Stone*. That suggests to us that it was about fossils, but we do not know, since Diogenes' text is corrupt and an alternative reading is *On Burning Stones*, which is presumably about coal or volcanoes.

Perhaps, then, it is not the fossils that are missing, just the texts. Alternatively, perhaps, Aristotle set aside reports of desert and mountain clams as fantasy. (Did not Herodotus also say that Egypt contained necropoli of winged serpents – that he had seen them?)* Or perhaps, to continue the excuses, Aristotle simply never got over to the far side of Lesbos. The hills were hot; he was a bad sailor; Theophrastus forgot to tell him about the stone trees. All this is possible. But I wonder whether he chose, deliberately, to ignore the reports or even the evidence of his own eyes. After all, if you believe in the eternity and immutability of organic kinds, it is just possible that you might dismiss a forest as a field of stones.

---

* Herodotus' winged serpents, it has been suggested, were amphibian fossils from Makhtesh Ramon in the Negev. Others moot that they originated in descriptions of *Spinosaurus* deposits in the Western Desert, the depictions of serpents with feathered wings that can be found on Egyptian sarcophagi (cf. the British Museum) or cobras whose hoods that Herodotus mistook for wings.

XCV

THAT THEOPHRASTUS MAY have written a book about fossils
tantalizes. That is because he took the road that his teacher did
not.

The first steps are small. Discussing differences between cultivars –
Thracian wheat, Egyptian pomegranates, Apulian olives and the like –
Theophrastus recognizes that a plant is shaped by both what the seed
gets from its parent *and* its environment. That's quite conventional.
But then he goes on to explain that when a cultivar is transplanted
from one region to another it acquires, within just a few generations, a
new nature:

> From this second source [differences in environment], moreover, arise
> peculiarities within kinds (that is to say varieties); and we often find
> that what was contrary to nature has become natural once it has
> persisted for some time and increased in numbers.

This is very un-Aristotelian. It allows the boundaries of formal natures
to shift. It also mingles the formal and material causes that Aristotle
strives so hard to separate. But Theophrastus doesn't let it rest there
for he also argues that the cultivars found in different countries are
'useful'. He means that Thracian wheat sprouts late because Thracian
winters are harsh and that if you plant a seed in a new country it will
eventually change *to meet the challenge*. Theophrastus' plants aren't
perfectly adapted; they can improve. His vision of the world is also
teleological, but where Aristotle's world is frozen perfection,
Theophrastus' is contingent and in flux.

He's so modest, so plodding, so reluctant to propose big theories, that
it's easy to miss his most radical claim of all. Up till now, Theophrastus has
just been talking about the origin of new varieties of wheat and grapes. If
that's evolution, then it's evolution of a pretty paltry sort. But what about
the origin of *species*? Can one *kind* of plant transform (*metaballein*) into
another? Yes, says our botanist, looking up from the ground, it's rather
marvellous when it happens, but it definitely can.

Wheat can transform into *aira*. These cereals, he says, are different kinds; you can tell them apart by their leaves. Some people doubt that one transforms into the other; they say that *aira* just happens to grow in wheat fields during especially rainy years. But, Theophrastus continues, the 'best authorities' agree that many people have sown wheat but reaped *aira*.

Well, maybe. It's not that Greek farmers didn't sometimes sow wheat and reap *aira*, they probably did, but the explanation for this on the face of it remarkable event isn't some lightning-bolt transformation. *Aira* is, as Theophrastus says, a totally different species, it's a weed called darnel (*Lolium temulentum*) and the reason that a farmer might find his fields full of it is that its seeds look a lot like grains of wheat.* The transformation of wheat into darnel is, then, just the report of a farmer who failed to sort his seed stock and who, confronted with a field of toxic cereal, had to explain the fact away.

But there *is* an unwitting truth to Theophrastus' transformist claim. Darnel doesn't mutate instantly into wheat, but the reason its seeds look so much like wheat grains is that they have evolved that way. Its history is written in the archaeology of the Levant. It's been a weed since before Babylon; farmers were sorting it from their seed stock in the Neolithic. But sorting is selection and selection is, given heritable variation, evolution. Over millennia the weed has evolved to mimic the grain the better to escape the farmer's sieve; by the fourth century BC it was a cuckoo infesting the seed banks of Europe. It took modern chemical herbicides to kill it off.

Would Theophrastus have bought this evolutionary tale? Probably – after all, he accepts that transformation can happen in a single season. True, he's uneasy about his wheat/darnel (it's one of several 'problems' that he considers about the generation of plants), but, having convinced himself of the fact, takes it in his theoretical stride. He discusses the origin of the transformation and concludes that some sort of 'corruption' in the seed

---

* The difference between wheat and darnel becomes obvious only if you turn them into bread. A fungal symbiont soaks darnel seeds with a cocktail of psychotoxic alkaloids and indolediterpene neurotoxins that causes dizziness, coma or death. In Attica *aira* was the drug of the Eleusinian rituals; in medieval Europe it was used to get a religious high. Its habit of sneaking into seed stock has made it a metaphor for false belief. It's the 'tares' of Matthew 13:24–30 ('Nay; lest while ye gather up the tares, ye root up also the wheat with them. Let both grow together until the harvest: and in the time of harvest I will say to the reapers, Gather ye together first the tares, and bind them in bundles to burn them: but gather the wheat into my barn'). In the seventeenth century it was a symbol of subversion and the Pope.

must cause the 'starting point' of the embryo to be 'mastered'. This, he continues, is analogous to what happens when a female (animal), or something even more unnatural, is produced, for we must think of the 'earth as a female'.

He's simply appropriated Aristotle's theory of monstrosity to explain the evolution of one natural kind into another – and it *is* evolution even if it is still far from Darwin's vision of a great tree of life. So often, when reading Aristotle, we sense the pressure of transformism. It is then that we should suspect that we are merely reading our own evolutionary preoccupations into texts that are, in fact, devoid of them. But the pressure must have been there, for Theophrastus, once his student, then his colleague, eventually his successor and, for more than twenty years, his friend, yielded.

XCVI

WILLIAM OGLE, WHO loved Darwin and Aristotle both, wished they could have met in person. In his letter to Darwin he imagines the Greek arriving at Down House. Aristotle considers Darwin with suspicion. He scans, as authors do, the study's bookshelves for his own works. He is astonished, as authors are, to find them not there – as, indeed, they weren't, Darwin having, by his own admission, long forgotten what little Greek he ever knew. He would also, continues Ogle, be astonished to discover that his views were of only antiquarian interest and that his old foe Democritus had triumphed; had, indeed, been reincarnated in Darwin. 'I have, however, such faith in Aristotle as an honest hunter after truth', writes Ogle, 'that I verily believe that, when he heard all you have to say on your side, he would have given in like a true man, and burnt all his writings.'

That seems optimistic. Aristotle would surely have scornfully pointed out that Democritus was oblivious to the appearance of design in nature, and – making the priority clear – have congratulated Darwin for placing final causes at the centre of his theory. He would have dismissed pangenesis as warmed-up Hippocratic theory and natural selection as a new label for Empedocles' maunderings. He would have been right about the first and wrong about the second. He would have been enchanted by the biota of the New World and impressed by the fossils. (You can't ignore a *Megatherium*.) Perhaps, upon reflection, he would even have granted that species evolve, that his grand vision of life's order had been subsumed by a grander one. I like to think so.

Were he to do so, he'd have to discard some of his metaphysics but, insofar as the two can be severed, not very much of his science. Theodosius Dobzhansky famously remarked, and evolutionary biologists endlessly repeat, that 'nothing in biology makes sense except in the light of evolution'. The sentiment is a fine and ringing one, always handy if there's a Creationist about, but it isn't really true, for quite a lot does.

Aristotle understands as Darwin did and we do that: (i) the complex morphologies and functions shown by living things require a primal source of order or information, his 'formal natures' or simply 'forms'; (ii) that these forms are dynamic, self-replicating systems; (iii) that they vary among kinds to give diversity; (iv) that they exert their power by modifying the flow of materials in development and physiology; (v) that organisms gain these materials from nutrition which is transformed internally; (vi) that this material is limited in quantity; (vii) that the manufacture of parts, production of progeny, indeed survival itself, all expend this material – that is, are costly; (viii) that these costs limit the forms and functions of organisms such that if they do or make one thing it is at the expense of not being able to do or make another; (ix) that these costs are not absolute: some organisms are more subject to them than others; (x) that these material constraints act in concert with functional demands to give the diversity of animals that we see in the world; (xi) that the parts of animals are suited to the environments in which they live, that they are, in a word, adaptations; (xii) that the functions of different organs depend on each other – that is, living things must be understood as integrated wholes. Much of modern evolutionary science is in this list – but evolution isn't.

You may object that these similarities are superficial. After all, evolution is a dynamical theory and Aristotle's world is static. But dynamics are difficult and so, when accounting for the features of animals, biologists often assume a world at equilibrium. What remains then for us, as for Aristotle, is an engineering problem, the search for the *optimal* solution in a set of possible solutions. 'Nature', he says, 'does that which, among the possibilities, is the best for the being of each kind of animal.' It is the engineer's credo and the starting point for biomechanics, functional morphology, sociobiology and all the other sciences of organismic design. It is surely no accident that Aristotle established this principle, and declared it fundamental, in a book on the locomotion of animals.

Although I have counterposed, as he does, Aristotle's use of teleological and material explanation, he clearly thinks that usually there is no conflict between them at all. When explaining the association of parts he sometimes appeals to functional harmony, sometimes to bodily economics, but often ecumenically opts for both. Rays have cartilaginous skeletons because they need to be flexible given how they swim *and* because, having spent all their earthy matter on their hard skins, they don't have anything left for the skeleton. Such double-barrelled arguments appear to be redundant, but in fact they're not; they are merely missing an additional premise. For Aristotle, functional demands and the allocation of resources are harmonized because 'nature does nothing in vain'. In their *Principles of Animal Design* (1998) Weibel and Taylor call this the 'principle of symmorphosis'.

The history of Western thought is littered with teleologists. From fourth-century Attica to twenty-first-century Kansas, the Argument from Design has never lost its appeal. Aristotle and Darwin, however, share the more unusual conviction that though the organic world is filled with design there is no designer. But if the designer is dead for whose benefit is the design? It's the prosecutor's question: *cui bono?*

Darwin answered that individuals benefit. Biologists have batted the question about ever since. The answers that they've essayed are: memes, genes, individuals, groups, species, some combination or all of the above. Aristotle, however, generally appears to agree with Darwin: organs exist for the sake of the survival and reproduction of individual animals. This is why so much of his biology seems so familiar.

Yet there *is* a deep difference between Aristotle's teleology and Darwin's adaptationism, one that appears when we follow the chain of explanation that any theory of organic design invites. Why does the elephant have a trunk? To snorkel. Why must it snorkel? Because it's slow and lives in swamps. Why is it slow? Because it's big. Why is it big? To defend itself. Why must it defend itself? Because it wants to survive and reproduce. Why does it want to survive and reproduce? Because . . .

Because natural selection has designed the elephant to reproduce itself. Darwin gave teleology a mechanistic explanation. He halted the march of *whys*. It is for this reason that Ogle celebrated Darwin as Democritus reincarnated. For, where Aristotle's organismal teleology is imposed upon recalcitrant matter, Darwin showed how, given a few simple conditions, it emerges from it. Darwin is an ontological reductionist; Aristotle is not.

Why, then, should Aristotelian animals strive to survive and reproduce? Aristotle can hardly invoke natural selection. (He's dismissed at least one version of it.) He could have said 'they just do', and left it at that, but then he would not be Aristotle, so he does have an answer, beautiful and a little mystical. Living things, he says, desire to survive and reproduce so that they can 'participate in the eternal and the divine'. When he asserts that living things desire to participate in the eternal he means that they are designed not to become extinct. *Cui bono?* It turns out that organismal design is not, after all, for the sake of individuals, for they always die, but to ensure that their forms/kinds, their species, persist for ever.

When Aristotle speaks of the divine he is not – the point must be made again – invoking a divine craftsman for none exists; rather, he is telling us that immortality is a property of divine things and that reproduction makes animals a little bit divine.

We are beginning to touch on Aristotle's theology, his ultimate explanation for why the cosmos is arranged the way it is and its relationship to an immortal God. Why should animal kinds be immortal? This is where we come to the end of explanation, to one of those indemonstrable axioms that lie at the bottom of every Aristotelian science, and from which all else flows, and it is simply this: it is better to exist than not to exist.

KOSMOS

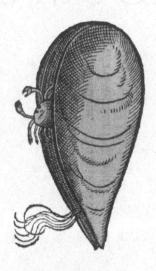

PINNOPHYLAX & PINNA — FAN
MUSSEL CRAB & FAN MUSSEL
— *NEPINNOTHERES PINNOTHERES*
& *PINNA NOBILIS*

XCVII

WHEN ARISTOTLE SPEAKS of 'perfection', it's often easy to understand him in simple zoological terms. More 'perfect' progeny are more fully developed at birth than less 'perfect' ones. And when he says that the arrangement of some organ is 'better' than another, he usually gives some quite ordinary functional reason for why it is so. But, as he reveals his vision of the order of living things, it becomes plain that another, metaphysical value system is also at work.

Aristotle's animal geometry (*above–below*, *before–behind* and *left–right*) does not map on to the geometry of modern biology (*anterior–posterior*, *dorsal–ventral* and *left–right*). That's fair enough: his geometry is trying to capture functional analogies, ours structural homologies. His valuation of the poles is more alien – when he tells us that *above* is more 'honourable' (or 'valuable') than *below*, *before* more honourable than *behind* and *right* more honourable than *left*. There is, apparently, some biological rationale for these assessments: sense organs are arguably more useful than buttocks or tails; eating is, for most people, more pleasurable than defecating; more people are right handed than left. Even so, one may wonder whether honour or value deserves a place in a functional biology – it doesn't have one in ours.

Yet his teleology is riddled with such value judgements. He says that the position of the heart in the middle of the body is dictated by its embryonic origins. But it is also located more *above* than *below* and more *before* than *behind*, 'For nature when allocating places puts more honourable things in more honourable positions, unless something more important prevents this' – the language suggests the seating plan at a dinner. One may wonder why, then, the human heart (actually its apex) is located on the inferior left, but Aristotle has inserted a caveat – 'when nature does nothing . . .' – and gives a patently *ad hoc* explanation that it's needed there to 'balance the cooling of things on the left'. He thinks, of course, that the right-hand side of the body, being more honourable, is hotter than the left, and that this is especially so in humans, and so the heart has to shift to compensate for the left's relative coolness.

Even when not speaking of honour, Aristotle appears to think that certain geometrical arrangements are simply 'better' than others, independently of their functional worth. He thinks it is better for organs to have a single origin. He likes symmetry. Given that the heart – the sensorium – is in the middle, it's 'best' that it have three chambers: the middle chamber is a single origin which the others nicely counterbalance. This is a murky side to his biology. One senses the influence of popular, Pythagorean or, most likely, Platonic notions of value. The biology – if one can call it that – of The Timaeus is not so much riddled with religious values as founded on them.

Plato's influence is most obvious when Aristotle considers man. He is explicit: man is his model not only because he's the animal we know best, but also because he is the most perfect animal of all. The axes of the body are most differentiated in humans; in other animals they're present but in a confused sort of way (in quadrupeds, recall, the above–below and before–behind are the same). In the same way, the characters of animals – courage, timidity, intelligence and the like – that are regulated by the sensitive soul are better developed in humans than in any other animal. For some of these features human exceptionalism is merely quantitative (we differ from animals by the more and the less); for others the difference is qualitative (we differ by analogy). There's a sense in which a swallow shows intelligence when it builds its beautiful little nest, but human intelligence is of an altogether different kind.

Since the capacities of the sensitive soul are most developed in humans, it is among humans that their variation is most obvious. You can see this in the difference between the sexes. Males are generally more courageous and faithful than females, but less compassionate, deceitful, shameless, jealous and depressive. Should a female cuttlefish be struck by a trident, Aristotle says, the male heroically sticks around to succour his mate; should the male be hit, the female just scarpers. It's like that in humans, just more so. It seems that Aristotle has, in general, quite a dim view of female character, that he thinks that women are less perfect than men. Actually, that's putting it rather gently for, in The Generation of Animals, he says that females are 'immature', 'deficient', 'deformed' and even 'monstrous'. Feminist scholars have made much of this.

As well they might. However, I don't want to put Aristotle in the dock for his gender ideology, but only for his science. It's not that he doesn't have his reasons – of course he does, he's Aristotle. He asks: why are the two sexes found in separate bodies? After all, it doesn't have to be that way;

it isn't so in plants; separate sexes need, then, to be explained.* The explanation is teleological: animals (most of them, at any rate) have distinct
sexes because it's 'better' that way. The reason it's better is quite abstract.
One of the ways in which Aristotle expresses la différence is by saying that
males contribute the moving cause and females the material cause to their
progeny. The moving cause is, he asserts, superior to the material since it
embraces the animal's definition and form. And, he continues, it's better
if superior things do not mix with inferior things. This is just axiomatic.
So it's better for males and females to exist in separate bodies than in a
single body.

The existence of separate sexes is, then, due to a division of labour
between the causal powers required for reproduction, with males having a
superior role. Superior? He actually says 'more divine'. Well, at least that
give females some purpose in life. The rest of his sexual biology is consistent
with this skewed assessment: girls are produced when the semen fails to
'master' the menses; males are hotter than females; semen is purer than the
menses; form is superior to matter and so on. He gives no empirical
evidence for any of this. On the other hand, eunuchs are mutilated and
feminized. The inference that females are defective is reasonable, even if it
doesn't really follow.

When he turns to humans as a species, his passion for connecting and
explaining, always ardent, becomes boundless. He links a long list of our
features together – libidinousness, volume of sexual secretions, fecundity,
posture, limbs, bodily proportions, hairlessness, blood type, heart structure, sociality and, above all, intelligence – in a complex causal web. One
can enter this web at almost any point, for example, at sex.

Aristotle thinks that we are exceptionally libidinous: only humans
and horses have sex during pregnancy. We are so because we produce
more seed, for our size, than any other animal.† Because women produce
so much menstrual fluid, they are also, for their size, unusually fecund.
Most large animals produce only one offspring, and so, usually, do

---

* Note that he is explaining the existence of separate sexes, not sexual reproduction per se.
So he's not interested in the modern question of the adaptive explanation for sexual
reproduction or recombination and its costs.
† That women menstruate more than any other species is true; that men produce more
semen, for their size, than any other mammal is not. Boars produce 250ml of semen/
ejaculate, a man about 2.5ml; given that they weigh about the same, ejaculate volume per
unit mass in man is far less. In fact, taking copulation frequency into account, men
produce less semen, per unit mass, than most farmyard animals.

women. But women also frequently bear two or three; he's even heard of quintuplets.

Why do we produce so much seed? Aristotle gives two answers, both based on his physiology. The first is that, of all animals, we have the hottest and most fluid bodies. The second is that we are naked. Unlike other animals, we don't have tusks, horns or even very much hair; since we don't expend nutrition on such things we can put our nutrition into seed. Aristotle is particularly insistent that hair is grown at the expense of semen. Eunuchs and women, he observes, don't go bald because they spend much less than men. On the other hand, bald men are exceptionally keen on sex. He also thinks that semen drains matter from the brain, which is why too much sex gives you sunken eyes.*

All this Aristotle explains in *The Generation of Animals*. But it is in *The Parts of Animals* that he gives the ultimate reason for human exceptionalism. That's where he explains why humans are naked. We are so, it turns out, because we alone have an ultimate weapon, one that can be turned into any other – a talon, claw, horn, spear or sword – as we please, namely, our hands; for our hands can make and grasp all of these; and following the principle of economy ('nature does nothing in vain') we therefore don't need any other.

Why do we have hands? Anaxagoras said that humans are the most intelligent animals because we have hands. That, says Aristotle, is to reverse the true direction of causality: we have hands because we are the most intelligent of animals (for only a highly intelligent creature would be able to use them). Moreover, we *can* have hands because, uniquely, we stand upright. So why do we stand upright? We do so because we grow that way. All animals are dwarfish, not only in stature, but in intellect, compared to us. And we grow that way because we are the hottest of all animals – which, along with our pure and thin blood, makes us the most intelligent of animals. So posture and intelligence are closely linked by material necessity. There's a final cause too, and here we come to the end of this long causal chain. We are upright and can reason not merely because we are the most perfect animal, but because we are the most divine. That is just part of the definition of our substance, not to be explained. Thus, it turns out, the reason that we are special in so many ways – even, piquantly enough, so rampantly libidinous – is because we, of all animals, are close to God.

---

* Since the brain is not the centre of higher cognition but a kind of radiator, this is not as deleterious as it may seem. It follows that while you can, in Aristotle's view, literally fuck your brains out, you cannot shag yourself senseless.

XCVIII

I N *HISTORIA ANIMALIUM*, DISCUSSING the various ways in which animals differ from each other, Aristotle distinguishes several levels of social organization. Most animals, he says, are solitary, some are gregarious, but a few are 'political' in that they work together for some common goal. Cranes, he thinks, are exceptionally intelligent birds in that they 'submit to a leader' who, with loud calls, keeps his flock in check in the course of their migratory flights.* His favourite political animal is, of course, the honeybee.

The intricate habits of bees obviously fascinate him. He records how they visit only one kind of flower at a time; how they recruit their fellows to a patch of blooms and how they waggle when they arrive at the hive carrying a load (but he does not know why).† While some workers busy themselves producing honey, others construct the comb and yet others collect water – a beautiful division of labour. The leader bee (his 'king', our queen) is also a specialist designed for only one purpose: the production of more bees. Honeybees have a collective goal: the maintenance of the hive. They keep the place spotless. They die in its defence. They ruthlessly regulate its internal economy and dispatch members surplus to requirements. Drones, those useless creatures, are particularly at risk.‡

It's all fascinating. Yet Aristotle's discussion of honeybee behaviour points to a glaring hole in his biology: behavioural ecology. He explains so much about animals, but not why they behave as they do. There is no *Habits of Animals* to set alongside *The Parts of Animals* and *The Generation of Animals*. As

---

* Eurasian cranes do co-ordinate their flights by bugling calls, but I have found no evidence that there is single leader. It is thought that flocks do not need centralized commands; at least they can be modelled by swarms of leaderless, interacting agents.
† Bees do indeed visit only one kind of flower at a time, a phenomenon known as 'single flower visitation'. The movement they make upon returning to the nest is the waggle dance studied by von Frisch between 1923 and 1947. Aristotle does not, however, say that it's a signal to the other bees. And workers do undertake specialized tasks.
‡ Curiously, Aristotle never explains what drones are good for. According to his model of bee generation, they're a reproductive dead-end and he says that they don't do any work. They seem to falsify his dictum that nature 'does nothing in vain'.

a result we don't know his answers to some extremely interesting questions.

How, for example, do bees regulate their affairs? In his *Oeconomicus* Xenophon gives one view. Ischomachus, a rather smug character, is telling Socrates how he instructed his young bride to manage their household. I told her, he says, about the queen bee. The queen bee (and she really is a queen rather than a king) instructs the workers what to do, parcels out the food, oversees the construction of the comb and the rearing of the young. You, my dear little wife, should do the same.

Xenophon's queen bee is the ruling intelligence of a command economy. Of course, his dialogue, written around the time that Aristotle was at the Academy, is no more a contribution to apiology than was Mandeville's *Fable of the Bees*, but it probably does reveal how an educated fourth-century Greek thought the hive ran. (More so since Xenophon was the sort of gentleman-farmer who could write an elegant little essay *On Hunting with Dogs*.) Xenophon's view, however, doesn't seem to be Aristotle's. His leader bee shows a dearth of managerial instincts: it just sits about generating more bees. The only time that it exhibits any initiative is when it zooms off with a swarm in tow to make a new hive. It also exists on sufferance. Workers, he says, often kill young leaders lest they lead to faction (multiple swarms) and so weaken the hive. And should two swarms meet and unite, one leader is eliminated. In Aristotle's hive the proletariat seems to be running the show.

But, as I said, it's hard to know exactly how, or to what end, Aristotle thinks the hive is organized since he didn't tell us. The absence of an ecological treatise is puzzling. The data are there. He saw that they could be built into a science for, in *Historia animalium*, he does venture a few ecological generalizations. In *The Parts of Animals* he also essays some ideas on how physiology affects animal characters. (Hot-blooded animals are courageous, cold animals timid – that sort of thing.) It's the teleology, the functional biology, that's missing. Perhaps there was a *Habits of Animals* long since lost; after all, only one-third of his works are extant. If so, he does not refer to it and the doxographers have left it off their lists. It's also possible that he felt no need to write one, having already written a treatise on the most social animal of all – the work that we call the *Politics*.

XCIX

'**M**AN IS, BY nature, a political animal' – it's his most frequently quoted apophthegm. It appears in Book I of the *Politics*. It is sometimes said to be Aristotle's definition of our species, but it isn't. If anything, his point is that we have quite a lot in common with some other animals. Aristotle's *politike episteme* – political science – is very sociobiological. Both sciences are rooted in animal behaviour; and both assume a strong view of human nature – that is, assume innate desires and capacities. Aristotle would agree with E. O. Wilson and Steven Pinker: humans are not born blank slates; they have an innate desire to co-operate.*

To illustrate this instinct, Aristotle gives a quasi-historical account of the origin of the state. It began with the formation of the household. The basis of the household was a union between male and female. This wasn't a reasoned choice, just the instinct to procreate. There was also a union between a natural ruler and subject who instinctively came together for the sake of protection. He means the domestic animals and slaves that nature has providentially provided for the Greeks. (Barbarians, being less *évolué*, do not distinguish between women and slaves.) Driving the point home, he adds that women and slaves are distinct since nature isn't some cheapskate coppersmith who makes a multi-purpose device. (If the analogy sounds familiar, that's because it appeared in his argument for specialization of insect organs.) It's understandable that Aristotle can't imagine a household without a woman. It's more striking that he can't imagine one without a slave or, at minimum, an ox.

The purpose of the family household, slaves and all, was to supply daily needs. Clusters of related households then formed multi-generational villages to supply non-daily needs. At first the villages were dispersed ('as was the manner in ancient times'), but then they formed denser associations for the sake of complete self-sufficiency. The city-state – the *polis* –

---

* Ironically, the image of the mind as *tabula rasa* has its origin in *The Soul* (*de Anima* 430a1). There, however, the image is merely used to explain the workings of the intellect and not the cognitive state of newborns. (The transition from potential to actual thought is compared to the act of writing on a tablet.) The modern use of the image is due to Avicenna, Thomas Aquinas and Locke.

was born. The ability, desire and need to live in a state are among the marks of humanity. Any man who, *by nature*, cannot live in the state is either a 'tribeless, lawless, hearthless' monster – he quotes Homer on the Cyclopes – or a god.

That most men and women have an instinct to procreate, or that domesticated animals have an instinct to serve humans, seems uncontroversial. Just such instincts result in households consisting of two parents, two children and a dog. Aristotle's account of the *genesis* of the state – social structures of increasing complexity driven by innate human desires for increasing economic capacities – also resembles many later evolutionary theories of the origin of the state.* But his story contains a less familiar element. Do some men have an instinct to be *ruled* by other men? Yes, says Aristotle, some men are 'natural' slaves:

> A human being who belongs, by nature, not to himself but to another is, by nature, a slave. One human belongs to another if, despite his being human, he is a piece of property. A piece of property is, as a distinct entity, a tool, suitable for action.

What, exactly, makes a man a natural slave? It's clearly not just the fact that he's owned, for Aristotle immediately notes that some men are 'legal' slaves; they're the plunder of war. Nor is a natural slave just a man who was born to slaves. Rather, he's one who is defective in some way and can't help but be a slave:

> People differ from each other as much as mind does from body and human from beast. Those whose function happens to be the use of their bodies (when this is the best that they can achieve) are slaves by nature.

Natural slaves are men so devoid of reason that they are basically animals.

Aristotle prized the life of the mind above all else; even so, this is quite extreme. In fact he quickly acknowledges that natural slaves are men and so have, at the very least, the ability to follow commands even if they can't think for themselves. The natural slave is, then, a barely sentient tool that nature has provided for the use of men capable of reason. He also suggests

* To give but one example, the account of state formation that Francis Fukuyama puts forward in *The Origins of Political Order* (2011) is sociobiologically inspired and also has a strong Aristotelian flavour.

that nature has made the bodies of natural slaves stronger and less erect than those of freemen, but he concedes that nature doesn't always get it right and sometimes gives a freeman the body or soul of a slave. (He avoids the concomitant, that a slave may have the soul of a freeman.)

This is not an attractive theory. Unsurprisingly, Aristotle has often been accused of defending the injustices of the society in which he lived by appealing to nature – that is, of committing the 'naturalistic fallacy': the derivation of an 'ought' from an 'is'. (It is an accusation also frequently levelled at sociobiologists with far less justification.) That may or may not be so. The more interesting question, however, is: does it contain some truth?

Set the question of ownership aside, and the difference between a free-man and a slave is, in Aristotle's view, the ability to exercise reason. To place it in a modern setting, the difference is between senior management and the workers that they control at, say, a Fulfilment Centre of the sort that mail-order firms run. For a senior manager, control is a monthly report to the board; for a 'picker', control is a handheld device that instructs him what to pick off the shelves and where, that plots an optimal path for him, and relays real-time efficiency data on his movements to roving 'control-lers'. It's a job that a robot could do were robots cheaper. Indulging in a whimsical thought experiment Aristotle says that if we had lyres that could play by themselves or automatic looms, we'd have no need of either serv-ants or slaves. How little he knew.

In the Fulfilment Centre, Aristotle's theory of *natural* slavery amounts to the claim that some men are naturally suited to be managers, just as others are naturally suited to be pickers. An objectionable doctrine? No, says the head of the hiring committee, having dismissed nine out of ten management trainees for want of 'natural leadership': that's just the way it is. Moreover, Aristotle would say that, given that men differ by nature in their ratiocinative abilities, it is better for *both* the master and the slave to have the relationship they do, and our managers would surely agree. The pickers might too?

It is not my intention to defend either Aristotle's theory of natural slav-ery or corporate hiring practices. I wish merely to show that his theory of natural slavery is not a pathological product of fourth-century Greek slave-owning society, but a general theory that speaks to the socio-economic structure of any state-level society, including our own. Indeed, it may be said that all modern battles over inequality ultimately turn on the question of whether 'natural slaves' exist and, if so, how to distinguish them from 'legal slaves'. This is most obvious in the extreme. I grew up in Apartheid South Africa, a state founded on the notion that Africans were, by nature, incapable of running anything just as Europeans, by nature, were so capable. There are hints in the *Politics* that Aristotle, too, believed that barbarians were, in general, natural slaves; he even suggests that slave-raiding wars are naturally just. Aristotle's word for a master's activity, *despotike*, doesn't really have an English equivalent. But *baasskap* – 'boss-ship' – works very well in Afrikaans.

C

THE GREEKS, SAID Plato, cluster around the Mediterranean like frogs around a pond. From Sicily to Asia Minor, and beyond to the southern Black Sea, there were more than a thousand Greek city-states. They came in a multitude of political flavours. By the time Aristotle was born, Athenian democracy had been running for more than a century. It wasn't the only democracy, just the most famous, powerful and extreme. Other states were run by aristocracies; many were ruled by kings. Some kings were good, others grotesque. Phalaris of Acragas, who had ruled that Sicilian statelet in the sixth century, was still remembered, though not fondly, for roasting his political opponents in a bull made of bronze – and because he, in turn, got roasted too. It is said that Aristotle collected 158 accounts of the Greek city-states, all of which are now lost apart from *The Athenian Constitution*, which was recovered from Egypt's sands in the late nineteenth century.* These accounts are the true subject of the *Politics*.

His explanatory system pervades the book. The state has a final cause: it exists for the sake of some purpose as surely as does the shell of a snail. Its formal cause is the constitution – not just a written document, but its whole economic, legal and political structure. The 'lawgiver', or rather his craft, is the efficient cause. By 'lawgiver' he means a man such as Solon of Athens (*fl.* 590 BC) or Lycurgus of Sparta (*c.* 800 BC) who moulded his city

---

* A fragment of *The Athenian Constitution* was found among the papyri of an ancient rubbish tip in Oxyrhynchus, south-west of Cairo, in 1879; more was later found in a tomb in Hermopolis.

and citizens into what they were. The state's people and territory are the brute matter from which it is formed.

All this sounds very biological and, like the *Meteorology*, the *Politics* is rich in organismal metaphors. The state not only has an origin, development and purpose, but an optimal size and self-maintenance mechanisms. It is composed of many interdependent, functional units, but it is also a whole. Its constitution holds it together, as an animal's soul unifies its parts. Inverting Heraclitus' metaphor of the river into which you cannot step twice, he likens the constitution to a river that retains its identity even though the waters that flow through it — the citizens — are ever changing. It can even decay or, at least, be transformed into something else. The *Politics* is, inescapably, political science written by a biologist.

We should not lean on metaphors too hard. As with his model of the physical processes of the sublunary world, they remain just metaphors. Humans, in Aristotle's view, may be political animals, but we are *more* political than any other. We are the only animal capable of moral reasoning, and the only one that can articulate its results by language. Hobbes, Hegel and Spencer — to name but three — would compare the state directly to an organism. Aristotle, the only biologist among them, does not. He also never says that the state has a *physis* — a nature, an internal principle of change — that all natural entities do. That is because the state, although a 'creation of nature', is not, in his view, a purely natural entity since it is also shaped by human agency. It is an organic-artefact hybrid — you could call it a cyborg state. 'Everyone has, by nature, an instinct for society. But the man who first instituted this performed the greatest service.' We have gone from the state as the product of herd instinct to the state as the product of some constitutional genius without pausing for breath. Philosophically, this is tricky; scientifically, it's unavoidable. Any human society is, inevitably, constructed from the desires, innate or not, of individuals *and* the laws of the land. 'Go tell the Spartans, stranger passing by, that here, obedient to their law, we lie,' runs Simonides' epitaph to the fallen heroes. But even a Spartan would rather be at home siring sons than generating flies by the Thermopylae pass.

Laws are necessary. There is a conflict between the true purpose of human life and our innate ability to achieve it. Humans, Aristotle says, should aim at happiness — *eudaimonia* — by which he means the active exercise of virtue according to reason. That, however, can be achieved only by submission to the state. He has a very dim view of human nature. True, we may have an innate capacity to co-operate and engage in moral reasoning, but without the rule of law we are the 'worst of animals'. We are savage, unholy, lustful and gluttonous.

If Aristotle's political science began with sociobiology it has now left it far behind. Indeed, his *politikē epistēmē* isn't a natural science at all, but a practical one: its purpose is to advise rulers. Should the philosopher speak to power, he may even be able to do a little political engineering: Plato tried it in Sicily; Aristotle may have tried again at Assos. Like Socrates–Plato, he has a vision of the ideal state. It's one that maximizes the number of its citizens who can live the good life, who can achieve *eudaimonia*. That sounds lovely, but – and here's the catch – in his state citizenship requires freedom from menial work, so tradesmen, craftsmen and labourers need not apply. (That women, children and slaves can't be citizens goes without saying.) It's a state in which the middle class is numerically predominant (a top-shaped income distribution) and a vaguely delimited, but apparently quite high, property-bar to citizenship. His state is designed to allow gentlemen to cultivate their souls. It is the sort of state that existed in England when Hanoverians sat on her throne and the landed gentry sat in her Parliament. It would be quite a hard sell today.

Aristotle's dislike of democracy is not just a wealthy philosopher's snobbery, but also a reaction to the Athenian way of government. Public life in fourth-century Athens was squalid. Every citizen could go up to the Pnyx and vote on the legislation of the day. Many did – if only for the sake of the three obols they got for attending. The result was institutionalized mob rule. Trained by sophists in the art, demagogues roused the rabble. *Sykophantai** – informers, blackmailers and slanderers – infested the legal system. A man could find himself in court on trivial or trumped-up charges, his fortune, home or life forfeit. Elected officials deposed each other by lawsuits. Brave military commanders, who had the misfortune to lose their battles and survive, suddenly saw the virtues of discretion and stayed abroad rather than return and argue for their lives. In 406 Athens executed six generals who had, so the accusation ran, failed to rescue the survivors of a naval engagement. Bribery and corruption were endemic. In his *Ecclesiazusae*, first performed in 392, Aristophanes has the women take over the government since the men are making such a mess of things. The farce

---

* A word that has not just a Greek provenance but an Athenian one, meaning 'a prosecutor of fig smugglers' – this is about trading with the enemy during the Peloponnesian War. It is the origin of the English 'sycophant' which, however, means something rather different, though equally unsavoury.

is crude but pointed: things were that bad. Even a philosopher, remote from public affairs, could be denounced and hauled before a court. Aristotle never forgot Socrates' fate.

No wonder Aristotle thought that he could do better. But he is no utopian. Rather little of the *Politics* is about the ideal state. Nearly all of it is about real states in their inexhaustible variety. Passionate for order, he tries to sort them out. Animals are classified by the variety of their organs and their relations to each other. States, he says, can be classified in the same way. The state's functional units are its classes: farmers, artisans, traders, labourers, military, the rich, public servants, administrators and judges. Their relationships – who rules whom – and the quality of their rule tell you what kind of state you have. The result is a complex taxonomy of power and virtue.

Aristotle's political pragmatism is reflected in his explanation of diversity. The main reason, he says, that there are so many different kinds of states is that people seek happiness in different ways and so make different ways of life and forms of government for themselves. The parallel with his teleological account of animal diversity is obvious. Here too, however, his teleology is not heedless: material necessity constrains constitutions. Oligarchies form in the plains where power rests on cavalry and hence accrues to the rich; democracies rise from arable soil where many people work their own farms. The character of its people also shapes the state. Europeans are spirited but not very bright and so organizationally useless; Asians are clever but supine and so tend to wind up as slaves. This is a consequence of their climate. The temperamentally middle-of-the-road Greeks ('courageous *and* sensible') have, of course, the best sort of character for good government. And they're free. Honesty, however, compels him to admit one weakness. If the Greeks could but *agree* on a single constitution, he says, they'd rule the world. If . . .

As Aristotle describes them, the most striking aspect of the Greek states is their fragility. Athenian democracy, it's true, was quite old. But across the Aegean monarchies, oligarchies and democracies alike lived mayfly lives. The picture he gives is of polities riding waves of scarcely controlled chaos. Much of the *Politics* is devoted to the causes and cures of instability. Since it isn't a purely natural entity, Aristotle does not give the state a life cycle, but not one of the constitutional forms he considers is immune to revolution (*metabole*).

Analysing the causes of constitutional change, Aristotle speaks of the desire of men for honour, money, power and justice. All of them lead to faction. He also speaks of how apparently trivial events – a squabble over a provincial heiress, say – can bring down the state. He touches on social and demographic factors, and points to the destabilizing effect of immigration even though – or is it because? – in Athens he's a resident alien and can't even own a house. But again and again it is to the malign effects of inequality that he returns. A sudden increase in the poor or rich or powerful will destroy or transform the state just as a monstrously hypertrophied body part will destroy an animal. No wild-eyed social reformer, he wants to know how to keep a lid on things – there are chapters full of tips for tyrants. But there are also arguments against manifestly mad laws. In *The Republic* Socrates–Plato, those dreamy utopians, had argued that women should be shared communally. For various quite cogent reasons Aristotle thinks that this is a bad idea. (The desires, much less rights, of the women in question are not among them.)

Although the state is, at least in part, an artificial construct, it is one of the instruments that allows humans – or those few lucky enough to be citizens – to manifest their full potential. In *The Parts of Animals*, describing the order of the living world, he expressly says that we are the one species capable of the good life. Like our arms, erect postures and reasoning minds, the state is an instrument of our divinity.

That is why, for all its flaws, Aristotle loved the *polis*. Correctly constructed, it could be the home of happiness itself. And yet the *Politics* is an essentially nostalgic work. By the time he wrote it, the age of the independent Greek city-state was past, and the age of empire had arrived. The conquerors were his friends; he was practically one of them. When Macedon made of proud Athens a vassal, Aristotle was still teaching Alexander at Mieza. The ironies haunt the book.

CI

THE EAGLE, SAYS Aristotle, is at war with the *drakōn*, which it eats. Elsewhere he says that the *drakōn* strikes, and destroys, the catfish in shallow waters. Although our 'dragon' descends from the *drakōn* via a long and complex transmutation, Aristotle means by it only a large serpent, probably the water snake, *Natrix tessellata*, which also went by the evocative name *hydros*. At the Vouváris' mouth, you can sometimes see them slipping into the water and swimming away.

Eagles do eat snakes, and snakes do eat catfish, but, like so much of Aristotle's ecological data in Book VII of *Historia animalium*, the first of these claims has the whiff of folklore, even myth, about it. In *Iliad* XII there's an aerial battle between an eagle and a monstrous, blood-red snake. The snake writhes free and falls among the Trojans as they make ready to attack the Achaean fleet. The Trojans take the fallen snake as an ill omen – as it turns out, rightly so. That particular mythological element is easily traced,* but the origin of Aristotle's belief that the dragon snake also sucks the juice of the *pikris*, a species of daisy, is more obscure.

Whatever their source, many of the dozens of competitive and preda-tor–prey relationships that Aristotle describes are, at least, plausible. The real weakness, again, is that it's data without explanation. Just as there is no zoological *Politics* to explain the habits of particular species, there isn't one to explain how and why different species interact as they do. Aristotle has the ingredients of community ecology in his hands but does not use them.

There is, however, one passage – precious and tantalizingly cryptic – that seems to reveal his views on the position of living things in not just the sublunary world, but also the cosmos. It appears in the twelfth book of his *Metaphysics*. Like Socrates and Plato before him, Aristotle believes that the constitution of the universe is good. In *Metaphysics* λ, 10 he attempts to identify the way in which it is so. One way in which it is good is that, like an army or a household, it has a hierarchical structure:

We must consider also in which way the nature of the whole possesses the good and the best – whether as something separated and by itself, or as its arrangement. Or is it in both ways, like an army? For an army's goodness is in its ordering, and is also the general. And more the general, since he is not due to the arrangement, but the arrangement is due to him. All things are in some joint arrangement, but not in the same way – even creatures which swim, creatures which fly, and plants. And the arrangement is not such that one thing has no relation to another. They do have a relation: for all things are jointly arranged in relation to one thing. But it is as in a household, where the free have least licence to act as they chance to, but all or most of what they do is arranged, while the

---

* In 1939 Rudolf Wittkower argued that the eagle–snake motif originated in Babylon four millennia ago whence it spread as far as Japan and the Aztec empire. Cultural diffu-sionists were bolder in those days.

slaves and beasts can do a little towards what is communal, but act mostly as they chance to. For that is the kind of principle that nature is of each of them. I mean, for example, that at least each of them must necessarily come to be dissolved; and there are likewise other things in which all share towards the whole.

Aristotelians often speak with admiration of the Philosopher's prose. They commend his ability to compress so much meaning into so few words. But, in truth, the pleasure that they derive from unravelling his tortured syntax and recondite metaphors is the pleasure of tackling a cryptic crossword. He is often shockingly opaque.* Were he not, classical philosophers wouldn't still be hacking at his texts more than two millennia after they were composed and fewer of them would have jobs. I have before me three monographs and one paper published within the last ten years. Each is by a gifted scholar and each analyses this one, metaphor-laden passage with an acuity, even brilliance, that I cannot hope to match. To varying degrees they all disagree about what it means. And I don't quite agree with any of them.

Rewritten into plainer English I think that the passage says this. 'What makes the cosmos good – even the best possible? An army or a household has an organizing principle (the general/master) and its members have an ordered set of relationships to each other. Does goodness depend on the organizing principle or on the ordered relationships? The answer is: on both, but mostly on the former since that dictates the latter. Like an army or a household, the organisms that inhabit the world are connected to each other by a set of ordered relationships. And, like them, that order is due to some organizing principle, not a man, but a common goal. ["For all things are jointly ordered with respect to one thing."] But not everyone in an army or a household contributes equally to that common goal. Senior members (officers/masters/higher animals) contribute more than junior members (troopers/slaves/plants); that's just their nature. Although all of the world's creatures are necessarily individual entities (and so have their own goals), they also all contribute to the common goal.'

Aristotle's household analogy is both beautiful and familiar. It appeared, albeit less explicitly, in his discussion of bodily economics, as the underpinning of what I called his ancillary teleological principles. Here he invokes it

---

* In the Renaissance, Aristotle's humanist critics often compared him to a cuttlefish hiding behind his own ink. That's funny but unfair; I think that he always *tries* to be clear; it's just that he often fails.

to explain the structure of the cosmos itself. But, of course, it is familiar for another reason. When, in 1866, Ernst Haeckel coined *oekologie* to describe the new science of the economics of nature, it was from *oíkos* – Greek for household – that he did so. The coincidence is testament to the metaphor's power. But it also makes us wonder: are all the different kinds of animals in the world truly like a household insofar as they are subject to some common organizing principle; or are they more like the residents of a hotel who just happen to find themselves under a single roof? On this question much of the history of modern ecology turns. It is a question that might be asked of Aristotle too.

Aristotle's claim that organisms are related to each other as members of an army or a household is, as I read it, a frank anomaly. It invokes a higher, inter-species level of organization, a common cosmic purpose or a global teleology. In the *Politics* Aristotle makes clear that a properly functioning household is not just an assemblage of self-interested individuals. Rather, it is a collective of co-operating members, directed by the household's master, whose collective purpose is procreation and protection. Yet, when speaking of animals, he hardly ever describes the kind of co-operative, even altruistic, inter-specific behaviour that we might hope to find in an army or a household. True, he claims that the *karidon* (or *pinnophylax*), a little symbiotic shrimp or crab, benefits the *pinna*, the giant fan mussel, in which it lives, but he makes nothing of it. And when he explains some animal's features in functional terms he almost invariably speaks of its benefit to that particular kind of animal. If all the species in the world share some organizing principle or common purpose beyond a desire for their individual survival, his zoology does not tell us what it is or how they accomplish it. Aristotle's forms are selfish forms.

Moreover, there is one kind of global teleology that he explicitly rejects. The strongest form of global teleology would be one that postulates that the world, perhaps even the whole cosmos, is a single super-organism. Such a world would be one over which Gaia reigns with a power far beyond James Lovelock's wildest imaginings. It would be a world like James Cameron's Pandora, one whose inhabitants are all connected by a vast signal transduction network, whose predators are not so much the ecological equivalent of jackals and hawks as phagocytes coursing through the planetary circulation, and whose animals rise as one in response to the planetary spirit's anguished call to arms. Or, since we are not in the twenty-second century AD but in the fourth century BC, it would be a world like the one that Plato describes in *The Timaeus*. Plato's perceptible cosmos is a copy of a single form, the Intelligible Living Creature. The name says it all. The cosmos has a 'soul'. Designed by the Dēmiourgos it is also designed *for* the Dēmiourgos. Even our bowels are arranged so that we can think about Him. But Aristotle is clear: the cosmos does not have a soul. (Though, as will become apparent, the celestial realm is not devoid of life either.)

It is for these, and other, reasons that most recent interpreters of *Metaphysics* λ, 10 have read the household analogy in a very weak sense. They argue that when Aristotle says that living things are 'jointly ordered with respect to one thing' he is just saying what he has so often said: that they all aspire to eternity. And I would agree but for three reasons. The first is that this reading renders the analogy otiose. Why even bother giving it? The second is that Aristotle *does* describe some altruistic species. Oddly, they include sharks.

He is explaining why sharks (and dolphins) have the faces they do. They have narrow snouts and their mouths are slung under their heads. These features, he thinks, make them inefficient predators since they cannot open their mouths very wide and, while seizing their prey, have to turn belly up which lets the little fish escape. He explains these awkwardnesses in two ways. One is that it prevents sharks from gorging themselves. That's quite consistent with Aristotle's usual style of explanation. He often argues that animals have built-in limits to the amount of food that they can eat, or the number of eggs that they can lay, or the quantities of semen that they can produce; and such limits, he invariably goes on to explain, are the consequence of some other feature that benefits the animal. They are, as we would say, functional trade-offs. It's his other explanation that is startling. For he *also* says that sharks have narrow gapes and under-slung mouths so that they don't devour all their prey ('nature appears to do this for the sake of the preservation of other animals'). Shark faces, it seems, are designed for the sake not only of sharks but of sardines too.*

The story of the shark's face is so strange that it's tempting to dismiss it as an un-Aristotelian interpolation. That seems unlikely since it appears in both *Historia animalium* and *The Parts of Animals*. So defenders of individual teleology sometimes say that Aristotle is just relating a popular notion – the sort of thing a fisherman might say. Or else, more subtly, that a good-for-sardines face is just an incidental benefit of its true,

---

* There is one other case where Aristotle suggests that a predator may have features that benefit its prey: at *Historia animalium* 563a20 he says that 'it is said' that nesting eagles abstain from food, and that their talons get turned so as to avoid harassing the young of wild animals. The information is weak, the reading is dubious and he makes nothing of it.

good-for-sharks design. I am less sure. A shark face that helps sardines to survive is just the sort of feature that we would expect if, as *Metaphysics* λ, 10 says, the world were like a household. In fact I believe that this passage solves a deep and hidden problem in Aristotle's ecology.

That is my third reason for taking Aristotle's household analogy seriously. Most scholars agree that Aristotle believes that: (i) organisms are designed to survive and reproduce; (ii) animal kinds are eternal. They have, however, failed to note that these two beliefs are, in general, incompatible. That is because, in a world in which organisms interact with each other, in which they compete and prey upon one another, there is no reason to suppose that all will persist for ever. Animals and plants often extirpate their competitors; predators eat up all of some prey species and then turn to eating something else. At least so it is in our world. In Aristotle's world, however, extinction is not an option; his metaphysics demands a balance of nature. Aristotle, I propose, grasps that such a balance does not emerge automatically from any self-interested assemblage of organisms, but must be designed by nature.

The evidence for thinking this is admittedly indirect. To begin with the evidence closest to the subject, he must have had – if I am right – some grasp of the fragility of ecological communities. In *Historia animalium*, speaking of fish, he says that if 'all their eggs were preserved they'd be infinite in number'. He's also impressed by the extraordinary fecundity of mice and speaks of how sometimes they multiply so rapidly that their predators cannot make a dent in their numbers, that they devastate entire crops, and that they then, suddenly, disappear again, but no one knows why.* His description suggests that he's relating an unusual phenomenon. That is indeed so. In the *Nicomachean Ethics*, discussing 'incontinent appetites' in humans – he's pretty severe on them – he asks whether animals can have them too. His genial reply is that since animals can't reason we don't generally speak of them as being 'temperate' or 'self-indulgent' except in a metaphorical way. Yet, he continues, some *kinds* of animals exceed others in 'wantonness, destructiveness, and omnivorous greed' – he's surely thinking of those pernicious mice – and that they are a departure from 'what's natural, as, among men, madmen are'.

Such passing comments certainly do not amount to an ecological theory. Yet they do tell us that he has a sense of the normal relationship of animal

---

* In his 1942 classic, *Voles, Mice and Lemmings: Problems in Population Dynamics*, Charles Elton notes that this passage contains the essence of the problem of the regulation of population numbers.

populations to their food and that, sometimes, this relationship goes awry. More generally, introducing his soothsayer-derived account of animal conflict in *Historia animalium*, he says: 'A state of war exists between animals which occupy the same place and get their livelihood [*zoë*] from the same sources.' It is one of his few explicit ecological principles – but it is a deep one.* He never says that animals might go extinct for want of food. On the other hand he sees that an adequate supply of food is not *automatically* guaranteed for any given species and that, faced with limited resources, individuals and species compete.

Passages that speak of ecological instability are, however, rare in Aristotle's works. He seems to believe that nature usually ensures that there's enough food for everyone. In the *Politics*, discussing the various ways in which men and animals make a living, he says that:

> *Nature seems to provide a basic livelihood to all when first born and when fully mature* [italics mine]. Some animals (e.g. larva-bearing and egg-laying animals) bring forth alongside their offspring enough food to last until they can provide for themselves. For a limited period live-bearing animals have food for the young inside them called milk. We should similarly infer that after birth plants exist for the sake of animals and other animals for the sake of human beings, the tame for service and food, and most wild ones for food, clothing and other uses. If nature makes nothing without purpose or in vain she must have made all animals for the sake of human beings.

Some have read this passage as implying that Aristotle's teleology is purely anthropocentric; that, like Xenophon before and the Stoics after, he sees the whole world, and all the animals in it, as existing *just* for the sake of man. He really can't mean that since the rest of his teleology is, as I've said, overwhelmingly directed at the survival of individual animals. But at the

---

* The claim here is not that Aristotle understands Gause's Competitive Exclusion Principle or Lotka–Volterra predator–prey dynamics; nor does he need to, for the idea that animals are designed to promote a balance of nature was likely commonplace in ancient Greece. Herodotus, for example, appears to claim that predators have fewer offspring than they might in order to not consume all their prey: 'Of a truth Divine Providence does appear to be, as indeed one might expect beforehand, a wise contriver. For timid animals which are a prey to others are all made to produce young abundantly, so that the species may not be entirely eaten up and lost; *while savage and noxious creatures are made very unfruitful* . . . [italics mine]'.

very least this passage does point out that plants, animals and men are connected to each other by a chain of trophic relationships; that they depend on each other, and that this isn't just an accident, but that nature has arranged matters this way. There is, then, a sense in which more perfect creatures typically use less perfect creatures as instruments of their survival since they eat them. But whose nature is at work here? When Aristotle says that 'nature' does this or that, he nearly always means the formal or material nature of some particular animal. Here, however, nature appears to refer to a higher level of organization. It appears to be the nature of the cosmos itself to ensure an adequate supply of food for all the animals in it.

The cosmos is a *holon*, a whole. As such it is like a soul, a household, a state or even a tragedy – Aristotle applies the term to them all. By a 'whole' he means a complex object that is something more than the sum of its parts, a system. But Aristotle is acutely aware of the fragility of wholes. His theory of the nutritive soul is, ultimately, an account of the material flows and regulatory devices that keep animals alive; his account of death is an account of how they fail. Much of his political theory is about the conditions that ensure the stability of the state. It would be strange indeed if he did not see that what is true of these wholes is also true of the greatest, most complex whole that he knows: the cosmos itself.

This, I believe, is the force of his household analogy. It is a statement that, if the components of the sublunary world – all its plants and animals forms – are to survive for ever, then their relationships must be arranged just so. Sharks *must* control their appetites for if they did not sardines would go extinct, and if sardines went extinct that would be bad for sharks. In the *Nicomachean Ethics* he reinforces this. He's speaking of the difference between true wisdom and mere political ability or 'prudence', by which he means the ability to manage a household or a state. He says that prudence for humans and prudence for fishes are quite different. That's unarguable, but raises the question: how can a fish be prudent at all? None of his fishes are 'political' by any definition. I suggest that he means that a fish – a shark – is prudent as a man is, by managing its income, by eschewing unnatural gluttony, by preserving its *oîkos* – its home – and so itself. He even suggests that they have foresight. Animals are indeed designed to promote their own interests, but not so much that they jeopardize the existence of other kinds, for that would jeopardize their own.

The hierarchical dimension of the household analogy is very unclear, but I believe Aristotle to be claiming that humans and animals have more diverse ways to realize their goals than, say, plants whose only function is to reproduce. Whether or not this is so, his household clearly invokes a far weaker form of global teleology than Plato's cosmic super-organism in which all interests are subordinate to those of the *Dēmiourgos*. It is more – to appeal to another organizational analogy – like the kind of mutual self-interest that exists among industrial corporations and the myriad companies that supply the components they need, all of which seek one thing – 'the most excellent thing there is' – in their case, profit.*

This vision of a cosmic teleology has one further benefit. It suggests a solution – albeit a frankly speculative one, for which there is no direct textual evidence – to the mystery of why spontaneous generators exist. Aristotle would deny Macbeth's dismal claim that a man's life is mere sound and fury signifying nothing. Speaking as a biologist (rather than a political scientist), he would say that the reason that he is born, grows to maturity and battles the vicissitudes of the world is so that he can reproduce his form. Not so the oyster. Its life, by his account, really does seem to be devoid of purpose for it perpetuates nothing. But perhaps this view is too narrow. For the oyster, and all its fellow spontaneous generators, do have this in common: they are eaten by other things. Most of them are at the bottom of the food chain. Perhaps, then, the purpose of spontaneous generators is to ensure the survival of the creatures that feed on them. They exist, as all living things do, to keep their world intact.

I should love Aristotle's teleology to be entirely aimed at the survival of individuals. That would make him seem quite modern – Darwinian, if not Neo-Darwinian. But if the picture that I have given of Aristotle's world is accurate, then it is a very different one from ours. In our world natural selection maximizes short-term reproductive success and is indifferent to eternity: '[Natural selection] does not plan for the future. It has no vision, no foresight, and no sight at all. If it can be said to play the role of watchmaker in

---

* Or is Aristotle's global teleology even stronger than this? Could it be that, to extend my metaphor, our companies are not merely bound together in a co-operative web in search of profit, but are explicitly directed to do so by some greater power to some greater goal? That would be more like the way in which, in the 1980s, the Japanese Ministry of International Trade and Industry (MITI) directed the *keiretsu* conglomorates for the sake of national economic growth. There's a spectrum of possibilities ranging from rampant individualism to superorganism status, and it's hard to say where, exactly, Aristotle thinks the world is located along that spectrum.

nature, it is the blind watchmaker.'* Just so. In our world, therefore, species drive each other extinct. Around AD 1280 the Maori brought the Polynesian rat to New Zealand. It ate up five species of native birds and three frogs as well as a variety of lizard, insect and land-snail species. The Maori themselves ate up nine species of Moa. Now imported European predators – Norwegian rats, black rats, stoats, weasels and cats – are working their way through what is left of the native fauna. If, in our world, there is a 'balance of nature', it is only the temporary truce of evenly matched opponents who, having battled in the theatre of ecological war, stand exhausted among the corpses of the less fortunate and well equipped. Aristotle's world is not a kinder one, for in it there are no truces, just battles that go on, without respite, for ever.

AKANTHIAS GALEOS – SPINY DOGFISH – SQUALUS ACANTHIAS

* Aristotle's shark is apparently a 'prudent predator'. The phrase was coined by Lawrence Slobodkin in Growth and Regulation of Animal Populations, 1961, but it was V. C. Wynne Edwards in Animal Dispersion in Relation to Social Behaviour, 1962, who argued that prudent predators could evolve by group selection. More generally, Wynne Edwards argued that ecological communities should be viewed as homeostatic systems, and he interpreted innumerable aspects of animal behaviours as such. George C. Williams, Adaptation and Natural Selection, 1966, demolished this view. He pointed out that group selection was a very weak force and that nearly all adaptations, predation behaviour included, were better interpreted as being the result of individual or genic selection. The recent revival of group selection notwithstanding, this conclusion remains sound.

CII

Experts all agree that the world had a beginning, but some [Orpheus, Hesiod and Plato] claim that, once begun, it is everlasting; others [Democritus] that, like any other natural artefact, it is subject to decay; and others [Empedocles, Heraclitus] that it alternates, being one moment as it is now, another moment changing and subject to decay, and that it is this process which continues without ceasing . . . Well, the idea that it has a beginning but is everlasting is quite impossible . . .

S O ARISTOTLE IN HIS cosmological treatise, *The Heavens*. He says that he wants to give the various theories a fair trial, but he's out to argue for his own. This rests on the claim, a novel one, that the universe is eternal, that it had no beginning and that it will never end. Since the forms of living things are eternal he needs, of course, an eternal cosmos to house them in. Even so, he gives a series of independent arguments for one.

Some of his arguments for the eternity of the cosmos are purely semantic; most are tortuous. The most lucid of them appears in the *Physics*. It focuses on the necessary existence not of cosmic matter but of cosmic change. Change is the object of his science and, since all natural entities have an internal principle of change − a *physis* − to prove the eternity of change is to prove the eternity of the objects of change as well.

Aristotle's proof rests on the necessary existence of prior causes. The argument is abstract, but a concrete example, Aeschylus' fate, will make it clear. For Aeschylus to be killed by a falling tortoise − so Aristotle would argue − both playwright and the immediate cause of his death, a tortoise travelling at speed, must first exist. That seems obvious enough. For the tortoise to fall, some existing thing must have changed: an eagle that opened its talons. For the eagle to have opened its talons, some existing thing must have changed: the eagle's sensitive soul − the cognitive-motor system that perceived Aeschylus' head, considered its goals and desires, fired its *pneuma* and sprang its talons wide. For the eagle's soul . . . but the point is clear: no matter how far back you go, any observed change necessarily implies the existence of a previous change and the existence of objects and subjects of change – ergo change is eternal.

Aristotle's argument is a generalization of the argument for the eternity of forms/kinds – organismal generation being a special kind of change. It's a good one if your physics are fully deterministic. Given a deterministic cosmos filled with change, change must have existed for as long as time has existed; and Aristotle has another argument to show that time has neither a beginning nor an end. We might expect that this would be the end of the matter, but it isn't. 'Eternity's a terrible thought. I mean, where's it going to end?' said Tom Stoppard's Rosencrantz; Aristotle, by contrast, worries that it might.

He worries that the causal chain might break. That's because his physics is based on the common-sense idea that an object in motion will eventually, naturally, come to rest. Before it does so, the object may contact another object and so set it in motion, but eventually the force will dissipate as, when you throw a stone in a pond, the ripples eventually fade to nothing.* So to keep the world in motion Aristotle needs a continuous source of change, a cosmic engine. To find one he looks to the heavens. The Egyptians and the Babylonians, Aristotle says, have been watching the heavens for generation upon generation, and their movements never vary.† If anything can guarantee eternal movement on Earth, then the stars can.

When Aristotle does biology one senses his solitude. Of course he could talk to Theophrastus and, later, his students; but who among his contemporaries cared about sponges and suchlike – cranky old Speusippus? Maybe. Astronomy was different. By the mid-fourth century there was a network of mathematical astronomers that spanned the Hellenic world.‡ Two of

---

* Galileo, by constrast, would argue that an object in motion comes to rest only when opposed by an equal force. This is his principle of inertia, codified by Newton in his First Law of Motion. Aristotle does not have the principle of inertia.

† Aristotle knows about comets and meteorites, but thinks they are sublunary phenomena. Greek astronomers apparently did not record any novae or supernovae, though early Chinese astronomers did.

‡ Consider Eudoxus' career. Born in Cnidus, Asia Minor, c. 390, as a young man he travelled to Athens to study, briefly, at the newly founded Academy. He then went to Heliopolis, Egypt, to learn astronomy and also apparently to Italy to study with Archytas, a friend of Plato's, and with Philistion of Locri, a philosophical physician. He was apparently very poor; friends and a passion for his subject kept him going. After further travels he returned to the Academy where he met Aristotle. By this time he had students of his own, among them Callippus, who later joined Aristotle at the Lyceum. Eudoxus finally returned to Cnidus, where he built an observatory and spent the remainder of his days looking at the stars, lecturing and doing legislative work for the city.

them, Eudoxus of Cnidus and Callippus of Cyzicus, were at the Academy
with Aristotle. The former was a first-class mathematician who had Archytas
of Tarantum, said to be the founder of mathematical mechanics, as a teacher.

Aristotle is uncharacteristically generous towards them. This is the
verbal theorist's deference towards colleagues who can actually do the
maths. (How well I know it.) In any event, when Aristotle needs a geometrical model of the cosmos, he just elaborates theirs. This model postulated
a spherical Earth* located in the middle of a series of concentric spheres in
which the heavenly bodies were embedded. The system (or, rather, systems
since Callippus improved, or at least modified, Eudoxus') was complex and
designed primarily to explain the retrograde motion of the 'wanderers'
(planetai) – the peculiar way in which they danced across the night sky
instead of progressing across it as the regular stars did.†

The details need not concern us; as far as Aristotle is concerned it isn't
natural science at all. The models produced by the mathematical astronomers may describe heavenly events; they may 'save the appearances'
(phainomena) – the phrase is attributed to Plato – and, while that's important, it's not enough. The stars aren't just mathematical constructs, they're
natural bodies; natural bodies are the objects of natural science; and natural
science needs causal explanations. What are the heavens made of? Why do
they rotate? It's not just that the astronomers had no answers to such questions; they didn't even think to ask them.

Of all the natural entities in Aristotle's cosmos, the celestial bodies –
the moon, sun, planets and, most especially, the stars – are most perfect
and divine. They are, he admits, the hardest to study: they are so far away
and we know so little about them, but that should not stop us from trying

---

* Aristotle says that the circumference of the Earth has been estimated (by whom he does
not say) as 400,000 stades. There is much uncertainty about the length of the ancient
stade, a foot-race distance: estimates vary between 150 and 210 metres, but taking the
median, 180 metres, Aristotle's number gives 72,000 kilometres, or 1.8 × greater than the
actual equatorial circumference. A generation later, Eratosthenes would estimate the
circumferance as 250,000 stades or 45,000 kilometres, 1.2 × actual. I'm impressed.
Aristotle adds some biogeographic evidence for the sphericity of the Earth: there are
elephants in Africa and Asia, so perhaps those who claim the existence of a continual
western landmass between the Pillars of Hercules and India are right. Just so Alfred
Russel Wallace and Alfred Wegener used biogeography to argue for (prehistoric) connections between landmasses.
† Retrogradation was explained by Copernicus by abandoning the ancient geocentric
cosmos. If, explained Copernicus, Earth is a planet that, like all planets, orbits the sun,
then our position relative to the other planets will shift in complex ways so that they will
sometimes appear to reverse direction relative to the stars.

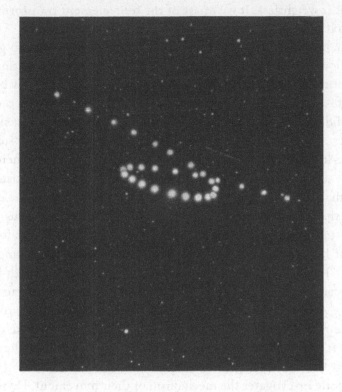

RETROGRADE MOTION OF MARS AGAINST THE STARS, AUGUST 2003

to understand them. When we tackle hard problems we should be content with even modest results. We find more delight in even a partial glimpse of a beloved's face than the plain sight of commonplace things.

Aristotle tackles the celestial bodies in *The Heavens*. He claims that the heavenly bodies, indeed the spheres in which they are embedded, are made of a unique substance – the 'first element' – *to prōton stoicheion*, traditionally called *aithēr*. This brings the total number of Aristotelian elements to five. Just as the four sublunary elements have a natural principle of change and rest, *aithēr* does too. Here, as always, when Aristotle wants eternity he looks for a circle; he thinks it's the simplest of all movements.* So he postulates that *aithēr*'s natural motion is circular but that it has no natural place of rest. Since it moves in a circle (rather than up or

---

* Here is another of the fundamental differences between Aristotelian and Newtonian physics: in the latter, movement in a straight line is the simplest possible motion; movement in a circle requires an additional centripetal force.

down) it is weightless. It isn't part of the four-element transformational cycle, so it's indestructible too.

Elemental *aithēr* was controversial stuff. In *The Timaeus* Plato gave the conventional view that the stars were made of fire. Proclus, writing in the fifth century AD, said that the Platonists thought *aithēr* positively barbaric. Some of Aristotle's Peripatetic successors abandoned it too. (It caught on in the Middle Ages.) Yet his reasons for dreaming it up were cogent. Were the stars made of some combination of the conventional sublunary elements it would be hard to explain the beautiful regularity of their movements – *aithēr* does so effortlessly. *Aithēr* also gives eternal existence. It means that the stars don't have to contend with the internal elemental turmoil that ultimately destroys everything on Earth, including us.

The strangest aspect of Aristotle's cosmology is not, however, its chemistry, but his application of teleological – functional – reasoning to the heavens. To say that the celestial bodies rotate about the Earth because they're made of *aithēr* is merely to give the material and moving causes. But, as always, Aristotle also wants a final cause. The celestial bodies rotate for the same reason that animals and plants reproduce: for the sake of being eternal. It's an odd claim – why should the stars move for the sake of anything? Odd it may be, but it's just a start.

Aristotle investigates the mechanism of their movement. He wants to show that each star is not motoring along under its own steam, but that they're being collectively transported in a single, rotating *aithēr* sphere. He provides some evidence for this: they all move in synchrony, so transportation seems like the most economical explanation – he compares them to ships being carried in a stream. Moreover, if they did move themselves, we'd see them rolling; but the moon doesn't roll since we can always see its 'face'. And here's another argument: if they moved themselves, then they should have locomotor appendages – feet, fins or wings – but they don't. (Aristotle does not bother to say that no one has seen wings sticking out of the moon. I suppose it's obvious.) But it cannot be, he continues, that nature has just neglected to give them locomotor appendages; after all, the celestial bodies are *perfectly* designed to do what they do – far better (he asserts) than any animal is. So their method of locomotion must be of a sort that does not require appendages: transport in a crystalline sphere of *aithēr*.

It is easy to translate Aristotle's teleological explanations of animal bodies into the adaptationist design-talk of modern biology. But celestial bodies? Planetary science tells us that the moon is round and orbits the

Earth because brute physics made it so; the fact that it lacks wings doesn't even arise. But that's the point. For us, the sun, moon and stars are inanimate; for Aristotle they are alive – as alive as a bee, an elephant or you. In a sense they are *more* alive; they are the most perfect of all living things. The cosmos as a whole may not have a soul – but a star does.

Aristotle's celestial biology is a little vague – how could it be otherwise? – for sometimes he suggests that it's not the stars (or planets) themselves that are alive, but the spheres in which they're embedded. Stars or spheres, there's definitely Life Out There. Is this another of Aristotle's cosmological novelties? 'But we think of them only as bodies, and units that have an order, but nevertheless are wholly without soul. However, they have to be accepted as possessing life and activity.' He even extends his zoological ladder-of-perfection upwards. The stars, or their spheres, are by virtue of their motions – and hence the means by which they achieve their goal – the most perfect; the planets, sun and moon are, in decreasing proximity to Earth, less so. The motionless Earth has no goal at all.

Like any plant or animal, the celestial bodies are, in general, designed to fulfil their own goals. But they are not entirely indifferent to sublunary affairs. The stars, unvarying in their orbits, have a very simple motion, but the other celestial bodies do not. The planets retrograde and the sun not only moves east to west but also has a secondary, west-to-east movement along the ecliptic. These are precisely the movements that the mathematical astronomers' models sought to describe. Aristotle, however, wants to give these more complex movements a purpose too. The secondary motions of the sun and the moon drive the seasons on Earth, and so the sublunary elemental cycle that stops the world from turning into an onion. But, he seems to suggest, this isn't just the consequence of material necessity, it's the *reason* that these secondary motions exist. Note the direction of causality. The sublunary elements do not merely cycle because the sun has secondary motions; the sun has secondary motions in order to make the sublunary elements cycle. Aristotle appears to be invoking the principle of conditional necessity from his zoology, which claims that the features of a living animal are designed to fit with each other, to the cosmos as a whole. The household analogy of *Metaphysics* λ, 10, then, is not just about how sublunary organisms depend on each other for their existence, but how they depend on the actions of the creatures orbiting the Earth. Entities both higher and lower in the chain of being are connected to each other by a criss-crossing network of benefits. Aristotle's ecology is literally cosmic in scope. That is why, in *Generation & Corruption*, he calls the sun 'the generator' and why, in the *Physics*, he amends his usual slogan 'a man gives rise to a man' to 'a man *and the sun* give rise to a man [italics mine]'. This isn't just a statement about the connectedness of all things; it's an assertion of cosmic purpose.

The whole scheme is magnificently absurd. Setting aside the claim that the celestial bodies are *alive*, even the weaker claim of a designed universe strikes us as strange. No astronomer believes that any of the universe's features – moons, planets, stars, nebulae, black holes, supernovae, galaxies – show evidence of design. Outside biology, teleology has no place in the scientific explanation of cosmic order. The universe just *is*.

Or is it? There is, at the heart of modern physics, a deep mystery. The Standard Model of particle physics and the λCDM cosmological model* that explain our universe so well – at least between scales of 10^{-21} and 10^{25} m – contain some thirty input parameters, for example, the masses of the elementary particles and the strengths of the three fundamental forces (electroweak, strong nuclear and gravity). Many are dimensionless and take apparently arbitrary values – except that, were they to vary from observed, the universe as we know it would not exist. To give but two examples, the cosmological constant, λ, is approximately the mass-energy density of one hydrogen atom per cubic metre. Quantum theory says it should be much greater, but if it were our universe would have expanded so quickly that neither galaxies nor we would be here. Again, neutrons are ~0.1 per cent heavier than protons: were the reverse true, protons would decay into neutrons, hydrogen would be unstable and conventional chemistry would not exist. This is known as the 'fine-tuning' problem.

Some physicists have tried to explain cosmic fine-tuning away by invoking the 'weak anthropic principle' – the idea that, were the physical constants of the universe not such that stars, planets, life and intelligent life could form, we wouldn't be here to wonder at the fact that they have. That is true, but does not solve the problem. If there is only one universe, and only a few solutions in the parameter space consistent with the evolution of sentient life, the odds against nature having got everything right must be, literally, astronomical. As Aristotle said, attacking Democritus *et al.* – 'They [the materialists] assert that chance is not responsible for the existence or generation of animals and plants . . . and yet at the same time they assert that the heavenly sphere and the divinest of visible things arose spontaneously, having no such cause as assigned to animals and plants.' The

* Lambda Cold Dark Matter.

empirical regularities that puzzle Aristotle and modern cosmologists are different: the root problem is the same.

Let's accept, *ex hypothesi*, that the cosmos shows the signature of purposeful order, the hallmark of design. Where does it come from? There are three, and only three, possible answers. The first is to appeal, as Plato did and Christians still do, to a beneficent creator who arranged things just so. The second is to appeal, as Democritus did and Epicurus would, to an infinite universe – infinity solving all low-probability dilemmas. The first can be discarded; the second I have no views on, though some cosmologists believe it to be true.* As a biologist, however, I like the third for it depends on the only known mechanism capable of creating order from disorder: natural selection.

This is the reasoning behind Cosmological Selection Theory, which proposes the existence of a population of universes – a multiverse; that these universes reproduce, that they do so with unequal success, and that they transmit their physical constants, allowing for some mutation, to their progeny universes. It's nothing more or less than cosmic Darwinism and an easy route to a universe which, depending on the fitness function, has any non-lethal combination of parameter values that you please. One multiverse theory holds that universes give birth to baby universes via black holes. In that case, the number of black holes in our universe (millions) would be a design feature. The plausibility of the physics need not detain us; but, granting them, or some similar scheme, it is clear that natural selection among universes will work. If it has, far from being the mere product of brute material necessity, some of the features of the universe would be as teleologically explicable as the parts of an elephant.

Such a cosmos would have a purpose. In such a cosmos, Aristotle would surely be at home. And yet he would reject a selectionist explanation for its origins, just as he would reject a creator and chance. Ask Aristotle, how did the cosmos come to have its purposeful features?, and he would say: that's a meaningless question since the cosmos *didn't come to be*. It just is and was and always will be, forever and forever and forever. Of all his theories this, I think, is the one that we most struggle to understand.

---

* To explain fine tuning, an infinite universe would have to contain an infinite number of local variations in the physical parameter values, rather than just those seen in the observable universe. In effect, this invokes a multiverse.

CIII

W E ARE APPROACHING GOD. I have mentioned, in passing, how Aristotle thinks that stars, humans and even bees are 'divine', but perhaps you took that merely as a *façon de parler*, Aristotle's way of describing beauty, high intelligence or a complicated social life. God has even cropped up once or twice by name, but perhaps you thought that He was just a metaphor for something like Plato's absolute good. If so, that is doubtless my fault. I have kept Aristotle's *theos* in the shadows. It may even be that I have done so deliberately; that I have been reluctant to reveal the degree to which my hero's scientific system is riddled with religion. Yet it is. In truth, God has been with us all this time.

Why does Aristotle think that the stars are alive? He certainly doesn't give any evidence for it. In *The Soul* he says, 'By life we mean the capacity for self-nourishment, growth and decay' – but the stars, he also says, do nothing of the sort. They neither have nor need organs. True, he says that they have souls and that they're enjoying themselves up there, but how does he know? It's mere assertion. It's a bizarre stance to take – at least it is for a scientist who's so cautious about the reproductive habits of bees. He seems to be star-struck:

> The reasons why the first body [*aithēr*] is eternal and not subject to increase or diminution, but unageing and unalterable and unmodified, will be clear from what has been said to anyone who believes in our assumptions. *Our theory seems to confirm experience and to be confirmed by it. For all men have some conception of the nature of the gods, and all who believe in the existence of gods at all, whether barbarian or Greek, agree in allotting the highest place to the deity* [italics mine], surely because they suppose that immortal is linked with immortal and regard any other supposition as inconceivable.

Which shows that he's God-struck too.

There is a passage in the *Metaphysics* where Aristotle undertakes some religious archaeology. Our remote forefathers have, he says, handed down to us a tradition that the heavenly entities are gods and that the divine encompasses all nature. Later, however, mythical elements were added on

– the zoomorphic and anthropomorphic gods of popular religion. But these were merely invented for 'the multitude' and because they were 'useful'. (I take this to mean that the mob needed pretty statues to worship, and the state needed religion to control the mob.) But, he continues, we should separate the original 'divine utterances' from these later accretions . . .

Aristotle has invented a new theology that seamlessly interweaves prehistoric superstition with cutting-edge science. He repeatedly refers to some antique belief about the gods – that they are located in heaven, are immortal or unchanging – and then shows, with a flourish, its consistency with cosmological theory. The reason, then – the only reason as far as I can tell – why Aristotle thinks that the celestial spheres are alive is because he thinks they're gods.

Aristotle says that theology is not a branch of natural science. It is, in his terms, *first* philosophy, a branch of our metaphysics; natural science is *second* philosophy. Different domains of knowledge, he also says, should be kept firmly apart because they depend on different first principles. Yet in his relentless drive for intellectual *Lebensraum* he marches across any disciplinary frontier he pleases. He discusses the nature of the gods not only in the *Metaphysics* but also in *The Heavens, The Soul, Generation & Corruption* and the *Physics*. They even appear in *The Movement of Animals*.

It can hardly be otherwise since the functioning of the world depends on them in so many ways. The rotations of celestial spheres are the ultimate *moving* cause of all natural sublunary change. As moving causes the celestial spheres can keep the cosmos going whether they're dead, alive or divine. But Aristotle also relies on them to give its creatures their goals; and for that they must be gods. Aeschylus died as the unwitting instrument of an eagle's ends. Her goal (we may suppose) was merely to feed the tortoise to her ceaselessly ravenous brood. In doing so she sought to emulate the eternal motion of the stars – not the petrified perfection of crystalline spheres, but the immortality of living deities. All she sought was a little slice of for ever. So, looking to our mates, do we all.

What do the motions of the god-spheres depend on? It may be thought that they depend on nothing: they're gods, after all. Besides, they're built of *aither* that just naturally moves in a circle. That, indeed, seems to be the theory that Aristotle gives in *The Heavens*; in the *Metaphysics* and *Physics*, however, he gives another, or perhaps it's just a shift in emphasis. In this second version celestial spheres do not move themselves. Their divinity is downgraded. *Aithēr* fades into the background and a host of mysterious entities appear on stage: the 'unmoved movers'.

The unmoved movers are the new deities. The spheres were almost mundanely comprehensible. Any science-fiction fan will take AU-scale sentient entities composed of exotic matter in his stride. The unmoved movers, by contrast, are gratifyingly abstract and paradoxical. There are said to be fifty-five of them, but also just one. They power the celestial spheres yet are utterly static themselves. They do this despite being indivisible, devoid of parts or even bodies. They are, in fact, immaterial. Aristotle means that they're not made of *any physical substance*.

The reason why there are so many unmoved movers is because each of them powers just one of the spheres in Aristotle's geometrical model of the cosmos. All the stars, having a simple, unitary motion, can be assigned to one sphere. The moon, the sun and five planets, with their more complex motions, require another fifty-four. (Parsimony was never a strength of geocentric cosmologies.) The unmoved movers have their origin in Aristotle's mature theory of motion. In this theory, still to be formulated when he wrote *The Heavens*, every motion requires a previous motion. Now *aithēr* spheres can no longer move themselves, they need to be kept in motion. But he doesn't want an infinite regress of movers driving every sphere, so he gives each one an unmoved mover. The evolution of his physical theory, cosmology and theology are all inextricably intertwined.

The unmoved movers obviously don't pull or push their spheres around since then they'd be *moving* movers; besides, they're immaterial so they can't. Instead each drives its sphere by being its object of love and desire. This sounds odd, but it's a different kind of moving cause from simple physical causation for it depends on cognition. Aristotle says that the unmoved movers 'touch' the celestial spheres, but are not touched by them. We are to understand this not as a literal, physical touch, but as a psychological alteration – the sort of thing we mean when we say 'I am touched by your solicitude' or 'I am moved by her beauty.' Even animals which can move themselves ultimately depend on objects of desire to prod them into action. Love, it turns out, really does make the world go round.*

---

* Aristotle's physics is devoid of the concept of a non-contact-dependent force. This is his best stab at one. Delbrück pointed out that the notion of an unmoved mover is inconsistent with Newton's Third Law (When one body exerts a force on a second body, the second body simultaneously exerts a force equal in magnitude and opposite in direction to that of the first body) – that is, the celestial spheres must be exerting an equivalent force upon the unmoved mover. Others have claimed that, considering certain physical

Aristotle's cosmos is now looking very busy. Add up all the spheres and their unmoved movers and he has 110 entities of varying degrees of materiality and divinity in Earth orbit. The reason that he can claim, within just a few paragraphs, that there are so many unmoved movers and also insist that there is only one is that, like so much in Aristotle's world, they come in a hierarchy with one at the top. This, of course, is the unmoved mover responsible for the outermost stellar sphere. The 'first unmoved mover' has, in some sense, control of all the others. It may be their ultimate object of love and desire. It is Aristotle's ultimate God.

In the *Metaphysics*, Aristotle reveals, in sweeping periods, the purpose and nature of this entity:

On such a principle [the first unmoved mover], then, depend the heavens and the world of nature. And it is a life such as the best which we enjoy, and enjoy for but a short time (for it is ever in this state, which we cannot be), since its actuality is also pleasure. (And for this reason are waking, perception and thinking most pleasant, and hopes and memories are so on account of these.) And thinking in itself deals with that which is best in itself, and that which is thinking in the fullest sense with that which is best in the fullest sense. And thought thinks on itself because it shares the nature of the object of thought; for it becomes an object of thought in coming into contact with and thinking its objects, so that thought and object of thought are the same. For that which is capable of receiving the object of thought, i.e. the essence, is thought. But it is active when it possesses this object. Therefore the possession rather than the receptivity is the divine element which thought seems to contain, and the act of contemplation is what is most pleasant and best. If, then, God is always in that good state in which we sometimes are, this compels our wonder; and if in a better this compels it yet more. And God is in a better state. And life also belongs to God; for the actuality of thought is life, and God is that actuality; and God's self-dependent actuality is life most good and eternal. We say therefore that God is a living being, eternal, most good, so that life and duration continuous and eternal belong to God; for this is God.

models of the celestial motions in *The Movement of Animals* 3, Aristotle appears to hint at Newtonian laws of motion. Such considerations are, however, beside the point: Newtonian mechanics simply cannot apply to his final cosmological model since unmoved movers are immaterial and therefore have no mass.

This is what God does: he thinks. Normal thinking isn't good enough for him, so he spends his time 'thinking . . . thinking of thinking' – *noēsis noēseōs noēsis*. This is a God who knows neither love nor hate, who neither creates nor destroys, who does not save, condemn or even judge; this is a God utterly indifferent to Earthly affairs, yet upon whom, ironically, the very existence of the universe depends.

In the *Nicomachean Ethics* Aristotle discusses the best sort of life to lead. The good life is obviously one of active virtue, and there are many ways in which virtue can be achieved – in politics or in the army, say. But the virtue that derives from such things is entirely utilitarian. The best way that a man can spend his life is in contemplation for that has no utilitarian goal; it's pleasurable in itself. Elsewhere he relates a story. Someone asked Anaxagoras what was the point of being born, to which the great *physiologos* replied: 'to study the heaven and order of the whole cosmos'. The answer rang true to Aristotle; he told the story at least twice. But he warns that none of us can ever achieve a life of pure contemplation. There are so many things, the mundane things of every-day life, and the human things – the sense is disparaging – that distract us from the divine life of the mind. Nevertheless we should 'strain every sinew' to ignore them and devote ourselves to pure reason. That is where true happiness lies.

His works, pitiless in their detachment, show us what he means. He was, after all, a man who wrote of the press of sexual desire but not of Pythia, his long-dead Asian love; of the rise and fall of states but not of Alexander, the conquering boy whom he unleashed upon the world; of the very structure of reality while scarcely mentioning by name his teacher whose life's work he effortlessly assimilated, appropriated and then destroyed. *This* is the life of reason – the scientific life – and, when you contemplate the rank of squat volumes, the shelves of papyrus scrolls that they once must have been, and the relentless march of his arguments page after page after page, you cannot help but feel that he has confused the causality; that he did not so much search for God as reconstruct Him in his own image.

Yet there is another side to Aristotle's God. For it is not only philosophers and scientists that can, indeed *must*, strive to be like Him; every natural thing partakes, to howsoever humble a degree, of His qualities. Indeed, it is only now that we can truly understand the meaning of the words with which Aristotle must have begun his great course, and with which I began this book:

So we should not, like children, react with disgust to the investigation of less elevated animals. There is something awesome in all natural things. Some strangers, so the story goes, wanted to meet Heraclitus. They approached him but saw he was warming himself by the stove. 'Don't worry!' he said. 'Come on in! *There are gods here too.*'

Even a cuttlefish is, in some way, divine. It is a sweet and solemn thought. Had I a God – *had* I a God – it would be Aristotle's God.

THE STRAIT
OF PYRRHA

ARISTOTELIS STA-
GIRITAE DE HISTO-
RIA ANIMALIVM LIBER
PRIMVS, INTERPRETE
THEODORO GAZA.

❧

In quibus animalia inter se differant, quibúsue conue-
niant, eorundemq́; naturæ diuersitae. Cap. I.

NIMALIVM partes aut incompositæ sunt,
quæ scilicet in similes sibi partes diuiduntur,
vt caro in carnes,& ob eā rem similares appel
lentur:aut compositæ,quæ aptè secari in par
tes dissimiles possint,nō in similes:vt manus
nō in manus secatur,aut facies in facies:qua-
propter eas dissimilares nominemus. Quo in genere partium

DETAIL FROM THEODORE GAZA'S TRANSLATION OF
ARISTOTLE'S ZOOLOGICAL WORKS, 1552

CIV

IN 340 BC, SWAYED by Demosthenes' nationalist oratory, Athens
allied with Thebes against Macedon. Philip, goaded into action,
marched south. In August 338 the armies met at Chaeronea. Philip
was victorious but merciful. He neither enslaved the survivors nor occu-
pied Athens, but returned the bones of her thousand dead for burial. When
he was assassinated two years later the Athenian mob celebrated. The new
king, they said, was just a boy. They forgot that Alexander was a battle-
hardened twenty-year-old. In 335 Thebes revolted. Alexander levelled it.
Athens submitted. That was the year that Aristotle returned. He had been
away for twelve years. He was nearly fifty.

Although a friend to the conquerors, he was not unwelcome. The city
remained divided between Demosthenes' nationalists and pro-
Macedonian aristocrats. Now the former were chastened (Demosthenes
narrowly escaping being fed to Alexander to appease his wrath), the latter
ebullient. And Aristotle had a close friend in Antipater, soon to become
Alexander's European viceroy. The new philosopher in town was probably
quite fashionable.

He rented some buildings at the Lyceum and began to teach. It is said that
he gave his more difficult, technical lectures in the morning and public
discourses in the afternoon. Senior colleagues – Theophrastus, Callippus –
would have lectured too. His students came from across the Hellenic world.
He set them to work. The data in *Historia animalium* could, perhaps, have been
collected by one man; but that *and* the 158 constitutions of the Greek states
and the list of victors of the Pythian games *and* his records of the dramatic
works performed at Athens? Other encyclopaedic projects are hinted at too.
Their scale suggests that the Lyceum was not merely a gathering of philosophically minded friends, or even a school, but a research institute.

We may wonder what, exactly, Aristotle taught. Superficially it's clear
enough. The works of Aristotle that we have all appear to be lecture notes
or, at least, unpublished manuscripts, and all derive from the Lyceum's
library. They are a curriculum. But matters cannot be so simple. In many
ways, large and small, the texts appear to contradict each other. The small
contradictions can be dismissed as the errors of scribes, the insertions of

successors, the places where Aristotle changed his mind about, say, the octopus' brain. The large contradictions are less easily explained. There are two approaches you can take to resolving them. First, you can try to show that the apparently inconsistent texts are, if read rightly, consistent after all. Second, you can allow that Aristotle changed his mind on important matters too; that by the time of the Lyceum some of the texts were out of date and languishing in the stacks, while others represented his current thought. That seems reasonable. Who, after all, philosophizes for forty years and doesn't?

There are fashions in these things. In 1923 Werner Jaeger, a young German philologist, published *Aristoteles: Grundlegung einer Geschichte seiner Entwicklung.** It gave a vision of Aristotle as he metamorphosed from a young man under Plato's influence, through several stages, to the mature, empirically minded philosopher of the Lyceum. Jaeger believed that he could order the composition of particular parts of the *Corpus Aristotelicum*, that he could show that *Metaphysics* Books A, B, M 9–10 and N were written in Assos against Speusippus, while Z, H and θ are part of an entirely separate, later enterprise; or that *Politics* II–III, VII and VIII, so Platonic in tone, were written before the empirical IV–VI.

His scheme, brilliant and a little mad, enchanted Aristotelians until the 1960s. Since then it has been unpicked so that little of it remains. These days, perhaps still in reaction to Jaeger, classical philosophers often emphasize the unity of Aristotle's thought. Points go to those who can show that the seemingly irreconcilable is, in fact, not. To admit that Aristotle might have changed his mind, or the meaning of his technical terms, seems to be held as an admission of defeat.

Yet this way of reading Aristotle conceals as much as it illuminates. After all, two facts are indisputable: that he began his intellectual life as a student of Plato, writing Platonic dialogues on Platonic themes, and that he ended it having developed a system of thought that, whatever its debt to his predecessors, contained the elements of natural science. It would be astonishing indeed had this transformation not left its mark on his works. For me, this transformation manifests as two Aristotles. We can, crudely, call them the philosophical and the scientific Aristotles. By this I do not mean Aristotle's distinction between first and second philosophy or *theologikē* and *physikē*; rather I mean *our* distinction, the distinction that

---

* Published in English as *Aristotle: Fundamentals of the History of his Development*, 1934, by the Clarendon Press, Oxford.

we see at a time when philosophy and science are very different enterprises.

It's partly a matter of style. On the one hand, there are the *a priori* arguments of the *Metaphysics*, the *Organon* and even the *Physics* and *The Heavens*. On the other, there are the arguments of the zoology, *Meteorology* and *Politics*, based on, or at least tempered by, data. Here, from *The Heavens*, is an example of the former. Aristotle is explaining why there cannot be more than one world:

> We must now proceed to explain why there cannot be more than one world – the further question mentioned above. For it may be thought that we have not proved universally of bodies that none whatever can exist outside our universe, and that our argument applied only to those of indeterminate extent. Now all things rest and move naturally and by constraint. A thing moves naturally to a place in which it rests without constraint, and rests naturally in a place to which it moves without constraint. On the other hand, a thing moves by constraint to a place in which it rests by constraint, and rests by constraint in a place to which it moves by constraint. Further, if a given movement is due to constraint, its contrary is natural. If, then, it is by constraint that earth moves from a certain place to the centre here, its movement from here to there will be natural, and if . . . [etc.]

I shall not try to explicate. It is an edifice of pure *a priori* reasoning, a series of claims that are taken to be self-evidently true or else derivable from other self-evidently true claims. It illustrates the truism that, although any science needs basic principles to get off the ground, it can't get far on them alone. Here, by contrast, is an example of the more empirical Aristotle. In *The Generation of Animals* he is explaining how animals nurture their embryos:

> As previously stated, in live-bearing animals the embryo achieves growth through the umbilical cord. In animals the soul has a nutritive power (alongside the others) so it sends this cord like a root into the uterus. The cord is made up of blood vessels in a sheath, more of them in larger animals such as cattle, a single one in the smallest and two in those of middle size. The embryo gets its nourishment in the form of blood through this cord: for many blood vessels terminate in the uterus. All animals without teeth in the upper jaw . . . [etc.]

The zoological works, too, are rich in long, deductive chains of reasoning, but supporting data are usually close to hand. That is the difference. To Aristotle both *The Heavens* and *The Generation of Animals* were *physikē*; to us one is cosmic philosophy, the other reproductive biology.

It's a matter not just of style but also of substance. There is often a conflict between theory and practice. There is a conflict between the syllogistic theory of demonstration of the *Posterior Analytics*, with its austere programmatic certainties, and how Aristotle actually does science.* In his empirical works he invokes other modes of demonstration but is vague as to what they are. Often he offers only dialectical plausibility: here are some explanations, here are some arguments against them; this one seems best. There is a conflict between his insistence that each domain of knowledge should be kept distinct and how he then ignores their boundaries. There is a conflict between the taxonomic essentialism of the *Categories* and the pragmatic casualness of his animal classification in *Historia animalium*. There is a conflict between his insistence in the *Metaphysics* that artefacts and animals are utterly distinct, that the former, indeed, aren't even to be granted the ontological status of 'entities' (*ousiai*), and the mechanistic flavour of his explanations in the zoological works. I shall return to this. There is a conflict between his simple male/female :: form/matter dichotomy and the complexities of his theory of inheritance. There is a conflict between his anti-materialism – that is, his entire causal theory – and his belief in spontaneous generation. Some scholars see one or more of these conflicts as irreconcilable; others see them as the same thoughts expressed in different ways. To me, they collectively speak of a philosopher who has been mugged by empirical reality – or at least what he takes it to be.

It is tempting to suppose that these two Aristotles belong to different times of his life; that there is an early, philosophical and a later, scientific Aristotle. The first sat under a tree in the Piraeus picking holes in Plato's theory of definition by division, the second on a quay in Lesbos prodding a pile of fish. I think that there is much truth to this, but have

---

* Jim Lennox and Allan Gotthelf have shown, in many finely argued papers, how the theory of demonstration given in the *Posterior Analytics* filters through *The Parts of Animals*, and, following them, I have tried to explain how. But I am also struck by the fact that the *Posterior Analytics* contains not a single example of syllogistic demonstration drawn from the zoology. Aristotle's examples are all about geometry and eclipses (well, there's one involving leaves). That's why, when I wanted to illustrate the method using a zoological example, I had to appropriate those modern sticklebacks.

neither the courage nor the expertise to separate them and certainly not to impose some chronology on the texts.* Besides, once you start down that road, it's hard to know when to stop. All students of the *Metaphysics* agree that it is a disparate series of treatises cobbled together by a later editor. But what about apparent contradictions within *The Generation of Animals*? It seems to be a single, if imperfectly unified, work. We would murder to dissect.

For this reason I, too, have tried to present an Aristotle who does not disagree with himself. And if, on one or two occasions, I have allowed that the texts are not consistent, that once he thought one thing and later another, it has been only as a final expedient, as an exegetical sword to be swung *in extremis* when all others have failed to cut the knot. This much is also true: cease rootling about in the texts, step back and view the *Corpus* from a distance, and a grand unity becomes apparent. Whatever its imperfections and inconsistencies, it offers a system of awesome *completeness*. Much of this is due to the biology. Of all the things in the world that he might have studied, that he might have devoted his life to, Aristotle identified living things as most worthy of his attention. Nearly all the rest – his metaphysics, system of causal explanation, physics, chemistry, meteorology, cosmology, politics, ethics, even his poetics† – bear the mark of that decision.

As to why the *Corpus* shows such varied styles of argument, the explanation is easily to hand. In our day philosophers and scientists are distinct academic castes with distinct ways of arguing. But who is to say that, more than two thousand years ago, a man could not be both at once? That the scientist might not displace the philosopher, but be added to him? Such a man, I take it, was Aristotle when, walking along the winding paths of the Lyceum's gardens, he began to teach.

---

* Though most scholars agree that the *Organon* is Academic; Guthrie, at least, thought that *The Heavens* is early and many suppose that *The Generation of Animals* is late.
† Artists often express disappointment upon reading his *Poetics*. Why, they cry, does Aristotle not tell us wherein beauty lies? He doesn't because it is not a treatise on aesthetics written by another poet, but a treatise on how plays *work* written by a biologist.

CV

H E TAUGHT FOR TWELVE years or so. But then Alexander died in Babylon. The Athenians celebrated once again. The anti-Macedonian party turned nasty. They accused Aristotle of praying to Hermias, his friend, dead so long ago, and charged him with impiety. Politics were certainly at play. Had they wanted to get him for heresy they should have attended his lectures on astrotheology. The Delphians had honoured him (and Callisthenes) for recording the victors of the Pythian Games. Now they revoked the honours and smashed the tablet on which they had been proclaimed. He decided to leave. 'I will not allow the Athenians to commit a second crime against philosophy.' He was thinking of Socrates.

He went to Euboea, the large island that is separated from the Attic mainland by only a narrow strait, another *euripos*. His maternal family had an estate there in Chalcis. Fragmentary letters from that time speak of solitary calm. He writes to Antipater that he regrets the revoked honours, but does not regret them very much. Another: 'The more alone I am, the more fond I am of myths.' Within a year he was dead.

His will, which has been preserved, begins: 'All will be well, but in case it is not . . .' He names Antipater as executor. He gives his daughter in marriage to Nicanor, once his ward, now an officer in Alexander's army. To Herpyllis – a slave?, a concubine?, a second wife?, in any event a woman who shared his bed – he bequeaths real estate, silver, furniture and slaves. He disposes of about a dozen slaves. Favoured slaves are to be freed, given money and subordinate slaves. A student is sent home. He commissions statues to the memory of his parents and guardians. Other statues are to be erected at the shrines of Zeus and Athena the Preserver to give thanks for Nicanor's safe return from the East. He asks to be buried next to Pythia 'as she wanted'. It isn't much. It's the only real glimpse that we have of Aristotle the man rather than Aristotle the mind.

Theophrastus became head of the Lyceum. Diogenes Laertius says that two thousand pupils attended his lectures. Presumably that's a cumulative total, but even so it shows that the school flourished. Theophrastus' last will and testament depicts it as having a temple to the muses, a museum

containing maps and a bust of Aristotle and, of course, a garden. He left it all
to his fellow philosophers so that they might reside there 'on terms of familiarity and friendship'. Strato 'The Physicist' became head of the school. It
continued, but without distinction, until 86 BC when Sulla pulled it down.*

At the foot of Lycabettos, the Hill of Wolves, lies a patch of scrub and some
ruins. They are no more than foundations, stone blocks arranged in grids, of
the sort that only archaeologists can understand. Flimsy structures covered in
plastic sheets show that the archaeologists themselves were once there, but the
dig has long been abandoned and now has a desolate air. On Rigillis Street,
along its ranks of purple-bloomed jacarandas, a high fence keeps you out, but
from the grounds of the adjacent Byzantine Museum you can get a decent view
through the wire. An irritable guard will accost you; but explain what you're
after, why you're there and he'll walk with you and – a true Athenian – have a
cigarette while you look at where Aristotle once taught.†

THE LYCEUM, CENTRAL ATHENS, JULY 2011

* It was refounded in the first century AD.
† The archaeologists think that they have uncovered a *palaestra*, a gym, but the remains
could well be a Roman villa; if so, then whatever's left of Aristotle's buildings is under
some apartment block or the National War Museum. There was talk of turning the site
into a park, but nothing has come of that nor will it, I now suppose, for many years.

Strabo tells the story that Theophrastus bequeathed the Lyceum's library to one Neleus who took them to Skepsis, a mountain village that lies interior to Assos on the Turkish shore; and that, for nearly two centuries, the scrolls lay rotting in a cave until they were bought by that Athenian bibliophile, then looted by Sulla and taken to Rome. It's probably true. Those are the books that, in the first century AD, were edited and arranged by Andronicus of Rhodes into the form that we have today. They can't, however, have been the only copies. Within a century of Aristotle's death, the Ptolemies began to build their great library at Alexandria. It certainly held Aristotle's and Theophrastus' works. Alexandria became the centre of scientific research. Mechanics, astronomy and medicine flourished. Many philosophers there called themselves 'Peripatetics' in honour of Aristotle's school.

Amid all this new science there is, however, a curious gap: biology. There were scientifically minded physicians (Herophilus, Erasistratus) and, later, the natural-history-minded encyclopaedists (Pliny), poets (Oppian) and paradoxographers (Aelian) of Rome. There was also the greatest physician-scientist of them all, Galen of Pergamon. But there was no one who tried to *explain* living things, in all their diversity, as Aristotle had. No one who did zoology or botany. No one who saw, as he saw, that each creature reveals to us 'something natural and something beautiful'. No one would do so for a thousand years and more.

CVI

QUESTION LINGERS. If, as I have claimed, Aristotle was indeed such a great biologist; if, as I have claimed, there is hardly a facet of our science that he did not illuminate; if, as I have claimed, many of our theories are built upon his, then why has his science been forgotten?

Of course, his neglect is not absolute. The authors of biology textbooks occasionally register dutiful obeisance ('Aristotle was the father of . . .') before passing swiftly on. Classical philosophers still study him as they always have and always will. But to modern biologists he is a void attached to a name. His scientific works and the system that they contain have been lost to common knowledge as surely as if they had been eaten by moths complete. And even when one chances across a scientist who, unaccountably, claims to know something of Aristotle's work, the assessment is more likely than not to be irrationally harsh, even unseemly: 'a strange and generally speaking rather tiresome farrago of hearsay, imperfect observation, wishful thinking and credulity amounting to downright gullibility' – so Peter Medawar, essayist, scientific statesman, Nobel Laureate in physiology and medicine on the books that contain the origin of his science.

Medawar wrote these lines in 1985.* Their tone, however, is pure seventeenth century. It's the tone of the early Royal Society of London, the association of scientists of which Medawar was rightly proud to be a Fellow. The anachronism explains all. Medawar's abuse was aimed not at Aristotle the father of science but at Aristotle its greatest foe. He was, indeed, re-enacting, for a new generation, the origin myth of modern science; the myth in which Aristotle was the giant who had to be slain so that we could pass through the straits of philosophy to reach the open sea of scientific truth that lay beyond; the myth in which Aristotle is little more than an endlessly fecund source of empirical, theoretical and methodological error; the myth that explains his absence from the scientific pantheon next to Linnaeus, Darwin and Pasteur; the myth that explains why not one scientist in a thousand can name, much less articulate a single result from, his scientific works. I say it is a myth and, insofar as history matters at all, it is certainly a pernicious one for it omits all that we owe him. But it is a myth that has this much truth: that Aristotle's science was the principle casualty of the Scientific Revolution. It may even be said that modern science was built on its ruins.

CVII

IN THE TWENTY-THREE CENTURIES since his death, Aristotle's works have been lost and found many times. In early medieval Christendom his oblivion was nearly complete. Bits of the *Organon*, relics of Byzantium, still circulated, but the *Metaphysics*, *Poetics*, *Politics* and natural science were all effectively extinct. The recovery of his works

---

* In *Aristotle to Zoos: A Philosophical Dictionary of Biology*, co-authored with his wife, Jean.

was in large part due to the Christian reconquest of Moorish Spain. In 1085 Toledo, that jewel of Al-Andalus, fell to Alfonso VI of Castile. Among the treasures contained within the city was most of the *Corpus Aristotelicum* preserved in Arabic along with paraphrases and commentaries by Avicenna, a Persian, and Averroës, an Andalusian, Muslims both. Translated into Latin by Michael Scotus, Aristotle's works began to circulate throughout Europe.

Two dates, appealing for their symmetry, capture the flux of his fortunes over the subsequent four hundred years. In 1210 the University of Paris banned the teaching of Aristotle's natural philosophy in the Faculty of Arts on pain of excommunication. In 1624 the Parliament of Paris, urged on by the Faculty of Theology, banned the teaching of any doctrine opposed to his on pain of death. The significance of the dates lies in the truism that authorities only issue bans when they sense the wind blowing against orthodoxy and that, by the time they get around to doing so, it's always far too late.

Aristotle's attractions proved irresistible to medieval scholars. Even the Parisian ban extended only to the Faculty of Arts; theologians could still read him and did. In 1245 Albert Magnus, a Dominican appointed professor at Paris, began a vast paraphrase and commentary on Aristotle's works based on Scotus' translation. A few decades later his student, Thomas Aquinas, began to construct his equally ambitious synthesis of Aristotelian metaphysics and Christian theology. Thomas abolished Aristotle's division between first and second philosophy – easily done since it was blurred by Aristotle himself – and turned natural philosophy into a branch of theology. Thomas's God, the *primum movens immobile*, is Aristotle's unmoved mover; the teleology of his ethics is Aristotle's too.*

The triumph of the Thomist synthesis rendered Aristotle's philosophy supreme. In *Inferno* IV, published around 1317, Dante called Aristotle 'the master of those who know'. The cost of philosophy was science. Following Thomas, the schoolmen of Oxford, Coimbra, Padua and Paris toyed endlessly with substance, potentialities, form-and-matter compounds, categories and all the other cogs in the Philosopher's metaphysical machine. Their method

* It was also Thomas who, suspecting that the Arabic text might be corrupt, commissioned William of Moerbeke to translate Aristotle's works from various Greek texts of Byzantine origin. Derived from Andronicus' edition, they are the basis of our own Greek text. The oldest-known Greek Ms. of *Historia animalium* is a ninth-century fragment of Book VI from Constantinople (Parisinus suppl. gr. 1156, Bib. Nat., Paris). Most of the other extant Mss. are twelfth to fifteenth century.

was disputatious, their factions innumerable, their writings interminable and their conclusions stultifying. Much of it wasn't very Aristotelian at all. They reigned over Europe's universities for three centuries.

There were, of course, deviations from Thomist orthodoxy. In the 1500s various thinkers, mostly extramural, critiqued the schoolmen on Platonic, Epicurean, Stoic, materialist or entirely novel grounds. In Warmia, Copernicus proposed a new cosmic geometry; in Calabria, Telesio sketched a new, materialist cosmogenesis. Given the intimate tie between natural philosophy and theology, such novelties were risky. The Neapolitan monk Giordano Bruno developed a comprehensive pantheistic cosmology and, for his efforts, was tried by the Inquisition for heresy and, in 1600, burnt at the stake.

Galileo captured the mood. In his *Dialogo sopra i due massimi sistemi del mondo*, 1632, he argued his physical system as a dialogue between three men: Salviati (Galileo's champion), Sagredo (a persuadable cipher) and Simplicio (an Aristotelian). Who, asks Simplicio, will guide us if we abandon Aristotle? Anyone with eyes in his head and his wits about him can serve as a guide, replies Salviati. But eyes were precisely what Simplicio lacked. He is said to have been modelled on Cesare Cremonini, Professor of Natural Philosophy at Padua, who – the story is so delicious that one suspects it to be apocryphal, but it nevertheless appears to be true – refused Galileo's invitation to look through a telescope at the mountains of the moon since if the moon were not a perfect sphere then it must be corruptible and Aristotle said it wasn't. How very Aristotelian – and, as Galileo observed, how utterly unlike Aristotle.

CVIII

ARISTOTLE'S PHYSICAL SYSTEM suffered grievously at the hands of the new scientists. By the middle of the seventeenth century his cosmology and theory of motion were obsolete. His chemistry took longer to kill. His biology, rich in empirical data, fared best. Even in the thirteenth century Albert Magnus drew from it the right conclusions. 'The aim of natural science', he wrote, 'is not simply to accept the statements of others, but to investigate the causes that are at work in nature.' And: 'Experiment is the only safe guide in such investigations.' He accordingly added much new animal lore, some of it first hand, some borrowed from other sources, to his synopsis of Aristotle's zoology. Compare Albert's use of Aristotle to Thomas's and it is hard to resist the conclusion that the eclipse of the former by the latter retarded the development of natural science by centuries.

This thought gains additional force from the fact that in the sixteenth century Aristotle's biology helped to break the hold of Thomist scholasticism. In 1516 Pietro Pomponazzi, professor at Bologna, published *Tractatus de immortalitate animae*, in which he counterposed the Thomist doctrine of the immortality of the soul, established as dogma by the Fifth Lateran Council of 1512, against Aristotle's arguments for its mortality. The book was burnt at Venice. Powerful friends and a cautious defence preserved its author from the same fate. In 1521 he published *de Nutritione et augmentatione* – on nutrition and growth – based on Aristotle's *Generation & Corruption*. And then he taught a course on *The Parts of Animals* – the first since antiquity. 'I don't want to teach you,' said this delightful man. 'I came here not because I am more learned, but because I am older. The love of science pushed me, therefore I am ready to be whipped and want you to teach me' – his words, to his students, faithfully recorded by one. It wasn't a zoology course, but Pomponazzi didn't hesitate to contradict Aristotle on empirical grounds. Discussing the (accurate) account of the avian nictitating membrane in *The Parts of Animals* II, 3, he lamented that he had dissected a chicken but failed to find it. 'I wasted my hen and I have found nothing!'

But Pomponazzi, unusually for a schoolman, had a medical degree from Padua. Within a few decades, the anatomists of Padua and Bologna's medical faculties – Vesalius, Fabricius, Falloppio, Colombo and Eustachi – were dissecting corpses. They were guided by that other great authority of antiquity, Galen, but did not hesitate to call Aristotle in their support. In 1561 Ulisse Aldrovandi became the first Professor of Natural Science at Bologna (*lectura philosophiae naturalis ordinaria de fossilibus, plantis et animalibus* was his splendid title). He established a botanical garden and a museum, and began to collate and rearrange Aristotle's zoology, as well as any other material he could find, into a vast encyclopaedia. Naturalists such as Salviani, Belon and Rondelet went down to the markets of Rome and Montpellier and sorted out the fish. This wasn't a rejection of Aristotle's science; it was its rediscovery and revival.

The anatomists and naturalists of the sixteenth century left Aristotle's explanatory theories largely intact. Harvey's demonstration of the circulation of the blood, 1632, and ovular embryology, 1651, cut deeper. But Harvey was a man who could love both Aristotle and the evidence of his eyes.

> For although it be a more new and difficult way, to find out the nature of things, by the things themselves; than by reading of Books, to take our knowledge upon trust from the opinions of Philosophers: yet must it needs be confessed, that the former is much more open, and lesse fraudulent, especially in the Secrets relating to Natural Philosophy.

How very true. Yet he also told John Aubrey that he'd be better off reading Aristotle than the new 'shit-breeches'. He meant by that, *inter alia*, Descartes.

Aristotle's empirical findings may have formed the foundations of modern biology, but his explanations of how animals actually work were vulnerable to assaults on his physical theory. The fascination of Aristotle's natural science is precisely the extraordinary way in which it interlocks. I said that Aristotle is no ontological reductionist; that he would never say that a child or a cuttlefish is *just* the stuff from which it is made. That's true: forms are, for Aristotle, more fundamental than matter. He is, however, a theoretical reductionist for he does believe that higher-level phenomena are explicable in physical terms. A son resembles his father because his father's form shaped him in the embryo. It sounds mysterious but can be explained in terms of the physical action of *pneuma* and the heating and cooling of material substances as evidenced by seminal foam. Very well, but dispense with *pneuma* and the whole account falls apart. Destroy Aristotle's theory of motion and much of *The Movement of Animals* no longer makes sense; deprive the elements of their 'natures' and the physiology of *The Length and Shortness of Life, Youth & Old Age, Life & Death* and *The Parts of Animals* ceases to work; revive atomism and the elements of the *Generation & Corruption* no longer cycle; set the Earth spinning about the sun and the celestial engine of *The Heavens* fails. Deprive the world of its eternity and you strip every living thing of its reason to be.

Yet it was neither his association with scholasticism nor his zoological errors, nor even the falsification of his physical theories, that accounts for the oblivion of Aristotle's scientific thought; for the fact that, if he is remembered as a scientist at all, it is as a muddle-headed ancient (scarcely distinguishable from Pliny), rather than as the engineer of the greatest scientific structure ever built by one man, and the first to boot; rather, it was the belief, a foundation stone of the New Philosophy, that his explanatory

system was corrupt to its core. And here Medawar gets it right. For he credits – no, celebrates – one man for having done more than any other towards
the destruction of Aristotle's reputation. Enter Francis Bacon.

CIX

THE FUTURE LORD CHANCELLOR OF ENGLAND brooded over
Aristotle's works like a vulture over a kill. No scientist himself, he
was the New Philosophy's most ardent theorist and propagand-
ist. In the prolix periods of *The Advancement of Learning*, 1605, his hostility to
Aristotle is palpable:

> And herein I cannot a little marvel at the philosopher Aristotle, that
> did proceed in such a spirit of difference and contradiction towards all
> antiquity, undertaking not only to frame new words of science at his
> pleasure, but to confound and extinguish all ancient wisdom, insomuch
> as he never nameth or mentioneth an ancient author or opinion, but to
> confute and reprove . . .

Aristotle was, Bacon said, like an 'Ottoman Turk, in the slaughter of his
brethren and with success'.

That Aristotle is generous with criticism and parsimonious with praise
towards his predecessors is undeniable. But so what? It's a scientist's job to
disagree. Besides, the remarkable thing is precisely how each of his books
begins with a round-up of what his predecessors thought before moving on
to his own solutions. Aristotle's treatises have the structure that academics
have used ever since.* As Bertrand Russell said, Aristotle was the first man
to write like a professor.

Bacon, however, had a complex agenda. He wanted to paint the Philosopher
in the colours of the quarrelsome scholastics, contrast their intemperate
disputations with the new, civil kind of scientific discourse that he envisioned
(but that his own writings hardly exemplify) and indict Aristotle for injustice
towards the true scientific heroes of antiquity, the *physiologoi*.

---

* The first two books of Strabo's *Geography* are, for example, devoted to defending his
heroes (Homer) and criticizing his opponents (Eratosthenes, Hipparchus, Posidonius).

He launched his attack from all sides. In the *Novum organum*, 1620, he accused Aristotle of pressing his facts to fit his theories:

> Nor is much stress to be laid on his [Aristotle's] frequent recourse to experiment in his books on animals, his problems, and other treatises; for he had already decided, without having properly consulted experience as the basis of his decisions and axioms, and after having so decided, he drags experiment along as a captive constrained to accommodate herself to his decisions: so that he is even more to be blamed than his modern followers (of the scholastic school) who have deserted her altogether.

The Royal Society's propagandists – Thomas Sprat (*History of the Royal Society*, 1667) and Joseph Glanvill (*Plus Ultra*, 1668) – echoed the charge. Glanvill was particularly caustic: 'he [Aristotle] did not use and imploy Experiments for the erecting of his theories; but having arbitrarily pitched his Theories, his manner was to force Experience to suffragate, and yield countenance to his precarious Propositions'.

Bacon's most serious charge was aimed at Aristotle's explanatory system. Of the four kinds of causal explanations that Aristotle insists natural science demands, Bacon ruled two – the formal and final – illegitimate. Natural philosophy should concern itself with the properties and movements of matter and them alone. Explanations such as 'the hairs of the eyelids are for a quickset and fence about the sight', or 'the firmness of the skins and hides of living creatures is to defend them from the extremities of heat or cold', or 'the bones are for the columns or beams, whereupon the frames of the bodies of living creatures are built' were no part of science, but should be left to metaphysics. They were 'remoras and hindrances to stay and slug the ship from further sailing'. They retard the search for the true, physical causes of things.

Bacon's attack on forms was subtler. It is, he said, futile to inquire into the form of a lion or an oak or gold or even water or air. To the degree that forms have a place in natural philosophy, they are only a list of the basic sensible properties of matter, heavy–light, hot–cold, hard–soft and the like. His formal properties were grounded in a particulate (in seventeenth-century jargon, 'corpuscularian') theory of matter. Heat, for example, is a type of motion found when particles are both set in motion and constrained in some way. He evidently envisioned a heat 'law' that relates particulate motion to temperature. Bacon sidestepped the question of how to get more complex objects – gold or lions – out of these basic properties; but, then, he was more concerned with providing general principles than usable theories. The thrust, however, was clear: a radical, un-Aristotelian, ontological reductionism in which there is room only for moving and material explanations. Bacon looked for a new philosophical champion in antiquity. Democritus. He would become the Attic poster boy of the new scientific age.

Bacon's aversion to Aristotle and Aristotelianism – he scarcely distinguishes the two – also stemmed from a particular vision of the purpose of science and its proper object of study. Its purpose, in Bacon's view, was not merely to understand the world, but to change it; its proper object of study, then, was the artificial rather than natural. Bacon was a technology enthusiast. Aristotle's philosophy, he said, was 'strong for disputations and contentions; but barren of the production of works for the benefit of man'. Bacon demanded a new, mechanistic natural philosophy underpinned by a unified physics that would explain the movements of both natural and artificial objects. Newton would provide one.

In biology, the cheerleader of mechanism was Descartes. Animals and plants, he declared, do not have souls – they are merely machines. This was the doctrine of the *bête machine* or beast machine. Descartes reduced the complex of Aristotelian changes to local motion alone, and founded his physiology on a corpuscularianism that he got from Gassendi and Beeckman. His mathematical physics was important, but his anatomy indifferent and he made no biological discoveries. (He contested with Harvey over the movements of the heart and lost.) His teleology was simply theistic. (Animals may be machines, but they are wondrous machines made by God.) But his explicit comparison of animals to automata resonated at a time when mechanical devices were proliferating. It did away with the obscurities of the Aristotelian nutritive and sensitive souls (rendered utterly opaque by the schoolmen) and gave a point of entry for experimental investigation. In 1666 the Danish anatomist Niels Stensen (Steno) wrote:

> No one but [Descartes] has explained all human function, and above all those of the brain, in mechanical fashion. Others describe for us man himself; Descartes speaks only of a machine, which at one and the same time shows us the inadequacy of others *and points out a method of investigating the function of the parts of the body* just as insightfully as he describes the parts of his mechanical man [italics mine].

The *bête machine* flexed its muscles and gave a lusty squall.

Such then, in brief, are the intellectual currents that destroyed Aristotle's science in the seventeenth century. His fortunes have varied since. Zoologists have always regarded him with affection. In the nineteenth century, Cuvier, Müller, Agassiz and many others even turned him into a bit of a cult.* To them, he was an illustrious forebear with a sharp eye for curious bits of zoology, an authority to wield against opponents, and even a fertile source of explanatory ideas – or so I have argued. In the eighteenth and nineteenth centuries, too, teleology was retrieved from the metaphysical and theological wastebasket to which Bacon and Descartes had consigned it. In some scientific circles, particularly German ones, final causes became respectable again. That was very Aristotelian. The consequences of that for his reputation were, however, ultimately malign. The association between teleology and vitalism revived and reinforced Bacon's old charge that Aristotle's science was unmechanistic. Hans Driesch, that errant embryologist, even wrote a history of vitalism that commenced proudly with Aristotle. Twentieth-century biologists would still be flogging vitalism long after it had expired. 'And so to those of you who may be vitalists I would make this prophecy: what everyone believed yesterday, and you believe today, only cranks will believe tomorrow' – Francis Crick in 1969. When, in 1954, Erwin Schrödinger published a little book on ancient science he simply stopped with Democritus. Why bother going further? Aristotle had nothing to say to modern science.

CX

BACON AND HIS successors said that Aristotle's methods were wrong and that his explanations were too. Both charges are grave, but are they just? Our ideas of what constitutes scientific explanation, and how to achieve it, are ever changing. It may be, then, that we can

---

* This was also a time when classicists with a gift for zoology and zoologists with a gift for classical studies such as C. J. Sundevall, H. Aubert and F. Wimmer, J. B. Meyer and W. Ogle studied his zoological works as zoology. In the twentieth and twenty-first centuries these works have been mostly studied for their philosophical insights, and I would say that D'Arcy Thompson's 1910 *Historia animalium* was the last edition in the zoological tradition except that W. Kullmann's great *The Parts of Animals*, 2007, is both deeply philosophical and will tell you the truth about the dolphin's respiratory tract.

see merits in Aristotle that our predecessors missed. Every generation must read Aristotle anew.*

That Aristotle made countless observations of the natural world is obvious to anyone who reads his books – even the men of the Royal Society conceded so much. Should you read Aristotle's biology, you may, however, wonder why Bacon and Glanvill keep going on about his 'experiments'. They said he did them but abused their results, using them merely to confirm what he knew or thought he knew. You, however, may be less dismayed by his abuse of experimental data than by its absence.

The difficulty is merely semantic. In the seventeenth century 'experiment' meant any investigation of a natural phenomenon that involved some sort of intervention. Aristotle's study of chick embryogenesis which involved finding eggs of just the right stage, carefully cutting open the shell and prodding the embryo to expose its heart is, in this sense, one. Starving and then strangling livestock to see their vascular systems is another. So are vivisecting tortoises and poking out the eyes of swallows. Aristotle sometimes indicates that he's actually tried it out by using the term *pepeiramenoi*: 'Saltwater, when it turns into vapour, becomes sweet; and the vapour does not form saltwater when it condenses again. This we know by *experiment*.' Or so *pepeiramenoi* is often translated.

A modern scientist would take a more austere view. Such manipulations, he would say, are just observations made using a fancy technique. Experiments are defined not by their technique but by their logical structure. A true experiment is the comparison of a deliberately manipulated situation to an unmanipulated control for the purpose of testing a causal hypothesis. And Aristotle's works, he would sadly conclude, are devoid of experiments of that sort.†

---

* '[Aristotle] must be detached from his historical roots and neutralized before he can become accessible to posterity' – Werner Jaeger, *Aristotle*, 1934.
† In the *Meterologica* Aristotle describes a variety of observations that are often said to be 'experiments'. The most curious of these also occurs in Book II, 3. He wants to make the case that seawater is a mixture of water and some earthy stuff, salt. He claims that if you make a sealed vessel of wax, and submerge it in the sea, it will fill with fresh water, the salt having been filtered out by the wax. The procedure, however, was not a true experiment because it was uncontrolled. A suitable control would have been the submersion of a similar vessel made of some impermeable stuff – glass, bronze – in salt water. The absence of a control was, presumably, the fatal flaw. We can be sure that wax will not filter seawater (if it did, the deserts of Arabia would have flowered long ago) so, if Aristotle carried out the procedure (and I rather doubt he did), then any fresh water that he found in the sealed vessel must have come from condensation as the vessel cooled in the sea. Had he done a proper control, he would have seen that he was not entitled to attribute any fresh

Why is this? Aristotle certainly understands experimental logic, for he repeatedly refers to what we would now call 'natural experiments'. The oysters that were taken from Lesbos to Chios did not breed in their new home: ergo the generation of oysters depends not on the presence of oysters but on the right kind of mud: ergo they are spontaneous generators. The inference is plausible but far weaker than Aristotle allows. Perhaps Chian waters were too cold for the oysters to breed; perhaps they did breed, but the infant oysters died undetected; perhaps . . . a dozen alternative explanations spring to mind. Ecologists and evolutionary biologists often speak of 'natural' experiments since it's hard to perturb the course of evolution or tweak whole ecosystems, but, as one of my colleagues, a famous ecologist himself, is fond of remarking, 'The thing about "natural" experiments is that they're not experiments at all'* – by which he means that in a true experiment the *only* variables that differ between control and treatment are those manipulated by the experimenter; and when you rely on nature to do your manipulations you can never be sure just what she's meddled with.

Theophrastus' reports of how wheat cultivars perform when grown in various places are better and, done deliberately, would be a reciprocal common-garden experiment of the sort that would fully justify his inference that wheat strains differ due to some inherited quality. But he didn't do them deliberately and so his inference, though very likely correct, is also weak. Who really knows what farmers get up to? If you believe them, you'll end up believing all sorts of things; you'll even believe that *aira* can evolve from wheat. Aristotle's version of a common-garden experiment is even more pleasing: to determine whether the infertility of a couple is due to a deficiency in the male, he says, let him copulate with women other than his wife, and see if he sires offspring with them. Now *that's* got the makings of a real experiment, and it would have been perfect had he also recommended, which he did not, the reciprocal treatment. But it's no more than a suggestion. To imitate my colleague: 'The thing about *thought* experiments is that . . .'

It's not as though the experiments were technically difficult. Do flies really spontaneously generate from rotten meat? All you need to test the idea are two jars, some fresh fish and a bit of fine cloth. That was the sum of

---

water found in the vessel to a marine origin and that his results shed no light on his original claim. The same kind of objection applies to his vivisections.
* Mick Crawley, who credits Nelson Hairston with having said it.

Francesco Redi's equipment. Does the embryo of a tetrapod truly emerge from a coagulum of semen and menstrual fluid? If so, then the coagulum should be visible in the dissected uterus of a freshly impregnated mammal; even a sheep would do. Aristotle didn't look; William Harvey did.*

Historians sometimes attribute Aristotle's failure to do experiments to his worldview. If you draw, as Aristotle did, a sharp distinction between natural and unnatural change then a manipulative experiment, which clearly involves the latter, can hardly shed light on the former. There may be something to this. In the centuries after Aristotle's death Greek technologists in Alexandria began to produce elaborate machines. In the first century AD Hero of Alexandria described a charming hydraulic device in which a cluster of bronze birds cease to sing as a bronze owl rotates to face them. The gadget-minded Alexandrians were also quicker than Aristotle to put their physical theories to the test. Hero's *Pneumatics* contains an account of an experimental programme that is almost worthy of Boyle.†

Perhaps, then, the reason why Aristotle did not roll a ball down an inclined plane as Galileo did is because the conceptual structure of his physics prevented him from doing so.‡ But does that explain why he did not ask a farmer to mate a white-fleeced ram to a black-fleeced ewe to see how the progeny turn out? It's not a particularly 'forced' intervention, he understood the logic (*vide* his discussion of that wayward woman of Elis) and the results of the experiment would surely have given him pause for thought when constructing his model of inheritance.§

---

* What is the earliest description of a true experiment? Perhaps it is Herodotus' account of one carried out by Psamtik, an Egyptian pharaoh of the seventh century BC. Psamtik, desiring to know the origin of humans, took two children and had them reared by goats in the absence of any human sound. (The controls are, implicitly, children raised by their parents.) The first comprehensible utterances of these goat-children resembled, it seems, the Phrygian word for 'bread'. Psamtik concluded, by a kind of recapitulationist logic, that the Phrygians were a more ancient people than the Egyptians. This is one of several manipulations involving humans that have come to be known as 'Forbidden Experiments', and is now quite unreproducible. Experimental science, it seems, was born in sin.

† Diels claimed that the opening pages of Hero's *Pneumatics*, which describes these experiments and was written in the first century AD, was largely based on one of Strato's lost works. This, if true, would imply a remarkable advance in experimental rigour at the Lyceum within just a few decades of Aristotle's death.

‡ That Galileo dropped cannonballs from the Leaning Tower of Pisa by way of experimentally falsifying Aristotle's theory of motion is apparently a myth. Others did though.

§ Black fleeces, we now know, are usually caused by a simple autosomal dominant at the MC1-R locus, so, by Mendel's laws, half the progeny would have been black and half of them white *or* all of them would have been black depending on the genotype of the black ewe. In either case, fleece colour would segregate independently of progeny sex. That

Indeed, had Aristotle done a few simple experiments, he would certainly have made fewer errors. But it's one thing to understand experimental logic, quite another to see it as the high road to truth. The question, however, is this: given that he didn't do experiments, does his method correspond to anything that we, today, might recognize as science? Plato's method plainly disqualifies his theories from being scientific ones — it was, after all, founded on a contempt for empirical reality. It's harder to be certain about the *physiologoi*'s methods — they are so diverse and we have so little idea what, exactly, they did. Aristotle, however, has a method for extracting truth from the empirical world, a very sophisticated one too. It is, I believe, very similar to one used today.

CXI

S CIENCE HAS ALWAYS embraced two very different styles of empirical investigation. The first is the style most familiar to us, in which causal hypotheses are tested by deliberate, critical experiments. It's the style adopted, and celebrated by, the founders of the Royal Society. The second is less familiar, but hardly less important. It's one in which data are amassed, patterns sought and causal explanations inferred from those patterns. It is the style that was once found only in the historical sciences — cosmology, geology, palaeontology, ecology and evolutionary biology — the sciences in which manipulative experiments are hard to do. That, however, is no longer true.

The first style dominated twentieth-century biology. First you identified the object of your study — some gene (say) in some creature that, for whatever reason, you thought particularly fascinating. You worked out a way to measure its activity. Then you manipulated it. You could 'knock out'

---

should have dented Aristotle's conviction that sons generally resemble their fathers and daughters their mothers. Of course, he seems to think progeny coat colour is influenced by the water that the parents drink — but *any* breeding experiment would have falsified that hypothesis. Incidentally, I asked two sheep farmers in Lesbos about the inheritance of fleece colour. One said that progeny fleece colour does generally take after the ram; the other denied it and just said that sometimes black × white matings give black offspring, sometimes white — which is true. Neither invoked Mendel's laws so I think that both men were reporting folk biology. The lack of consensus is, in my experience, quite typical.

your gene – kill it dead in its tracks – or 'over-express' it – turn it on in unexpected ways and places. You would see how your manipulation affected the gross phenotype of your creature or, perhaps, the behaviour of other genes – but not too many of them since each test was complicated, expensive and time consuming. All this would take years. When you were done you'd publish a paper like this one:

> Morita, K. *et al.* 2002. A *Caenorhabditis elegans* TGF-beta, DBL-1, controls the expression of LON-1, a PR-related protein, that regulates polyploidization and body length. *Embo J.* 21:1063–73

The authors of this paper – and there are thousands like it – compared long mutant worms with regular worms and so described the role of a few genes in controlling the length of their worm. They know full well that they've unravelled only a few links in a vast causal network that might influence worm length, but since they believe in their results, they rest secure in the knowledge that their causal claim is true; and that, modest though their discovery may be, when it is combined with a thousand others like it, something important will emerge.

That paper was published just over a decade ago. How dated it now looks. For, as the twenty-first century has progressed, the notion of studying just a few genes at a time has become quite passé. The problem is no longer how to find a gene of interest – a single, modern, genome-sequencing machine can pump out fifty-four gigabases of sequence per day.* That's around sixteen human genomes, each with 25,000-odd genes. An 'expression array' chip can show, for any tissue you can grind up and put on it, which of those 25,000 are active and to what degree. Other technologies will allow you to survey thousands of metabolites or proteins all at once. Biologists speak of the 'omics – genomics, transcriptomics, metabolomics, proteomics, but they all really mean just one thing: data – lots of it.

Here is a typical 'omics paper:

> Fuchs, S. *et al.* 2010. A metabolic signature of long life in *Caenorhabditis elegans. BMC Biology* 8:2

---

* Illumina HiSeq 2500 High Output run, 1 × coverage per genome. This number will be obsolete by the time this book is published and will seem quaint should someone read it in 2024.

The authors of this paper – and there are thousands like it – compared long-lived mutant worms with regular worms and described many differences in their metabolites. This paper (Fuchs) has much in common with the one above (Morita): same worm (*C. elegans*); same lab (mine); similar problems (growth *v.* ageing). But there's a fundamental difference in method. Where Morita studied a few genes in detail, Fuchs studied hundreds of metabolites superficially. The consequences of this for what they could claim were profound – and were reflected in the titles. Where Morita spoke boldly of 'control', Fuchs admitted only to having found a 'signature'. She found dozens of differences between their long-lived and short-lived worms, but had no idea which of them matter, which – if any of them – actually make the long-lived worms live long. Technology gave Fuchs comprehensiveness. It cost her causality. Of course Fuchs and her colleagues did not despair. They – we – trawled her data for patterns. We found them and, from them, wove a causal model that we fondly believe might contain some truth. But we'd be the first to admit that we really don't know.

Fuchs' paper is very much in the second style. It's the characteristic style of the age of Big Data, one that is spreading into sociology, cultural history, engineering and economics. Its method is always the same: hoover up all the data going, order it by some sort of classification, visualize its structure, infer a causal model. The tools – multidimensional scaling, network graphs, Self Organizing Maps and the like – are all new, but the style is old.

It's Aristotle's style. It's the style to adopt when you discover a new world; when, instead of worrying away at the phenomena your predecessors looked at, you chance across some vast new domain of things to study, splendid in their confusion, fascinating in their order and opaque in their causes. These days technology – better sequencing machines, faster computers, bigger telescopes – give us our new worlds. Aristotle needed none of them. He just had to walk down to the shore to find a whole domain of things that had never been studied before. *Historia animalium* is the Big Data repository of his day. Not so big, you say? Maybe – but his other Big Data project, the 158 constitutions he collected, would be impressive even now. They were the basis for his other great exercise in causal explanation, the *Politics*.

The two styles also have a very different relationship between theory and data. In the first, a specific hypothesis is tested. The result is either consistent with the hypothesis or not. In the second, a narrative is constructed. You let the data speak. What they tell you is, of course, strongly influenced by what you think and hope you'll hear. When Glanvill complained that Aristotle forces 'Experience to . . . yield countenance to his precarious Propositions', he identified the danger. We are much more keenly aware than Aristotle was that any given empirical pattern may be explained by several different models, yet even so remain susceptible to the same mistake. That is why both the styles of science that I have spoken of are needed. Data trawls and pattern analyses give you models; targeted experiments tell you whether or not they're true. Many scientists use them both.

Besides the lack of strong causal inference, the second – Aristotelian – style has another weakness. Lots of data always means low-quality data, especially when it's gathered from anywhere and everywhere.* Genbank – the vast database in which biologists deposit their DNA sequences – is notorious for its errors. This doesn't stop them from using it. They take the data, run what checks they can and hope that the errors cancel out and that truth will emerge in the aggregate. That, too, seems to be very much Aristotle's style. He makes hundreds of factual claims that he must have known were based on uncertain data. It was probably a deliberate choice. His data were grist for the empirical generalizations and causal theories that were his ultimate goal; and he seems to have felt that the risk of incorporating dubious data was worth the prize of finding them. The data from his soothsayer-derived passages on animal behaviour are feeble. But they do illustrate the different kinds of interactions among animals, some of them competitive and some of them predatory, and suffice for an important generalization about how agonistic interactions among animals increase when food runs short. In The Heavens he suggests that we should still theorize even when the evidence is very slight and the object of our investigations is far away. History judges such scientists bold when they're right – and rash when they're wrong.

* You could call this the Wikipedia Principle.

CXII

THERE IS A belief, and I think it is a very widespread one, that something is wrong with Aristotle's explanations; that they are, in some way, fundamentally unscientific. Sometimes it is said that his appeal to the 'natures' of things is circular. In *Le Malade imaginaire*, 1673, Molière's Aristotelian quacks explain that opium induces sleep because it possesses a sleep-inducing principle. Ever since, arguments of this kind have been known as *virtus dormitiva* explanations and rightly treated with scorn. At other times it is said that Aristotelian natures possess a 'creative impulse' or else 'occult forces'. Applied to his biology, these are polite ways of saying that he's a vitalist – which many have said too. And then there are those who have said that final or formal causes *are* those creative impulses and occult forces and have no place in modern science.

All of these charges, endlessly repeated, are echoes of the Scientific Revolution. Often they have been repeated by Aristotle's foes who knew little of what he said or did. Yet even those who have known Aristotle intimately, and loved him dearly, have sometimes thought his explanations bankrupt. William Ogle did. So too, remarkably, did D'Arcy Thompson. *On Growth and Form*, the strange and beguilingly beautiful book that he published just seven years after his translation of *Historia animalium*, is a paean to Democritus.

Over the last fifty years or so, scholars have explored, uncovered and displayed the explanatory wealth of Aristotle's biology as never before. I have tried to show something of what they have found. Their discoveries, and our ever-changing understanding of the natural world, demand that we reopen the investigation into these antique accusations.

The claim that Aristotle's explanations are not merely wrong but *unscientific* comes down to the claim – as old as Bacon – that they are *unmechanistic*. Let us accept its premise: that a scientific account, ancient or modern, of some phenomenon must give a mechanistic explanation for it, or at least permit the possibility of one. Most scientists will find this uncontroversial. The question is: what do we mean by 'mechanistic'?

The term is a slippery one. We can surely agree that a mechanistic explanation is, minimally, one couched in terms of a physical theory. Beyond that, views vary. Here are a few definitions that I think are wrong. Some philosophers and historians also demand that the physical theory in question be a *correct* one, or at least a *particular* one – Newtonian mechanics or atomism, say. Such restrictions are obviously ahistorical. Why should any particular physical theory be so privileged? Physical theories come and go: the discovery of subatomic particles may have rendered Dalton's atomic chemistry redundant or even wrong, but not unmechanistic, much less unscientific.

Mechanistic explanations are also sometimes held to be those that eschew any reference to final or formal causes. That seems wrong too. Certain kinds of complex phenomena demand final and formal explanations; mechanistic explanations are not excluded by, but rather complement, them. Other philosophers demand that mechanistic explanations contain explicit comparisons to machines – pulleys or clocks, say. This is also too restrictive. Ask a biologist how proteins are made in the cell, and he will tell you about the 'ribosomal machinery'. Ask him then what kind of artefact a ribosome resembles and he will say, well, it's a bit like a CD player since they both translate information encoded in one physical form into another, and it's a bit like a locomotive since they both travel down 'tracks' (mRNA). Probe a little further, however, and he will also acknowledge the vacuity of the similes, that humans have never built anything like a ribosome, nor indeed anything so clever, but that even so the physics do make sense.

I propose, then, that a mechanistic explanation is simply one that explains a phenomenon in terms of the physical theory *of the day*. Granting this definition, Aristotle biology is replete with them. They are two of the four planks of his explanatory scheme, the moving and material causes. He is always, to be sure, saying that it is the 'nature' of some animal to do this or that and, had he left it there, his explanations would indeed be vacuous or occult. He doesn't. He then explains how and why.

In this book I have sketched Aristotle's account of five interlocked biological processes: (i) the nutritional system by which an animal takes up complex matter from its environment, alters its qualities and redistributes it to its various tissues so that it can grow, thrive and reproduce; (ii) the thermoregulatory cycle by which it maintains itself and which, as it ages, falls apart; (iii) the CIOM system by which an animal perceives and responds to its environment; (iv) the epigenetic processes of embryonic development and its related spontaneous-generator version; (v) the inheritance system. All of these processes are underpinned by Aristotle's physical theory and are, as such, mechanistic. That the physical theory is wrong is irrelevant; in the long run, all physical theories are.

All these processes explain some part of the workings of the soul. But soul is not something superadded to them: they *are*, collectively, soul; more precisely, soul is the dynamic structure of these physical processes (or their result). Again, that Aristotelian souls run on an obsolete theory of motion, a defunct chemistry and an oft-erroneous anatomy is beside the point. Descartes, for all his *bête machine* rhetoric, had his animals move by means of 'animal spirits' percolating through their nervous systems – *pneuma* by another name. If Aristotle's biology becomes unmechanistic at any point, it's when he considers higher cognitive functions – *phantasia*, reasoning, desire. They're merely black boxes. But we can forgive him this – they are for us too.

Although mechanical similes are not needed for a theory to be mechanistic, they are often the sign of one. When explaining how animals work Aristotle incessantly invokes them. Bellows, irrigation ditches, porous pottery, cheese-making, toy carts and, of course, those enigmatic automatic puppets, all appear in his biology. For all that, he never draws the Cartesian comparison of a whole creature to a machine. Doubtless this is because the mechanical devices of Aristotle's day were so rudimentary.* We can see that his heart–lung cycle is a thermostat but he obviously didn't – he just said how he thought it all works.

This, then, is Aristotle's dilemma. He sees that artefacts and living things are both made of more basic stuff, that they change and that these changes must be explicable in terms of physical principles. Yet, when looking at his world, he also sees that there is no artefact remotely capable of doing what creatures so effortlessly do. His solution is to acknowledge the parallels but keep them firmly apart. The cybernetic properties of living things even cause him to give them the special ontological status of 'entities' – *ousai* – while denying that status to artefacts. He would surely have dismissed Descartes' talk of beast machines as empty rhetoric. In Descartes' hands it was. It wouldn't stay that way.

Aristotle's enemies (and some of his friends) have also made formal and final causes far more mysterious than they really are. Aristotle saw that complex objects – and nothing is more complex than a living thing – cannot assemble willy-nilly by chance but must be modelled on a

---

* But who really knows what devices were current in the fourth century? The most sophisticated mechanical device of Greek antiquity, the Antikythera mechanism, an analogue computer composed of at least thirty interlocking gears, designed to demonstrate celestial motions, was built in Rhodes around 87 BC, a few centuries after Aristotle's death. But, until it was hauled from the sea, no one supposed the Greeks remotely capable of building such a thing.

pattern located elsewhere. Long absent from science, molecular biology made form – *eidos* – respectable again. In *What is Life?* Schrödinger, quoting Goethe ('Being is eternal; for there are laws to conserve the treasures of life on which the Universe draws for beauty'), argued that the chromosomes, which he envisioned as aperiodic crystals, contain a 'code-script' and are 'the law-code and executive power – or, to use another simile, they are architect's plan and builder's craft – in one'. The last is one of Aristotle's similes too. It was Max Delbrück at Caltech who made the connection explicit. In his charming essay 'Aristotle-totle-totle' he told of how, in the course of a long correspondence with André Lwoff at the Institut Pasteur in Paris, he discovered the Philosopher's works. After quoting bits from *The Generation of Animals* he wrote, 'What all of these quotations say is this: The form principle is the information which is stored in the semen. After fertilization it is read out in a pre-programmed way; the readout alters the matter upon which it acts, but it does not alter the stored information, which is not, properly speaking, part of the finished product.' And then he suggested that, were Nobels handed out posthumously, Aristotle should get one for discovering the principle (if hardly the substance, much less the structure) of DNA. In 1969 Delbrück got one for his work on mutation.

Final causes, too, have been demystified. Aristotle saw that they are needed when the phenomenon to be explained appears to have a goal. They arise then as the answers to several related questions which he asked and which modern biologists do too. When we ask *why* do goal-directed entities exist, we give Darwin's answer: because evolution by natural selection produced them. That is shorthand for the whole edifice of population genetic theory that renders benevolent creators null and void. When we ask *what* their goals are, we answer by pointing to all the adaptive devices that allow them to feed, move, mate, defy their predators and, ultimately, survive and reproduce. It is Bacon's sneers at teleological explanations of this sort, those 'remoras and hindrances', that now look quaint. To argue, as he did, that the functional study of eyelashes, skin and bones should be no part of science is to betray a remarkable incuriosity about the point of one's own body.

We can also ask *how* goal-directed things, living or not, work. That is the most difficult kind of final explanation, and its answer lies in the beating heart of the science of complex objects. Cybernetics, General Systems Theory and Control Theory formalize the general principles; systems biology shows those principles at work in living things; synthetic biology how those same principles can be used to reshape them. In 2010 JCVI-syn1.0, the world's first artificial cellular life form, fired its molecular motors. The distinction between artefact and organism dissolved in a Petri dish.

Aristotle's answers to these questions, all of which are embraced by his final cause, are sometimes similar to ours and sometimes, but hardly surprisingly so, very different. That they are scientific questions and that he gave scientific answers to them cannot be denied; or at least that he did so until he looked to his God to give creatures, not least himself, their ultimate purpose in life.

What, finally, of Bacon's accusation that Aristotle's science was useless to man? It's the eternal *cri de coeur* of the science bureaucrat. (You scientists want all the money going, but what, exactly, do we get in return?) Neither complaint, Bacon's nor the bureaucrat's, is entirely baseless. But, just as few modern scientists are utterly indifferent to the utility of their work, neither was Aristotle. His father was a physician, so it's no surprise to find two books titled *On Medicine* listed among his lost works. And, although his books on ageing – *Youth & Old Age, Life & Death* and *The Length and Shortness of Life* – do not reveal what we can do to nurture the internal fire whose vitality dictates the length of our days, he does conclude the latter with this:

> Our investigation into life, death and related subjects is almost complete. On health and disease it is to some extent up to natural scientists as well as doctors to consider their causes. But it is important to note the differences between these two groups of investigators in how they treat different problems; for it is clear that to some extent they cover the same ground; *doctors who display curiosity and intellectual flexibility have something to say about natural science and declare that their theories arise from it and the best practitioners in natural sciences tend to end up with medical theories* [italics mine].

Think of it as the Invitation to Biomedical Science. 'This our science', wrote D'Arcy Thompson, 'is no petty handicraft, no narrow discipline. It was great, and big, in Aristotle's hands, and it has grown gigantic since his day.' Aristotle could not have conceived just how vast the science that he founded would become. Yet, as I contemplate the elaborate tapestry of his science, and compare it to ours, I conclude that we can now see his intentions and accomplishments more clearly than any previous age has seen them and that, if this is so, it is because we have caught up with him.

CXIII

A ND BECAUSE WE know, and know intimately, the one other scientist in history who resembles him more than any other.
They were so very much alike. Both were the sons of famous physicians, but both preferred to study nature. Both were voracious for facts. Both were ruthlessly, powerfully logical – and not much good at maths. Both were bold and rash in equal measure and, in being so, left us visions of life imbued with – there is only one word – *grandeur*. If there is a difference, it is only in the scale of their accomplishments. After all, Darwin didn't invent science itself from scratch; Aristotle did.

They also shared a scientific style. Seeking facts to support their theories, both cast their nets wide to catch them. Both interrogated farmers, fishermen, hunters and travellers – though Darwin could add pigeon fanciers to the list too.* Both papered over vast inferential cracks in their evidence – Darwin, *inter alia*, the mechanisms of heredity, gaps in the fossil record and the invisibility of natural selection. Both made voluminous, if often fleeting, observations. And both men occasionally made far too much of the facts they knew, or thought they knew.

In *The Origin of Species*, Darwin tells of a small rodent, the Tuco-Tuco, *Ctenomys*, which infests the Argentinean pampas. It lives in burrows and, so Darwin assures us, is frequently blind; indeed, one that he had kept alive while travelling with the *Beagle* 'was certainly in this condition, the cause, as appeared on dissection, having been inflammation of the nictitating membrane'. Such an inflammation, he continues, being injurious to the animal, would tend to select for eyeless Tuco-Tucos and so, eventually, result in something like a mole. It's a very reasonable argument and an important one too, for it's the only example that Darwin has of natural

---

* Darwin, of course, did do experiments – he bred pigeons, tested how long snails could survive in saltwater and how bees made their combs – but such was hardly the stuff from which the *Origin* was made. In 1860, Richard Owen wrote an anonymous review of the *Origin*. Eager to deny Darwin's originality, he focused on the 'direct observations of nature which seemed to be novel and original' in the work – the experiments. The extent to which he misunderstood Darwin's method and achievement would be funny were it not so plainly malicious.

selection, the driving force of evolution, in action. Unfortunately, it is almost certainly not true. Some years ago, following in Darwin's footsteps, I searched for teary-eyed Tuco-Tucos in Argentina and Uruguay in vain. I interrogated gauchos and scientists, and all denied that the animals have anything wrong with their eyes. A gaucho offered an explanation for Darwin's observation: 'Well, you know, when we catch the Tuco-Tuco we hit it with a spade. They move fast and they are fierce! Maybe this is why Carlos Darwin's Tuco-Tuco bleed from the eyes, eh?'

The lesson is one that every biologist, every scientist, knows or must learn: that the practice of science demands a peculiar intimacy with the object of your investigations. You must know its form, its foibles, its pretty little ways, for, if you don't, you'll make a mistake or else miss some astonishing thing that it does, and that is almost as bad. That is why Darwin spent eight years on barnacles. He sought, in Barbara McClintock's phrase, a 'feeling for the organism'. My own postdoctoral supervisor expressed the same sentiment, albeit more restrictively, when, on the first day in his lab, he told me: 'Know the Worm.' A Delphic utterance? Not at all. I knew exactly what he meant.*

Aristotle, I believe, would have too. Intimacy with the natural world shines from his works; it does from Theophrastus' as well. This intimacy allowed them, the men of the Lyceum, to begin the process of sieving the ocean of natural history folklore and travelogue for grains of truth from which to build a new science. Aristotle even said so:

> Failure to understand what is obvious can be caused by inexperience: those who have spent more time with the natural world are better at suggesting theories of wide explanatory scope. Those who have spent time arguing instead of studying things as they are show all too clearly that they are incapable of seeing much at all.

The passage comes from On Generation & Corruption. The argumentative types are the Platonists. Their obsessions — intangible Forms, numerology and geometry — caused them to deny the evidence of their eyes. They were blind to the structure of the world, this world. The passage is a prelude to the Invitation to Biology. For, when Aristotle said that we must attend to even the humblest creatures since there are gods there too, he was not only urging some students to pick up their cuttlefish, but he was arguing, as he would until

* Scott Emmons quoting Paul Sternberg.

the end, with Plato's shade. He was doing what every scientist who opens a new domain of inquiry must do: defend it before his peers. Of the whole, vast natural world, the Academy deemed only the stars worthy of study. But, and this is Aristotle's point, we do not live among the stars; we live here, on Earth.

Nor do we live just anywhere on Earth. If D'Arcy Thompson was right, as I believe he was, then this is what Lesbos and the lagoon at Pyrrha gave to Aristotle: a *place*, calm and lovely, where he could be among natural things. Lesbos was for him what Chimborazo was for Humboldt, the Malay archipelago for Wallace, the Amazon for Bates and a Berkshire wood for Hamilton. It was what the Atlantic rainforest of Brazil, the bleak pampas of Patagonia, the black volcanic rocks of the Galapagos and a field in Kent were for Darwin. Biologists often have such places. They need them, for ideas do not come from nothing, they come from nature herself.

CXIV

W HEN ARISTOTLE SPEAKS of Kalloni, it is always the *euripos Pyrrhaiēn* – the strait of the Lagoon – of which he speaks. It is through the *euripos* that the fishes funnel on their annual migrations. It is there that the scallops wax and wane and there that the bottom boils with starfish. I wanted to see it for myself.

The *euripos* is formed by a submerged reef that juts out from the north-west shore. Seagulls, fossicking among its hidden rocks, appear to be walking on water. Aegean currents are feeble, but the reef constricts the entrance to a gullet so that, at tide's change, the whole body of water attempts to traverse it in glissading cascades.

An oyster diver said he'd take us. We made a date for the waning moon, calculated for the slack, loaded the boat at Apothika and suited up *en route*. As we approached the dive site, a tuna leapt high and clear, blue against blue against blue. A negative buoyancy entry put us above a rocky bottom at seven metres. Pink and brown sponges squatted between eelgrass beds. Silver and black sea bream finned against the current. David K., who has an enthusiasm for sea slugs, disappeared to look for them. He later reported the aeolids *Cratena peregrina* (purple cerata, orange rhinophores) and *Caloria elegans* (black-tipped cerata against a white body) and the dorid *Discodorus atromaculata* (a Belgian chocolate, squashed and leaking marzipan).

Towards the drop-off, shoals of orange-pink *Anthías* and electric-blue juvenile *Chromís* fluttered between stands of gorgonians. These fragile zoophytes are usually found below thirty metres, but here, as on the undersea cliffs of Sulawesi, they live in the shallows. Their branches – some golden, others white – ramified in a curious reticular geometry. Clusters of translucent *Clavelína* ascidians draped from them like crystals on a chandelier. From beneath a ledge, a dusky grouper flashed away.

At ten metres sponges, hovering between form and formlessness, crowded upon each other. One looked like a weird desert succulent, another like a mutant hand, another resembled an engorged ear stuck unaccountably to a rock. Coralline algae draped from boulders in stalactites. An octopus prinked by.

The tidal currents were evidently responsible for these riches. Twice a day their nutrient and plankton-rich waters sweep through the *eurípos* fuelling an intensity of life that I have not seen elsewhere in the Aegean. And then, disorientingly, fifteen metres down, I came across a coral wall. It was as though I had unwittingly swum through Suez to the coral gardens of the Red Sea. Looking closer, I saw that my coral reef was in fact an enormous boulder that had been colonized by a solitary coral, *Parazoanthus axinellae*, but so densely that the illusion of a tropical reef was complete. Their golden cups, ringed with tentacles, radiated like a thousand small suns.

Parazoanthus axinellae. Strait of Kolpos Kalloni, Lesbos, August 2012

'It has been said', wrote Borges, 'that all men are born either Aristotelians or Platonists.' Philosophers may wince at the opposition, but I suspect it to be literally true. Plato invites us to the world of abstractions, Aristotle to the world of tangible things. You begin with particulars, a box of seashells, say; gather them together and rearrange them endlessly in order to apprehend their logic and order. This apprehension, Aristotle says, is the gift of reason and the beginning of science. It is also where true beauty lies. This was inarticulately obvious to me when I was ten.

As we age we become trapped by our habits of mind, by what we already know, as surely as fish are in the sea. Science, the glittering medium in which we swim, dictates what we see. That is how it should be and inescapably is, for no one sees the world unmediated by theory and expectation. Yet how we long to see it afresh. 'For as the eyes of bats are to the blaze of day, so is the reason in our soul [oblivious] to the things that are most evident of all' – *Metaphysics* 993b10. Aristotle, armed with the method that he discovered, that precarious combination of theory tempered by experience that is the essence of science, turned to a part of the world that no one had ever looked at before, described it, explained it and, as Thompson said, won for it a place in Philosophy. We can envy him for have doing so. Swept along in the seething currents of scientific progress we struggle to emulate him. But Aristotle shows us what we must do.

And why. When I found the *kēryx* it was lodged between two boulders. Its foot, as mottled as a leopard's pelt, spilled from beneath its shell. Its tentacles were zebra striped. Never before had I seen one alive. The thick shell was covered with a filigree of bryozoans and a patchwork of coralline algae; its apex was grey and worn. It must have been very old. The great snail's proboscis was stuck into a black sea urchin whose guts it was slowly rasping away. The sea urchin's spines waved a last, futile defence, but its systems were failing fast. This is the world that Aristotle gave us: the vividly perceptible world of living things, whole and at home; the world that he enjoins us to love and understand. Aristotle wrote thousands of sentences, but one, the first of his *Metaphysics*, defines him: 'All men, by nature, desire to know.' Not all forms of knowledge, however, are equal – the best is the pure and disinterested search for the causes of the things. And, he has no doubt, searching for them is the best way to spend a life. It is a claim for the beauty and worth of science.

GLAUX — LITTLE OWL — *ATHENE NOCTUA*
ACROPOLIS, ATHENS, 2013

GLOSSARIES

I. TECHNICAL GLOSSARY

aithēr	ether
anō	above
antithesis	opposite position (anatomy)
analogon	analogue
aphrodisiazomenai	highly sexed (women)
aphros	foam
apodeixis	demonstration
aristeros	left
arkhē	origin/principle
atomon eidos	indivisible form
automata	spontaneous/self-moving things
balanos	glans
basileia	queen
basileus	king
bios	lifestyle
delphys	uterine body
Dēmiourgos	Creator
dexios	right
diaphora/diaphorai	difference/pl. (in some feature)
dynamis	potentiality, potency, power
eikōs mythos/eikotes mythoi	likely or plausible story
ekhinos	omasum/hedgehog/sea urchin/wide-mouthed jar
eidos/eidē	form/pl.
emprosthen	before
entelekheia	actuality
epagōgē	induction
epamphoterizein	to dualize

epistēmē	knowledge
eurípos	strait
geēron	earth
genos/genē	kind/pl.
gēras	old age
gēs entera	guts of the earth
gonē	semen
hippomanein	they are 'stallion-mad'/nymphomaniacs
historia tēs physeōs	study of nature
historiai peri tōn zōiōn	*Historia animalium* [*Enquiries into Animals*]
holon	whole
hylē	matter
hystera	uterus/female reproductive organs
katamēnia	menses
katō	below
kekryphalos	reticulum
keratia	uterine horns
khelidonias	swallow wind
khōrion	amniotic sac
kinēsis/kinēseis	movement/pl.
kotylēdones	cotyledons/caruncles
limnothalassa	lagoon, lit. lake sea
logos	definition, essence
lysis	relapse/mutation
mathematikē	mathematical science
megalē koilia	rumen
metabolē	transformation
mētra	cervix
mixis	compound
myes	muscles
mythos	story
mytis	cephalopod 'heart' (i.e. its digestive gland)
neuron/neura	sinew/pl.
nous	reason
oikoumenē	known world
onta	things
opisthen	behind
organon	instrument/tool/organ
ornithiai anemoi	bird winds

ousia/ousiai	substance, entity/pl.
pepeiramenoi	having tried or tested something
peri physeōs	on nature
phainomena	appearances
phantasia	mental representation
phantasma/phantasmata	mental image/pl.
physis	nature
physikē epistēmē	natural science
physikos	one who understands nature
physiologos/physiologoi	one who studies nature/pl.
pneuma	pneuma
polis	city state
politikē epistēmē	political science
prōton stoicheion	first element
psychē	soul
sarx	flesh (i.e. muscles)
sōma	body
sperma	seed
stoma	mouth
stomakhos	oesophagus
symmetria	proportion
symphyton pneuma	connate breath
syngennis	kindred
synthesis	mixture, agglomeration of parts
ta aphrodisia	sexual intercourse
technika	skilled activities
telos	end
theologikē	theology
theos	god
thesis	position (anatomy)
to agathon	the good
to hou heneka	that for the sake of which
trophē	nutrition/way of life
tōn zōiōn	livelihood

II. ANIMAL KINDS MENTIONED

> Considering this mass of valuable information, one must particularly regret that
> the author [Aristotle] did not suspect that the nomenclature of his time might
> become opaque, and that he therefore took no precautions to ensure that the
> species he discusses are recognizable. This is the general defect of the ancient natu-
> ralists; one is almost obliged to guess the identities behind the names they used; the
> often changing tradition induces error; thus it is by arduous deduction, and bring-
> ing together features scattered among authors, that one gets a positive result for
> some species; but we are condemned to remain ignorant of the majority of them.
>
> Georges Cuvier and Achille Valenciennes,
> *Histoire naturelle des poissons* (1828–49)

The task of identifying Aristotle's animals started around 1256 when Albert
Magnus began to assemble his *De animalibus* based, in part, on *Historia animalium*.
Zoologically minded classicists and classically minded zoologists have been at it
ever since. They have had mixed success. Aristotle's descriptions of his animals are
often so thin as to defy identification. However, other classical texts using the same
or similar names provide clues, as do the vernacular names used by modern Aegean
and Adriatic fishermen and hunters. Biogeography helps too. Or one can simply go
to the Lagoon to see what's there. One scholar who did so plausibly identified
Aristotle's *kōbios* as any of three species of goby and his *phykis* as the blenny, *Parablennius sanguinolentus.**

Although generations of scholars have laboured to identify Aristotle's animals,
there is no recent, comprehensive list of them. For this reason I tabulate the
230-odd Aristotelian animal kinds mentioned in this book, along with my best
guess as to what they are. Scholars have varied in their willingness to pin Aristotelian
kinds down to Linnaean species. Some are enthusiastic while others think that it
can hardly be done at all. I have taken the middle road. After all, when Aristotle
says *hippos*, he must mean *Equus caballus*, that is, a horse – at least when he doesn't
mean the *hippos* crab or the *hippos* woodpecker. When, however, he says *kephalos* we
are less sure. He certainly means a grey mullet since that's what they're still called
in Greece today, but he could mean any or all of *Mullus cephalus* (flathead grey
mullet), *Chelon labrosus* (thicklip grey mullet), *Oedalechilus labeo* (boxlip mullet), *Liza
saliens* (leaping mullet), *Liza aurata* (golden grey mullet) or *Liza ramada* (thinlip grey

* Tipton (2006).

mullet), all of which are found in Greek waters and are notoriously hard to tell apart.* Moreover, Aristotle mentions at least four different fishes that are plausibly grey mullets, so it's likely that he, and fishermen, distinguished at least some of the six modern nominal species. But which of Aristotle's grey mullets correspond to ours must probably always remain a mystery.

There is also a trap for the unwary. Linnaeus and other early taxonomists often gave their European species classical names on the basis of ancient descriptions. Sometimes they were right to do so. Linnaeus' *Chamaeleo chamaeleon chamaeleon* – the European chameleon – is certainly Aristotle's *chamaíleōn* since it's the only lizard that answers to his detailed description.† Sometimes, however, they were on much less certain ground. Linnaeus thought that Aristotle's *rhinobatos* was the guitarshark so he called the guitarshark *Rhinobatos rhinobatos*; and since both the fish and what Aristotle says about it are interesting, it's nice to think that that is what it actually is, but we can't be sure since he doesn't say much.

My list is based on several editions of *Historia animalium* and *The Parts of Animals*‡ as well as monographs on ancient animals.§ I have tried to make ambiguity plain. In general, large mammals can be identified to modern species; birds to genus or not at all (*Historia animalium* contains a swathe of strange, possibly Egyptian or Babylonian, bird names); fish to species, genus or family depending on their prominence, uniqueness and depth of description; insects mostly to family or order; marine invertebrates anywhere from species to phylum. For a few of Aristotle's creatures, however, we can say little more than that they are probably animals and that they live in the sea.

---

* Koutsogiannopoulos (2010).
† True, the African chameleon, *Chamaeleo africanus*, occurs in Pylos in the Peloponnese, but that's thought to be a Roman introduction. Why the Romans should have carried chameleons around the Mediterranean basin is hard to say.
‡ HA: CRESSWELL and SCHNEIDER (1862), THOMPSON (1910), PECK (1965), PECK (1970) and BALME (1991). PA: OGLE (1882), LENNOX (2001a) and KULLMANN (2007).
§ KITCHELL (2014) on mammals and some other animals, THOMPSON (1895) and ARNOTT (2007) on birds, THOMPSON (1947) on fishes, DAVIES and KATHIRITHAMY (1986) on insects, SCHARFENBERG (2001) on cephalopods and VOULTSIADOU and VAFIDIS (2007) on marine invertebrates.

English name	Aristotle's name	Linnaean name

ANIMALS   *ZŌIA*   METAZOA

BLOODED ANIMALS   *ENHAIMA*   VERTEBRATA

man (humans)	*anthrōpos*	*Homo sapiens*
LIVE-BEARING TETRAPODS	*ZŌOTOKA TETRAPODA*	MAMMALIA (MOST)
ass, Asian wild (onager)	*onos agrios*	*Equus hemionus*
ass, Asian wild (onager)?	*hēmionos**	*Equus hemionus?*
ass, domestic (donkey)	*onos*	*Equus africanus asinus*
baboon, hamadryas	*kynokephalos*	*Papio hamadryas*
bear, Eurasian brown	*arktos*	*Ursus arctos arctos*
beaver, Eurasian	*kastōr*	*Castor fiber*
bison, European	*bonassos*	*Bison bonasus*
camel, Arabian (dromedary)	*kamēlos Arabia*	*Camelus dromedarius*
camel, Bactrian	*kamēlos Baktrianē*	*Camelus bactrianus*
cat	*ailouros*	*Felis silvestrus cattus*
cattle	*bous*	*Bos primigenius*
cattle, wild	*tauros*	*Bos primigenius (auroch)*
deer, red?	*elaphos*	*Cervus elephas?*
deer, roe	*prox*	*Capreolus capreolus*
dog	*kyōn*	*Canis lupus familiaris*
dog, Molossian	*kyōn en tēi Molottiāi*	*Canis lupus familiaris* (mastiff)
dog, Laconian	*kyōn Lakōnikos*	*Canis lupus familiaris* (hound)
dog, Indian	*kyōn Indikos*	*Canis lupus familiaris* (Indian pariah dog?)
dormouse	*eleios*	Gliridae
elephant, Asian†	*elephas*	*Elaphas maximus*
fox	*alōpēx*	*Vulpes vulpes*

* Aristotle also uses this term for the regular mule; its relationship to the onager is unclear; see KITCHELL (2014).
† Aristotle does not say where his elephant was seen; it is most likely the Asian elephant on the basis of its association with Alexander's expeditions alone.

gazelle, dorcas	*dorkas*	*Gazella dorcas*
unknown bovid	*pardion*	Bovidae
giraffe?	*hippardion*	*Giraffa camelopardis?*
goat, ram	*tragos*	*Capra aegagrus*
goat, ram	*khimaira*	*Capra aegagrus*
goat, ewe	*aix*	*Capra aegagrus*
hare, European	*dasypous*	*Lepus europaeus*
hare, European	*lagōs*	*Lepus europaeus*
hartebeest	*boubalis*	*Alcelaphus buselaphus*
hedgehog, northern	*ekhinos*	*Erinaceus roumanicus*
hippopotamus	*hippos potamios*	*Hippopotamus amphibius*
horse	*hippos*	*Equus caballus*
hyena, striped*	*hyaina*	*Hyaena hyaena*
hyena, striped	*glanos*	*Hyaena hyaena*
hyena, striped	*trokhos*	*Hyaena hyaena*
jackal, golden?	*thōs*†	*Canis aureus?*
jerboa	*dipous*‡	Dipodidae
leopard	*pardalos*	*Panthera pardus*
lion, Asian	*leōn*	*Panthera leo persica*
lynx, Eurasian	*lynx*	*Lynx lynx*
macaque, Barbary	*pithēkos*	*Macaca sylvanus*
macaque, Rhesus?§	*kēbos*	*Macaca mulatta?*
mole, Mediterranean¶	*aspalax*	*Talpa caeca*

* Beginning with WATSON (1877), there's a long, and incorrect, consensus that Aristotle's *glanos/hyaina* is the spotted hyena, *Crocuta crocuta*, but the mane alone identifies it as the striped hyena *Hyena hyena*. Furthermore, Aristotle's description of its genitals doesn't fit the massively masculinized genitalia of *Crocuta* females. I assume that the *trokhos* is the same animal, but that's less certain; see FUNK (2012). KITCHELL (2014) says that Oppian distinguished the spotted and striped hyena, so perhaps the former wasn't entirely unknown to the ancients.

† KITCHELL (2014) points out that this animal has a bewildering number of identifications. It may be jackal, civet or some sort of viverrid.

‡ This is the ancient Greek name for the animal. Aristotle does not actually use it, but just speaks of mice with long legs or that walk on their hind legs – clearly the jerboa.

§ Aristotle mentions three non-human primates: the *kynokephalos*, *pithēkos* and *kēbos* (excluding the textually dubious *khoireopithēkos* of HA 503a19). The *kynocephalos* is certainly the Egyptian baboon, *Papio hamadryas*, since it has a doglike face and no tail; the *pithēkos* is said to have a short tail and so is likely the Barbary macaque, *Macaca sylvanus*. The *kēbos* is said to have a tail, but the tailed African *Cercopithecus* are all sub-Saharan, so perhaps it's a report of the Asian rhesus macaque, *Macaca mulatta*, from Alexander's expedition. See KULLMANN (2007) p. 709 and KITCHELL (2014).

¶ The *aspalax* could be the naked mole rat, *Spalax*, of Asia Minor or the Mediterranean

mongoose, Egyptian	*ikhneumōn*	*Herpestes ichneumon*
mouse	*mys*	*Mus* sp.
mouse, field	*arouraios mys*	*Apodemus* sp.
mouse, spiny	*ekhinos*	*Acomys* sp.
mule	*oreus*	*Equus africanus asinus* (m) × *Equus caballus* (f)
mule	*hēmionos*	*Equus africanus asinus* (m) × *Equus caballus* (f)
mule (hinny)	*ginnos*	*Equus caballus* (m) × *Equus africanus asinus* (f)
nilgai	*hippelaphos*	*Boselaphus tragocamelus*
oryx	*oryx*	*Oryx* sp.
otter	*enhydris*	*Lutra lutra*
pig	*hys*	*Sus scrofa domesticus*
porcupine, crested	*hystrix*	*Hystrix cristata*
rhinoceros, Indian*	*onos Indikos*	*Rhinoceros unicornis*
seal, monk	*phōkē*	*Monachus monachus*
sheep	*krios*	*Ovis aries*
sheep	*ois*	*Ovis aries*
sheep	*probaton*	*Ovis aries*
shrew	*mygalē*	Soricidae
tiger	*martikhōras*	*Panthera tigris*
marten	*iktis*	*Martes* sp.
weasel	*galē*	*Mustela* sp.
wolf, grey	*lykos*	*Canis lupus*
CETACEANS	KĒTŌDEIS	CETACEA
dolphin	*delphis*†	Delphinidae
whale	*phalaina*	Odontoceti

mole, *Talpa caeca*. Both *Spalax* and *T. caeca* are blind and have eyes covered in skin, but the latter seems more biogeographically plausible. (*T. europea*, the common European mole, is found north of the Alps and is disqualified by its small, but externally visible, eyes.) THOMPSON (1910) n. HA 491b30 favours *T. caeca* simply because it is rather more common than *Spalax* in the areas that Aristotle knew personally; see KULLMANN (2007) p. 457.
* The *onos Indikos* is generally thought to be an Indian rhinoceros (OGLE 1882 p. 190, THOMPSON 1910 n. 499b10). LONES (1912) p. 255, looking at its feet, disagrees. Lones is right to say that the rhinoceros has three toes and the *onos Indikos* one, but the rhino's central toe is much larger than the others and so could easily be mistaken for a hoof.
† Likely the bottlenose dolphin, *Tursiops truncatus*, but Aristotle does not distinguish the several Delphinid spp. found in the Aegean.

BIRDS	*ORNITHES*	AVES
bee-eater, European	merops	*Merops apiaster*
blackbird	kottyphos	*Turdus merula*
bustard, great	ōtís	*Otis tarda*
chaffinch	spíza	*Fringílla coelebs*
chicken	alektōr	*Gallus domesticus*
chicken, Adrianic	adrianíkē	*Gallus domesticus*
cormorant, great	korax	*Phalacrocorax carbo*
crane, Eurasian	geranos	*Grus grus*
crow, hooded	korōnē	*Corvus corone*
cuckoo	kokkyx	*Cuculus* sp.
dove, turtle	trygōn	*Streptopelia turtur*
duck, teal?	boskas	*Anas crekka?*
eagle	aietos	*Aquila*
flamingo, greater*	phoiníkopteros	*Phoenicopterus ruber*
nightjar	aigothēlas	*Caprimulgus europaeus*
goldcrest	tyrannos	*Regulus regulus*
goose	khēn	*Branta* sp.
grebe, great crested	kolymbis	*Podiceps cristatus*
vulture	aigypíos	*Aegypius* sp.
hawk	hierax	Accipitridae, small
heron	pellos	*Ardea* sp.
hoopoe, Eurasian	epops	*Upupa epops*
ibis†	íbis	Threskiornithidae
jay, Eurasian	kissa	*Garrulus glandarius*
kestrel	kenkhris	*Falco* sp. *tinnunculus* or *F. naumanni*
kingfisher	alkyōn‡	*Alcedo atthis*
kite	iktínos	*Milvus* sp.
lark	korydalos	Alaudidae
nuthatch, rock	kyanos	*Sitta neumayer*
ostrich	strouthos Líbykos	*Struthio camelus*
owl, little§	glaux	*Athene noctua*

* Not mentioned by Aristotle, but now very common in Kalloni. The only references to a flamingo (or what might be one) in ancient Greece are in Aristophanes' *Birds*, 273 and Heliodorus.
† Either the glossy ibis, *Plegadis falcinellus*, found in Greece (Kalloni) or the sacred ibis, *Threskiornis aethiopicus*, found in Egypt.
‡ May also refer to a species of tern.
§ Athena's owl. The ancient proverb 'bringing owls to Athens' is the Greek equivalent of bringing coals to Newcastle.

owl, Ural?	aigōlios	Strix uralensis?
partridge	perdix	Alectoris or Perdix
pelican, Dalmatian	pelekan	Pelecanus crispus
pigeon	peristera	Columba sp.
pigeon, wood	phatta	Columba palumbus
quail	ortyx	Coturnix vulgaris
raven	korax	Corvus corax
seagull	laros	Laridae
sparrow	strouthos	Passer sp.
stilt, black-winged	krex*	Himantopus himantopus
stork, white	pelargos	Ciconia ciconia
swallow	khelidōn	Hirundo rustica
tit	aigithallos	Parus sp.
tit, coal	melankoryphos	Parus ater
turtle dove	trygōn	Streptopelia turtur
woodpecker†	dryokolaptēs	Dendrocopus sp.
woodpecker	hippos	Dendrocopus sp.
woodpecker	pipō	Dendrocopus sp.
woodpecker, green	keleos	Picus viridis
wren	trokhilos	Troglodytes troglodytes

* Traditionally identified as the corncrake, Crex crex; but this is dubious and the krex is mentioned by Aristotle as a long-legged waterbird with a short hind toe and a quarrelsome disposition (THOMPSON 1895 p. 103; ARNOTT 2007 p. 120) which does not fit the corncrake well, but does the black-winged stilt.

† Dryokolaptēs is a general name for woodpecker (literally 'tree-pecker'). Aristotle (HA 593a5, HA 614b10) speaks of at least four kinds of woodpecker as well as the hippos, some of which are easily identified, others not. When he refers to a small woodpecker with reddish speckles he must mean Dendrocopus minor since it is the only small woodpecker found in Greece that answers to the description. When he refers to a larger woodpecker that nests in olive trees he must mean D. medius since it is the only species to do so; interestingly it does so only in Lesbos (Filios Akreotis, pers. comm.). When he refers vaguely to a 'larger' species he could mean one of the three large Dendrocopus: the white-backed, D. leucotos, Syrian, D. syriacus or greater spotted, D. major, which are all about the same size (8–10 inches). Hippos may be a copyist's error for pipō. In addition to these Aristotle refers to a green woodpecker, clearly Picus viridis. See THOMPSON (1895) and ARNOTT (2007).

EGG-LAYING TETRAPODS	ŌIOTOKA TETRAPODA	REPTILIA* + AMPHIBIA
chameleon	*chamaileōn*	*Chamaeleo chamaeleon chamaeleon*
crocodile	*krokodeilos potamios*	*Crocodylus niloticus*
gecko, Turkish?	*askalabōtēs*	*Hemidactylus turcicus?*
lizard	*sauros*	Lacertidae
tortoise	*chelōnē*	*Testudo* sp.
terrapin	*emys*	*Mauremys rivulata?*
turtle	*khelōnē thallattia*	Cheloniidae

SNAKES	OPHEIS	SERPENTES
snake, water	*hydros*	*Natrix tessalata?*
snake, large	*drakōn*	Serpentes
Ottoman viper	*ekhidna*	*Vipera xanthina*

FISHES	IKTHYES	CHONDRICHTHYES + OSTEICHTHYES
blenny, rusty?	*phykis*†	*Parablennius sanguinolentus?*
blotched picarel	*mainis*	*Spicara maena*
catfish, Aristotle's	*glanis*	*Silurus aristotelis*
comber	*khannos*	*Serranus cabrilla*
comber, painted	*perkē*	*Serranus scriba*
eel, European	*enkhelys*	*Anguilla anguilla*
goby	*kōbios*	*Gobius cobitis?*
'goby, white'	*leukos kōbios*	unknown
gurnard	*kokkis*	Triglidae
gurnard	*lyra*	Triglidae
John Dory	*khalkeus*	*Zeus faber*
mullet, grey	*khelōn*	Mugilidae
mullet, grey	*kephalos*	Mugilidae
mullet, grey	*kestreus*	Mugilidae
mullet, grey	*myxinos*	Mugilidae
mullet, red	*triglē*	*Mullus* sp.
parrotfish	*skaros*	*Sparisoma cretense*

* Not a valid taxon; now the Sauropsida, which includes birds as a clade of dinosaurs.
† The *phykis* been variously identified as a goby (*Gobius niger*), a species of wrasse (e.g. *Symphodus ocellatus*), THOMPSON 1910 n. *HA* 567b18, THOMPSON (1947) pp. 276–8, or a blenny (*Parablennius sanguinolentus*), TIPTON (2006). It's hard to know since all of these are found in Kalloni or its surrounds and the description is vague and may be confused with other fishes.

pipefish	*belonē*	*Syngnathus* sp.
salema	*salpē*	*Sarpa salpa*
scorpionfish	*skorpaina*	*Scorpaena scrofa*
sea bass, European	*labrax*	*Dicentrarchus labrax*
sea bream, annular	*sparos*	*Diplodus annularis*
sea bream, gilthead	*khrysophrys*	*Sparus aurata**
sea bream, pandora	*erythrinos*	*Pagellus erythrinus*
sea bream, striped	*mormyros*	*Lithognathus mormyrus*
sea bream, white	*sargos*	*Diplodus sargus sargus*
sea perch, swallowtail	*anthias*	*Anthias anthias*
shad	*thritta*	*Alosa* sp. or another Clupeid
smelt, sand	*atherinē*	*Antherina presbyter*
tuna, blue fin	*thynnos*	*Thunnus thynnus*
unknown	*korakinos*	unknown
unknown, sardine-like	*khalkis*	Clupeidae
unknown, sardine-like	*membras*	Clupeidae
unknown, sardine-like	*trikhis*	Clupeidae

CARTILAGENOUS FISHES	*SELAKHĒ*	CHONDRICHTHYES
angelshark	*rhinē*	*Squatina squatina*
dogfish, smooth	*leios galeos*	*Mustelus mustelus*
dogfish, spiny	*akanthias galeos*	*Squalus acanthias*
dogfish, spotted	*skylion*	*Scyliorhinus* sp.
frogfish†	*batrakhos*	*Lophius piscatoris*
guitarfish?	*rhinobatos*	*Rhinobatos rhinobatos?*
ray, torpedo	*narkē*	*Torpedo torpedo*
skate or ray	*batos/batis*	Rajiformes
shark	*galeos*	Galeomorphi + Squalomorphi

UNCLASSIFIED BLOODED ANIMALS		
tadpole or eft	*kordylos*	Amphibia
bat	*nykteris*	Microchiroptera
fruit bat, Egyptian (flying fox)	*alōpēx*	*Rousettus aegyptiacus*

* Sometimes confused with *Chrysophrys auratus*, an Indo-Pacific fish, due to a complicated history of synonymy.
† *Contra* Aristotle, the frogfish is not a cartilagenous fish.

English name	Aristotle's name	Linnaean name
BLOODLESS ANIMALS	*ANHAIMA*	INVERTEBRATA*
'SOFT-SHELLS'	*MALAKOSTRAKA*	CRUSTACEA (MOST)
crab	*karkinos*	Brachyura
crab, fan mussel	*pinnophylax*	*Nepinnotheres pinnotheres*
crab, ghost	*hippos*	*Ocypode cursor*
lobster	*astakos*	*Homarus gammarus*
shrimp	*karis*	Nantantia + Stomapoda
shrimp, fan mussel	*pinnophylax*	*Pontonia pinnophylax* or similar spp.
spiny lobster	*karabos*	*Palinurus elephas*
shrimp, mantis	*krangōn*	*Squilla mantis*
'SOFT-BODIES'	*MALAKIA*	CEPHALOPODA
cuttlefish	*sēpia*	*Sepia officinalis*
octopus, common	*polypodōn megiston genos*	*Octopus vulgaris*
octopus, musky	*bolitaina*	*Eledone moschata*
octopus, musky	*heledōnē*	*Eledone moschata*
octopus, musky	*ozolis*	*Eledone moschata*
paper nautilus	*nautilos polypous*	*Argonauta argo*
squid, European	*teuthis*	*Loligo vulgaris*
squid, sagittal	*teuthos*	*Todarodes sagittatus*
'HARD-SHELLS'	*OSTRAKODERMA*	GASTROPODA + BIVALVIA + ECHINOZOA + ASCIDIACEA + CIRRIPEDIA
cockle	*khonkhos, rhabdōtos trakhyostrakos*	Cardidae
limpet	*lepas*	*Patella* sp.
mussel, fan	*pinna*	*Pinna nobilis*
oyster	*limnostreon*	*Ostrea* sp.
razorfish?†	*sōlēn*	Solenidae?

* Not a valid taxon.
† Aristotle says the *sōlēn* can't live if torn off a rock. Elsewhere, however, he says that it is free living and might be able to hear. One of these must be wrong. The *sōlēn* is

scallop	*kteis*	Pectinidae
sea urchin, edible	*esthiomenon ekhinos*	*Paracentrotus lividus*
sea urchin, long-spine	*ekhinos genos mikron*	*Cidaris cidaris*
sea squirt	*tēthyon*	Ascidiacea
snail, murex	*porphyra*	*Haustellum brandaris*
snail, murex	*porphyra*	*Hexaplex trunculus*
snail, trumpet	*kēryx*	*Charonia variegata*
snail, turban	*nēreitēs*	*Monodonta* sp.?

'DIVISIBLES'	*ENTOMA*	INSECTA + CHELICERATA + MYRIAPODA
ant	*myrmēx*	Formicidae
bee, honey (drone)	*kēphēn*	*Apis mellifera*
bee, honey (queen, lit. king)	*basileus*	*Apis mellifera*
bee, honey (queen, lit. leader)	*hēgemōn*	*Apis mellifera*
bee, honey (worker)	*melissa*	*Apis mellifera*
beetle, dung	*kantharos*	Scarabaeoidea
butterfly	*psychē*	Lepidoptera
centipede or millipede	*ioulos*	Myriapoda
cicada	*tettix*	*Cicada* sp.
clothes moth	*sēs*	*Tinea* sp.
cockchafer	*mēlolonthē*	*Geotrupes* sp.
flea	*psylla*	Siphonaptera
fly	*myia*	Diptera
fly, horse	*myōps*	*Tabanus* sp.
grasshopper	*akris*	Acrididae
locust	*attelabos*	Acrididae
louse	*phtheir*	Phthiraptera
mayfly	*ephēmeron*	Ephemeroptera
pseudoscorpion	*to en tois bibliois gignonmenon skorpiōdes**	*Chelifer cancroides*
scorpion	*skorpios*	*Scorpio* sp.
spider	*arachnē*	Araneae
tick	*kynoraistēs*	*Ixodes ricinus*
wasp	*sphēx*	Vespidae

traditionally identified as the razor-clam (*Solenidae*), a sand-burrower, and among the most active and perceptive of all bivalves.

* Literally 'The thing that looks like a scorpion that comes to be within books.'

wasp, hunting	*anthrēnē*	Vespidae
wasp, fig	*psēn*	*Blastophaga psenes*
wasp, parasitoid	*kentrinēs*	*Philotrypesis caricae?*

UNCLASSIFIED

fish louse	*oistros ō tōn thynnōn*	*Caligus* sp.
hermit crab	*karkinion*	Paguroidea
jellyfish?	*pneumōn**	Scyphozoa?
red coral	*korallion*	*Corallium rubrum*
sea anemone	*knidē*	Actinaria
sea anemone	*akalēphē*	Actinaria
sea cucumber?	*holothourion*†	Holothuria?
sponge	*spongos*	Dictyoceratida
sponge, black Ircinia	*aplysias*	*Sarcotragus muscarum?*
starfish	*astēr*	Asteroidea
worm	*helminthes*	Plathyhelminthes + Annelida + Nematoda, etc.
worms, tape	*helminthōn plateion genos*	*Taenia* sp.
worm, nematode ('round')	*strongyleion*	*Ascaris?*
worms, unknown	*akarides*	unknown

* VOULTSIADOU AND VAFIDIS (2007) identify this as the dead man's fingers sponge, *Alcyonium palmatum*. That's plausible too.
† VOULTSIADOU AND VAFIDIS (2007) identify this as the soft coral, *Veretillum cynomorium*. That's plausible too.

APPENDICES

Here I present some of Aristotle's data and models as he might were he writing now: in tables and diagrams. Such devices are not in principle un-Aristotelian since he clearly used abstract models to explain biological phenomena at least occasionally – for example, when he explains animal geometry in *PA* or perception and movement in *MA*.* Nevertheless, my justification for using them does not rest upon such examples, for my purpose is not to reproduce his methods, but rather to understand the strengths and weaknesses of his data and his explanations. The absence of data tables in his work is particularly painful: he can take a book (e.g. *HA* VI on avian life history) to explain patterns that would now be summarized in a single table in *Nature* – and in the Online Supplementary Information at that. In the same way it is also impossible to know whether the heart–lung cycle he gives in *JSVM* 26 really works as he says it does without building a control model or else a physical analogue – and the first seems a lot easier. Classical philosophers may shy at the resulting tables and diagrams; to them such devices may seem incongruously modern. I would ask them to view them merely as tools analogous to their use of modern symbolic notation to explicate and test the coherence of Aristotle's logic. Scientists will be less fussed; to them, the utility of such devices will seem obvious and they will only wonder how Aristotle got as far as he did using mere words. I would ask them to remember that, although he was smart, he did live a long time ago.

* NATALI (2013) ch. 3.3.

I. A DATA MATRIX FOR TWELVE ARISTOTELIAN KINDS AND SIX MORPHOLOGICAL FEATURES

This table displays some of the morphological features that Aristotle thinks some animals have. His information is not always correct. For convenience the feature states are first coded as integers. If Aristotle thinks an animal kind has more than one feature state this is indicated with a slash, for example 0/1; intermediate states are indicated as 0.5; no data as 'NA'. This table is based on the following sources. **Foot type**: lion, dog, sheep, goat, deer, hippopotamus, horse, mule, pig, HA 499b5.

Astragalus with foot type: lion, pig, man, cloven-hoofed animals, solid-hoofed animals, HA 499b20; human HA 494a15; camel HA 499a20. **Horns with cloven hoofs**: ox, deer, goat HA 499b15. **Tooth number and horns**: horned animals, camels, HA 501a7, HA 499a22. **Tooth type and horns**: pig, lion, dog, horse, ox, HA 501a15; elephant HA 501b30. **Stomach type and horns and tooth number**: HA 495b25; HA 507b30, human HA 495b25. The feature matrix shows a strong association between the various features that Aristotle describes. These associations then become the target of explanations. This table could be expanded to include more kinds and features, but I do not do so since for these either his data are incomplete or he makes little of them.

CODING

FEATURE	STATE	
tooth no.	teeth in upper jaw ≠ teeth in lower jaw	0
	teeth in upper jaw = teeth in lower jaw	1
tooth shape	flat	0
	saw	1
	tusks	2
stomach	simple	0
	complex	1
horns	absent	0
	present	1
feet	solid-hooved	0
	cloven-hoofed	1
	multi-toed	2
astragalus	absent	0
	present	1

MATRIX

	FEATURE					
KIND	tooth no.	tooth shape	stomach	horns	feet	astragalus
ox	0	0	1	1	1	1
goat	0	0	1	1	1	1
sheep	0	0	1	1	1	1
deer	0	0	1	1	1	1
camel	0	NA	1	0	1	1
pig	1	2	0	0	1/0	1/0
horse	1	0	0	0	0	0
mule	1	0	0	0	0	0
elephant	NA	0/2	1	0	2	0
lion	1	1	0	0	2	0.5
dog	1	1	0	0	2	0
human	1	1	0	0	2	0

11. RESOURCE (TROPHĒ) ACQUISITION AND ALLOCATION PATHWAY FOR A LIVE-BEARING TETRAPOD (A MAMMAL)

This diagram summarizes Aristotle's vision of the metabolic system, how nutrition is taken up, transformed and allocated to its various ends. The arrows represent material flows. Aristotle's 'uniform parts' are roughly equivalent to our tissues except that he is emphatic they have no microscopic structure such as atoms or cells. All uniform parts derive from blood, itself a uniform part. There are two great branches in the network, earthy uniform parts and fatty uniform parts, with flesh being at the terminus of a branch of its own. All reactions produce waste; and all uniform parts are broken down into waste and excreted, giving an open system. Some nutrition goes to fuel the internal fire. The nodes represent specific transformations of nutrient. The supporting statements for network are as follows. **Blood** is the final/universal nutriment: PA 650a34, PA 651a15. **Flesh** is made from the purest nutriment and bones, sinews, etc. are residues: GA 744b20. **Flesh** is concocted blood and **fat** is the surplus blood left over from this: PA 651a 20. **Fat** is concocted blood: PA 651a21. **Fat** can be soft or hard (suet or lard): PA 651a20. **Semen** comes from blood, specifically from the part that forms fat: PA 651b10; GA 726a5. **Marrow** is partially concocted blood: PA 651b20. **Hoofs, horns and teeth** are related to **bone**: PA 655b1, PA 663a27. **Bones** and **marrow** are made from a common precursor: PA 652a10. **Cartilage** and **bone** are fundamentally the same thing: PA 655a27. Deposits from the bladder and gut are **residues** of nourishment: PA 653b10. **Bile** is a residue of nourishment: PA 677a10.*

---

* See LEROI (2010) for further details.

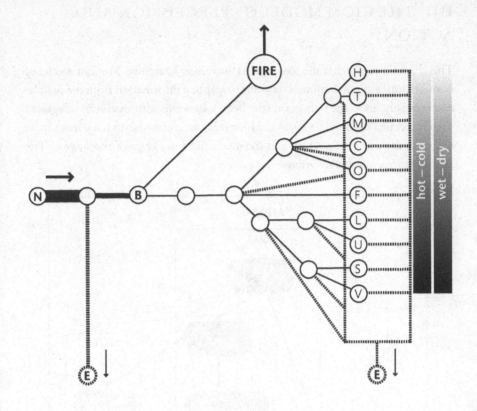

LEGEND

N   nutrition
B   blood
H   hooves, hair, nails
T   teeth
M   marrow
C   cartilage
O   bone
F   flesh
L   lard
U   suet
S   semen
V   vaginal secretions, menstrual fluid, milk
E   excreta: urine, bile, faeces

III. THE CIOM MODEL OF PERCEPTION AND ACTION

This diagram represents the Centralized Incoming Outgoing Motions model of how Aristotle supposes animals transmit perceptual information from the peripheral sense organs to the sensorium (the heart), how this information is integrated with respect to the animal's goals and how it is transformed into movement in its limbs via the action of pneuma and the mechanical workings of the sinews.* The arrows represent causal relations.

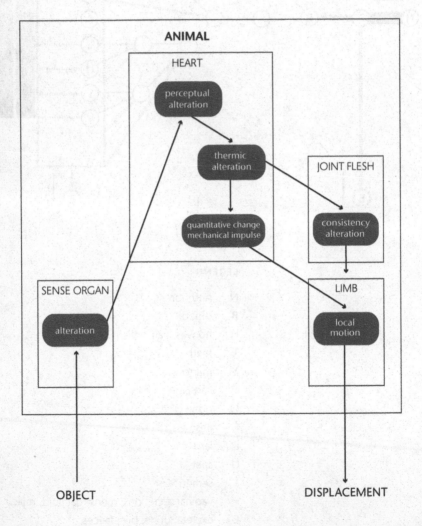

* GREGORIC and CORCILIUS (2013).

IV. CONTROL DIAGRAM OF ARISTOTLE'S HEART— LUNG THERMOREGULATORY CYCLE

This is the simplest of many possible models that could describe the heart–lung cycle that Aristotle sketches in *JSVM* 26.* The arrows represent control relations. To make Aristotle's model work we need various assumptions explicit that he does not. Here, we assume that the animal has an ideal 'reference' temperature, T_r. The goal of the system is to maintain the temperature of the heart, T_h, at that temperature. The system works in the following way. Nutrition enters the heart and is concocted. The temperature of the nutrition (now blood), T_n, rises above the reference temperature. If that increase in temperature is sufficient to exceed heat loss due to diffusion (see below), it will increase the heart temperature, T_h. Since lung volume is a function of the difference between T_h and T_r, lung volume increases. This results in an increase in the rate of air flow through the mouth, F_a. Since air temperature, T_a, is lower than the reference temperature, heart temperature declines and the lung contracts. The result is a negative feedback control system. Note that we allow for the constant loss of some fraction of heart heat by diffusion, perhaps via the brain that, in Aristotle's view, acts as a radiator. This will tend to damp the system making it less sensitive to increases in T_n and gives an equilibrium at T_r. This system will work only if air temperature is lower than the ideal reference temperature. If, however, $T_a > T_r$, then no amount of air will reduce T_h, the negative feedback loop will become an unstable positive feedback loop, and the animal's lungs will stay permanently open or permanently closed, either way extinguishing the fire (due to excess cold or consumption of all the nutrient), thus resulting in death. As described here, the system will tend to a stable dynamic equilibrium rather like a thermostat. However, if additional delays or non-linearities are included, it will produce the oscillatory behaviour that Aristotle supposed explained the lung's movements. The model was produced with the kind help of David Angeli, Electrical Systems Control Group, Imperial College London.

* King (2001) pp. 126–9.

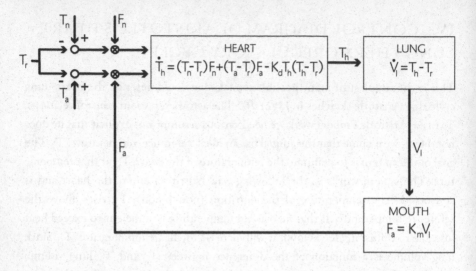

LEGEND

T_r reference temperature
T_h heart temperature
T_n nutrient temperature
T_a air temperature
F_n nutrient flow
F_a air flow
V_l lung volume
K_d heat diffusion constant
K_m air intake constant
O   sensor
⊗   multiplier
+   positive regulation
−   negative regulation

V. ARISTOTLE'S LIFE-HISTORY DATA: LIVE-BEARING TETRAPODS AND BIRDS

These tables summarize Aristotle's life-history data. His data are a bit more complex than the tables suggest and, again, are not always correct. Since Aristotle does not have descriptive statistics, he often says that something is 'generally' the case; if so, that is the value I give. If he gives a range, I report a median but ignore exceptional cases. When he says that he is uncertain (e.g. about the great lifespan of the elephant or the short lifespan of the sparrow) I have indicated this with a u. In some cases Aristotle does not explicitly say that a particular kind has some value for a given life-history variable, but just speaks generally about the megista genos – for example, 'very few birds propagate in their first year'. In such cases, I have indicated the value as belonging to all kinds within that greater kind unless noted otherwise; but in cases where he does not say explicitly that a value applies to a megista genos I have not assumed it. For example, he probably knows that most large live-bearing tetrapods (mammals) have one brood per year, but he does not say so. The exception to this rule is body size. Aristotle never reports quantitative data for body size, nor even whether an animal is big or small except in the context of a functional explanation. From such explanations, however, it's clear that he thinks a human or an ostrich is 'large', a pig or a chicken is 'medium-sized' and a cat or sparrow is 'small' relative to the megista genos to which each belongs; I have filled in appropriate body sizes accordingly. Most of these data come from HA V and VI; data on embryonic perfection come from GA IV. Aristotle argues correctly that multi-toed animals (fox, bear, lion, dog, wolf, jackal, etc.) have imperfect young; solid- and split-hoofed animals (cow, horse) have perfect young. The pig is an oddity, being split-hoofed and having relatively perfect offspring. Among the birds, Aristotle names ravens, jays, sparrows, swallows, ring doves, turtle doves and pigeons as having imperfect neonates – but doesn't name any perfect ones. He probably bases his generalizations on more data than he reports.

zoōtoka tetrapoda LIVE-BEARING TETRAPODS

KIND	adult body size	age at maturity (years)	lifespan (years)	broods per year	brood size	gestation time (months)	relative perfection
	FEATURE						
mouse	S				many		
hare	S				4		I/P
cat	S		6 I		<8		I
mongoose	S		6 I		<8		I
jackal?	S				4		I
goat	M	1	8 I		1.5	5	P
sheep	M	1	10 I		<4	5	P
pig	M	0.75	15 I		<20	4	P
wolf	M				<8	2	I
dog	M	0.9			<8	2	I
leopard	M		12 I		4		I
lion	M		5 I	1	4		I
bear	L				3.5	1/9	I
horse	L	3	37 I	1	1.5	11	P
ass	L	3	30 I	1		12	P
cattle	L	1.5	15 I		1.5	9	P
deer	L				1.5		P
camel	L	3	>50 I		1	10	P
human	L	<21	40f/70m rl		1	9.5	
elephant	L	7	250u I	1	1	30	

L: large
M: medium
S: small

I: simple lifespan
rl: reproductive lifespan

P: perfect
I: imperfect

ornis BIRDS

KIND	adult body size	age at maturity (years)	lifespan (years)	clutches per year	clutch size	time to hatching or fledging (months)	relative perfection
coal tit	S	>1*		1	>20		
tit	S	>1		1	many		
sparrow	S	>1	1u l	1	many		I
kingfisher	S	0.3		1	5		
bee-eater	S	>1		1	6.5		I
swallow	S	>1		2			
nightjar	S	>1		1	<3	0.5	
cuckoo	S	>1		1	<2		
jay	S	>1		1	many		I
pigeon	S	0.5	8 l	10	<3		I
turtle dove	S	0.4u	8 l	≤2	<3		I
wood pigeon	S	0.4u	30 l	≤2	<3		I
hen	M	>1		many	many	1.7	
partridge	M	>1	>16 l	1	many	0.6	
raven	M	>1		1	4	0.6	I
kite	M	>1		1	2	0.6	
hawk	M	>1		1			
kestral	M	>1		1	4		
Ural owl?	M	>1		1	4		
peacock	L	3	25 l	1	<12	1	
goose	L	>1		1		1	
bustard	L	>1		1		1	
vulture	L	>1		1	2		
eagle	L	>1		1	3	1	
ostrich	L	>1			many		

L: large
M: medium
S: small

l: simple lifespan
rl: reproductive lifespan

P: perfect
I: imperfect

* When Aristotle says that 'very few birds propagate in their first year' he certainly means that they propagate in their second, that is, the following breeding season, typically spring.

VI. RELATIONSHIPS AMONG SOME LIFE-HISTORY FEATURES, ILLUSTRATED USING MODERN DATA

In GA IV and LBV, Aristotle claims that various life-history features are associated with each other in certain ways. His claims are correct at least for placental mammals. Below, I illustrate four of these associations using data from the panTHERIA database of mammalian life history.* I exclude Orders not seen by Aristotle (e.g. Marsupialia) or else excluded from his tetrapods (Chiroptera, Cetacea), and then model the log-transformed data using linear regression. Four of Aristotle's claimed relationships are shown: brood size and adult body size (negative), gestation time and longevity (positive), adult body size and longevity (positive) and fecundity and adult body size (negative). Much more sophisticated analyses of this sort have often been published.† They usually aim to take various confounding effects into account and so reduce, but hardly eliminate, the difficulty of inferring causal relations from comparative data.

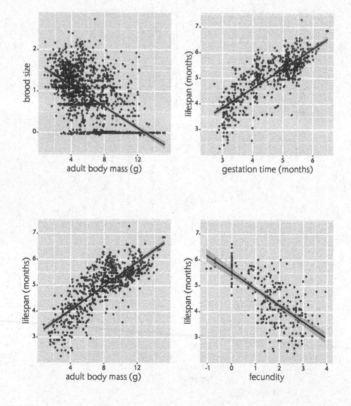

* JONES et al. (2009).
† For example, MILLAR and ZAMMUTO (1983), DERRICKSON (1992), STARCK and RICKLEFS (1998), BIELBY et al. (2007).

NOTES

THE EXEGETICAL LITERATURE on Aristotle's writings is ancient, disputatious and vast. Modern classical philosophers, working out what Aristotle was getting at in his *Physics*, often cite Alexander of Aphrodisias' commentary even though it was written in the second century AD. I, however, must eschew such erudition and the notes below have only two modest goals. The first is to guide you to Aristotle's texts. If you want to read for yourself what he said about the mole's eyes, the notes to Chapter LIV will tell you where to do so. The second is to give you an entrée to the most important, recent and accessible secondary literature. Unfortunately these qualities rarely coincide since Aristotelian scholarship is glacially slow and often appears in those Patagonias of academic publishing, *Festschriften* and conference proceedings. For the most part I have neither justified my readings by appeal to this literature nor attempted to adjudicate on disputes within it. If, on occasion, I cite scholars who offer different interpretations from my own, it is only to warn of important disagreements among experts or a small unorthdoxy of my own.

References to Aristotle's works are in the form of 'Bekker numbers', which refer to Immanuel Bekker's 1831 edition of the Greek text. They look like this: *HA* 608b20, where *HA* refers to the treatise, *Historia animalium*, and 608b20 to the line number. Any given work, e.g. *HA*, is also divided into 'books' and 'chapters' which I do not generally use unless citing an entire chapter, e.g. *HA* I, 1 – *HA* book I, chapter 1. Using these numbers you will be able to find any given text in any decent edition written in any language.

The Oxford *Works of Aristotle Translated into English*, 1910–52, edited by J. A. Smith and W. D. Ross, is available free online. It was revised and published in two volumes as *The Complete Works of Aristotle: The Revised Oxford Translation*, Princeton, 1984, edited by Jonathan Barnes. However, if you want a copy of *HA*, hunt down a second-hand paper copy of the 1910 Oxford edition: D'Arcy Thompson's. True, I am being sentimental, but it also has the notes that the Princeton and online editions don't. The Loeb editions, published by Harvard, are also invaluable and contain the Greek.

These editions have been partly superseded by the *Clarendon Aristotle* which gives an English text with important commentaries; however, the only biological work currently available in this series is Jim Lennox's *The Parts of Animals*, 2001. German readers will want *Aristoteles: Werke in deutscher Übersetzung*, Akademie Verlag; however, again, the only available biological works are Jutta Kollesch's *IA* and *MA*, 1985, and Wolfgang Kullmann's *PA – Über die Teile der Lebewesen*, 2007. The standard Greek edition of *Historia animalium* is David Balme's (with Allan Gotthelf) *editio maior*, Cambridge, 2002.

WORKS BY ARISTOTLE

Cat	Categories (Categoriae)
APo	Posterior Analytics (Analytica posteriora)
Top	Topics (Topica)
Phys	Physics (Physica)
DC	The Heavens (de Caelo)
GC	On Generation & Corruption (de Generatione et corruptione)
Meteor	Meteorology (Meteorologica)
DA	The Soul (de Anima)
PN	Small Treatises on Nature (Parva naturalia)
Sens	Sense and Sensible Things (de Sensu et sensibilius)
SV	Sleep (de Somno et viglia)
LBV	The Length and Shortness of Life (de Longitudine et brevitate vitae)
JSVM	Youth & Old Age, Life & Death, incl. Respiration (de Juventute et senectute, vita et morte, incl. de Respiratione)
HA	Enquiries into Animals (Historia animalium)
PA	The Parts of Animals (de Partibus animalium)
DM	The Movement of Animals (de Motu animalium)
IA	The Progression of Animals (de Incessu animalium)
GA	The Generation of Animals (de Generatione animalium)
DP	On Plants (de Plantis)*
Mirab	Marvellous Things Heard (de Mirabilibus auscultationibus)*
Prob	Problems (Problemata)*
Metaph	Metaphysics (Metaphysica)
EN	Nicomachean Ethics (Ethica Nicomachea)
EE	Eudemian Ethics (Ethica Eudemia)
MM	Great Ethics (Magna moralia)*
Pol	Politics (Politica)
Poet	Poetics (Poetica)
FR	Fragments (Fragmenta)

* Pseudo-Aristotelian.

WORKS BY THEOPHRASTUS

HP   *Enquiries into Plants (Historia plantarum)*
CP   *Explanations of Plants (de Causis plantarum)*
St   *On Stones (de Lapidibus)*

WORKS BY PLATO

Rep   *The Republic*
Tim   *The Timaeus*
Phaedrus *The Phaedrus*
Phaedo   *The Phaedo*
States   *The Statesman*
Laws   *The Laws*
Philebus *The Philebus*
Georgias *The Gorgias*

WORKS BY OTHER ANCIENT WRITERS

Athen   Athenaeus, *Deipnosophists*
DK   Pre-Socratic texts (Diels–Kranz number)
DL   Diogenes Laertius, *Lives of the Philosophers*
Econ   Xenophon, *Oeconomicus*
Herod   Herodotus, *The Histories*
Hesiod   Hesiod, *Theogony*
HM   Aelian, *Historical Miscellany*
Mem   Xenophon, *Memorabilia, or, Recollections of Socrates*
NA   Aelian, *On the Nature of Animals*
Paus   Pausanias, *Description of Greece*
Plin   Pliny, *Natural History*
Plut   Plutarch, *Life of Alexander*
Strab   Strabo, *Geography*
Symp   Xenophon, *Symposium*

I

Shells & snails. A. on shells, *HA* 528a20; on the internal anatomy of snails, *HA* 529a1. THOMPSON (1947) p. 113 doubts the etymology of *kēryx* from herald and suggests it's just an ancient name for snail.

II

The Lyceum. Sulla sacked Athens in 87–86 BC during the first Mithridatic War; see KEAVENEY (1982) p. 69. *Strab* XIII, 1, 54–5 describes how he took A.'s works to Rome. *Strab* IX, 1, 24 and *Paus* I, 19, 3 give late descriptions of the Lyceum; LYNCH (1972) discusses its topography and function. The sayings and description of A. are given in *DL* V, 1–2; *DL* V, 17–22 [trans. HICKS (1925)]. In some sources it's a shame to keep quiet while Isocrates (a sophist) rather than Xenocrates (a fellow Academician) speaks. Most scholars agree that A.'s extant works are lecture notes, e.g. ACKRILL (1981) p. 2, GRENE (1998) p. 32, BARNES (1996) p. 3, ANAGNOSTOPOULOS (2009b). Canguilhem's historiographical strictures were issued in *L'objet de l'histoire des sciences*, 1968, but I found them in PELLEGRIN (1986) p. 2. A. speaks of the study of nature at *PA* 639a13, *PA* 644b17, *PA* 645a6 and *DA* 402a7. He gives the curriculum of the great course at *Meteor* 338a20. The *Invitation to Biology* can be found at *PA* 645a15.

III

D'Arcy Thompson. A life of D'Arcy Thompson was written by his daughter, THOMPSON (1958). THOMPSON (1910) identifies the jerboa at *HA* 606b6, *n.* 1, and discusses the *Rhinobatos* and relations at *HA* 566a27, *n.* 6. THOMPSON (1910) p. vii makes the case for A.'s stay in Lesbos – more generally, the Eastern Aegean – as the period in which he did the bulk of his biological work. JAEGER (1948) ignored this in his chronology, but LEE (1948) argued for Thompson's view, while SOLMSEN (1978) criticized Thompson on the basis of the inauthenticity of the main passages in *HA*; LEE (1985) defended Thompson again. BALME (1991) p. 25 considered the period in Lesbos the 'likeliest' for the bulk of A.'s work on *HA*, but thought some of the other biological works were earlier, perhaps even Academic. KULLMANN (2007) pp. 146–56 reviews the arguments for the chronology of the zoology and concludes that 'The time at Lesbos is thereby the *terminus post quem* [i.e. earliest date] for the drafting of the zoological works. There is much to suggest that all the zoological works were conceived in the same period of A.'s life. Whether they were developed only later, we do not know' [trans. AML]. THOMPSON (1910) p. iv despairs of annotating A.'s natural history.

IV

Lesbos. For the birds of Lesbos see DUDLEY (2009); for its geology see ZOUROS *et al.* (2008); for its botany see BAZOS and YANNITSAROS (2000) and BIEL (2002). Makis Axiotis, a local doctor, naturalist and polymath, has also written several excellent books on the island's fauna and flora (in Greek) which you can buy locally.

V

At the lagoon. My Aristotelian synopsis of the animals of the Lagoon is compiled from the following passages: *HA* 621b13, *HA* 544a20, *PA* 680b1, *HA* 547a4, *HA* 548a8, *HA* 603a22, *GA* 763b1. THOMPSON (1913) adds *HA* 548b25 which refers to the sponges of Cape Malea; however, although there is a Cape Malea on Lesbos, there's a much more famous one on the Peloponnese so I exclude it. A.'s word for 'lagoon' is *limnothalassa* or 'lake sea', cf. *GA* 761b7, *HA* 598a20, which he does not specifically apply to Kalloni.

VI

Fish as food. Archestratus' foodie fragments are collected and translated by WILKINS *et al.* (2011). In his classic work on Greek consumption, DAVIDSON (1998) has much to say about the importance of fish.

VII

The Pre-Socratics. For accessible introductions to the thought of the *physiologoi* see LLOYD (1970) and WARREN (2007). BARNES (1982) and BARNES (1987) give a generous selection of the texts and commentaries that are both witty and illuminating. Barnes is, however, by his own admission, not very interested in the scientific theories of the *physiologoi*, so these excellent books must be supplemented by KIRK *et al.* (1983). For some, e.g. FARRINGTON (1944–9) and LLOYD (1970) p. 9, the *physiologoi* 'leave the gods out'; others, e.g. SEDLEY (2007), are more inclined to see the divine in their explanations. LLOYD (1970) p. 10 and BARNES (1982) ch. 1 argue that the *physiologoi* are characterized by debate or reason. Thales' account of earthquakes is due to *Aëtus* III, 15 and Seneca's *Naturales quaestiones* III, 14; 6.6. A. discusses Hesiod at *Metaph* 983b19; *Hesiod* 116–20. Heraclitus is cutting about his contemporaries and predecessors at *DK* 22B40. The reference edition of the *Corpus Hippocraticum* is Littré's Greek/French one: LITTRÉ (1839–61), but the corpus is available in Greek/English: JONES *et al.* (1923–2012) and LONIE (1981). 'Hippocrates' wants to 'explain how man and the other animals . . .': *Littré* VIII,

Fleshes, I [trans. modified from JONES *et al.* (1923–2012) vol. VIII]; speaks of the uses of oxymel: *Littré* II, *Regimen in acute diseases*, 16. A. mentions Hippocrates only once and not in a medical context: *Pol* 1326a15. Empedocles' quackery is recorded in *DK* 31B111; A. criticizes his style at *Metaph* 985a5.

VIII

A. arrives at the Academy. A.'s biography has been cobbled together, with subtle scholarship, from a variety of late and unreliable *vitae.* DÜRING (1957) was, for many years, the standard account; now NATALI (2013) has produced an excellent new analysis of them. A. speaks of the abandonment of natural science at *PA* 642a29. A list of Plato's students can be found at *DL* III, 46. Socrates' despair at his own muddle-headedness is recorded in *Phaedo* 99B. His anti-science is recorded by Xenophon in *Mem* I, 1.11–15. Cicero commends Socrates' ethical turn in the *Tusculan Disputations* vol. 10.

IX

Plato's anti-science. Speusippus' character is recorded in *DL* IV, 1. The dialogue between Socrates and Glaucon is from *Rep* 527C–531C.

X

The Timaeus. BURNYEAT (2005) analyses the meaning of *eikōs mythos.* Plato's numerological theory of the elements is given at *Tim* 54D–55C [trans. CORNFORD (1997)]. GREGORY (2000) and JOHANSEN (2004) give general accounts of Plato's natural philosophy. HAWKING (1988) looks for the mind of God. (He later gave up.) A. speaks of the conflict between love of truth and friendship at *EN* 1096a11; in the later tradition this often becomes 'I love Plato, but I love truth more.'

XI

At the Academy. The (fairly implausible) anecdote about how A. bullied the elderly Plato is recorded by Aelian: *HM* III, 19. We hear about Hermias and Assos in *DL* V, 3–9, who also records the inscription on Hermias' statue; cf. *Athen* XV, 696 and *Strab* XIII, 1, 57. ANDREWS (1952) discusses whether A. was politically involved in Hermias' court; Plato probably never met Hermias – at least so his *Sixth Letter*, about friendship, which is addressed to Hermias as well as to the Academicians Coriscus and Erastus, seems to imply, NATALI (2013). A. speaks of the optimal age of marriage at *Pol* 1335a27; he was around thirty-seven at the time, from which we infer (indirectly enough) that Pythia was eighteen. 'A spray of myrtle...': Archilochus [trans. BARNSTONE (1972) p. 29].

XII

Assos. A report of the excavations at Assos is given by CLARKE *et al.* (1882).

XIII

Theophrastus. The archaeology of ancient Eresos is recorded by SCHAUS and SPENCER (1994). T.'s life is given in *DL* V, 36–57. His botany can be read in the Greek/English editions of HORT (1916) and EINARSON and LINK (1976–90) which, however, have been superseded by the Greek/French editions of AMIGUES (1988–2006) and AMIGUES (2012), or will be, when the latter is finished. The rest of T.'s fragmentary writings have been collected and analysed in the long series of monographs titled *Theophrastus of Eresus: Sources for his Life, Writings, Thought and Influence,* and related volumes, edited by the late Robert Sharples, William Fortenbaugh and Pamela Huby of the great Theophrastus Project.

XIV

Lesbos. 'He will be a lucky...' is from THOMPSON (1913) p. 13, a tribute from one great zoologist to another.

XV

A. as a scientist. A. uses *physikē* [*epistēmē*] at *Metaph* 1026a6 and *physikos* at *Phys* 197a22. The term 'scientist' was defined by WHEWELL (1840) vol. I, p. 113, though used by him earlier.

XVI

Epistemology. 'All men... desire to know' at *Metaph* 980a21 [trans. ROSS (1915) modified], continuing in *Met* I, 1. The *Metaphysics* is a compilation of related texts. The fashion used to be, following JAEGER (1948), to analyse them into different layers of development, but this is now thought to be hard to do; see BARNES (1995b) for an introduction to their contents and relationships.

XVII

The source of empirical information. OWEN (1961/1986) and NUSSBAUM (1982) discuss what A. means by *phainomena*, but do not, I think, credit his empiricism sufficiently; see BOLTON (1987) for a corrective. For a clear statement on A.'s sense of empirical reality, and the primacy of observation, in doing science see *DC* 306a5 among others. For all that, A.'s inquiries into *phainomena* do often begin not only with his own observations but with 'reputable opinions' or 'the opinions of the many or the wise', what he calls *endoxa*, e.g. *Top* 100b21. 'Some animals are live bearing' is from *HA* 489a35. I estimate the number of empirical claims in *HA* from a sample of 1,500 words chosen at random from THOMPSON (1910) *HA*. Plato's acceptance of hieroscopy is apparent in *Tim* 71–2. We press on those in front at *DL* V, 20. BOURGEY (1955), PREUS (1975) and LLOYD (1987) discuss the sources of A.'s empirical data. A. talks about diviners and bird behaviour at *HA* 608b19. THOMPSON (1895), THOMPSON (1910) *n.* 609a4 and PREUS (1975) pp. 34–6; *ibid.* pp. 278 *n.* 113, 115, 116 suggest that a good deal of A.'s bird lore was astrological in origin; see also A. on the *alkyōn*, Ch. LXXX. PREUS (1975) p. 22 discusses A.'s use of 'mythos', while LLOYD (1979) ch. 3 discusses the relationship between Greek science and popular belief. For A. on the astragalus see *HA* 499a22, *HA* 499b19; gall bladders, *HA* 506a20; rejections of myths about cranes, *HA* 597a23; lions, *HA* 579b2; wolves, *HA* 580a11; talking heads, *PA* 673a10.

XVIII

A. and the fishermen. A. speaks of vocalizing fish at *HA* 535b14; ONUKI and SOMIYA (2004) describe the sounds that *Zeus faber* makes and the mechanism by which it makes them. Athenaeus waxes sarcastic at *Athen* VIII, 352. There is a sentimental idea that farmers and fishermen are unusually knowledgeable about the creatures that they see, but the evidence shows otherwise, e.g. THOMPSON (1998) on the seal folktales of the Scottish isles. A. discusses fellating fishes at *HA* 541a13, *HA* 567a32, *GA* 756a7; a story presumably due to *Herod* II, 93. It is often said that H.'s account refers to the mouth-brooding Tilapia (*Oreochromis nilotica*), but in both A. and H. the reference appears to be to a saltwater or estuarine fish whereas Tilapia are freshwater. A. presses the claims of expertise at *PA* 639a1 and *HA* 566a8.

XIX

Chameleons. On the chameleon, *HA* 503a15. This passage is unusual in that the animal in question has not been chopped up and distributed across *HA* system by system; instead it seems to be a preliminary summary of findings that await further analysis, BALME (1987a). LONES (1912) p. 157 says that the chameleon does have a spleen, but that it is small, about '0.11 inches long'.

XX

A. and Alexander. In his Life of Alexander, Plutarch, Plut 668, 7, 4 [trans. Dryden], relates Alexander's education at A.'s hands. NATALI (2013) doubts the whole Mieza story, but it's unclear why – he agrees that A. taught Alexander, and he must have done so somewhere. The story of Alexander's Iliad is also told by Plutarch; A. wrote books for Alexander about how to lead and how to run colonies but, fragments aside, they have been lost. LANE-FOX (1973) gives a life of Alexander. Pliny tells the story that Alexander funded A.'s research: Plín VIII, 44; Athenaeus, Athen IX, 398e, amplifies it. LEWES (1864) p. 15, OGLE (1882) pp. xiii–xiv, ROMM (1989) and most recent scholars deny Pliny's story, though JAEGER (1948) defended it since it fitted with his developmental scheme for the composition of A.'s works. LLOYD (1970) p. 129 points out that the idea that the state or a king would fund scientific research directly, rather than hosting scholars, as Hermias did, was probably alien to fourth-century Greece and that the first recording of state-funded research is the library at Alexandria, third century BC. Pliny (Plín VIII, 42) is usually credited with ex Africa semper aliquid novi, but A. tells us that the saying was old even in his day, HA 606b20. See Glossary II for a list of A.'s animals mentioned in this book and their identification.

XXI

Exotica. A. tells us about the martikhõras at HA 501a24, and goes on about Ctesias' unreliability on elephant sperm at HA 523a26 and India at HA 606a8. A. mentions the oryx at HA 449b20 and the onos Indíkos at HA 499b19 and PA 663a19. A. is ambivalent about the status of the 'so-called' Indian ass, and says that 'it is reported' that it is horned and has one hoof; if it is a rhino, then he's wrong, for it has three toes: see Glossary II. Herodotus speaks of trusting his eyes at Herod II, 99, II, 147, IV, 81, V, 59. Information that A. takes from Herodotus, but does not credit him with includes: the menopausal priestesses, HA 518a35/Herod I, 175; Herod VIII, 104; camels fight horses, HA 571b24/Herod I, 80; lions, HA 579b7/Herod VII, 126; cranes, HA 597a4/Herod II, 22; Egyptian animals, HA 606b20/Herod II, 67; flying serpents in Ethiopia, HA 490a10/Herod II, 75; camel knees, HA 499a20/Herod III, 103. Ethiopean sperm, HA 523a17/Herod III, 101. Herodotus talks of gold-digging ants at Herod III, 101–5. For his serpents with wings see also Ch. XCIV. Besides Ctesias' Persica and Indica and Herodotus' Histories, A. may have drawn on Herodorus of

Heraclea's *Heraclea* which he mentions, but there are many other histories that he does not mention, but that he might have drawn upon nevertheless, e.g. Heraclides of Cyme's *Persica* (mid-fourth century) and Damastes' *Periplus* (fifth century).

The elephant et al. There is a substantial literature on whether or not A. saw an elephant and, if so, whether it was an Asian or African one. I think he saw neither; however, PREUS (1975) p. 38 suggests that A. may have seen an elephant in a Macedonian zoo even though there is no evidence that the Macedonians ever had a zoo. ROMM (1989) discusses the issues and tries to disprove Pliny's story by arguing that A. saw an African elephant, while BIGWOOD (1993) concentrates on the possible literary sources of A.'s knowledge of the elephant. See Chs XXXVI and XLVII for more on elephants. A. gives his mostly inaccurate information about the lion at *HA* 579a31, *HA* 594b18, *HA* 629b12 and *GA* 760b23. His information about the European distribution of the Asiatic lion mostly comes from Herodotus, writing around 430 BC, but *HA* 629b12 also seems to rely on information from hunters and distinguishes two kinds of lion; Xenophon, writing around 380 BC, independently speaks of hunting lions in Macedonia; see BIGWOOD (1993) p. 236 *n*. 6 and SCHNITZLER (2011) on its historical distribution. A. talks about the ostrich at *HA* 616b5, *PA* 644a33, *PA* 658a10, *PA* 695a15, *PA* IV, 14, *GA* 749b15 and *GA* 752b30. He discusses the camel's toes at *HA* 499a23. What he says about them is a bit obscure and various interpretations turn on exactly what A. meant by 'back' and 'front', LONES (1912) pp. 191–2. I interpret 'back' and 'front' to mean 'hind limb' and 'forelimb'. If this is correct, then A.'s statement is correct since the cleft of the hind feet is indeed deeper than that of the forefeet. A. does not say exactly how many chambers the camel's stomach has, which is just as well: their number and relationship to ruminant stomachs has been the subject of debate for centuries, WANG *et al.* (2000). On projectile-defecating bison see *HA* 630b9, cf. *Mirab* I, and anti-predator behaviour in the Bovinae, ESTES (1991) p. 195; the same behaviour has been recorded in American bison.

XXII

The hyena. A. describes the hyena at *HA* 579b15; cf. *GA* 757a3. Many have seen in this account a description of the spotted hyena's (*Crocuta crocuta*) pseudo-hermaphroditism, but this is implausible, see Glossary II: *hyaina/glanos/trochos*. THOMPSON (1910) at *HA* 579b23 has a translation error – 'male' should read 'female' – A. is *not* saying the female has an organ resembling the male's. BIGWOOD (1993) postulates Callisthenes as a source of A.'s knowledge about exotic zoology; he also mentions Eudoxus of Cnidus, see Ch. CII. BROWN (1949) discusses the

NOTES TO PAGES 53-62

relationship between A., Alexander and Callisthenes, while ROMM (1989) talks about the reputation-boosting tradition. There may, of course, be more than one Unknown Collaborator; recall the immense network of correspondents that Darwin drew on.

XXIII

The anatomies. It's hard to know exactly how many different kinds of animals A. dissected, but LONES (1912) pp. 102–6 suggests forty-eight spp., which is surely too generous since he includes the elephant and other animals about which A. is pretty vague. A. talks about dissecting the mole or *aspalax* at HA 491b28; see Glossary 11. His *tour de force* of cuttlefish anatomy is given at HA IV, 1. A. refers to the diagram of the dissected cuttlefish at HA 525a8. He frequently appeals to diagrams or tables in his works as discussed by NATALI (2013) ch. 3, 3.

XXIV

Human internal anatomy. A. says that we should first understand the parts of humans at HA 491a20. LLOYD (1983) ch. I, 3 discusses man as a model and gives a list of features that A. claims are unique to humans, but notes that this list is qualified in various places, e.g. when A. considers apes. A. speaks of the obscurity of the internal anatomy of humans at HA 494b19. He does specifically refer to the shape of the human stomach and spleen, HA 495b24 and HA 496b22, but otherwise there is little evidence to suggest he dissected a human corpse. LEWES (1864) pp. 160–70 discusses whether A. dissected a man and, p. 157, unfairly dismisses A.'s dissections by arguing that his skills are inferior to those of a modern-day anatomist; COSANS (1998) gives a more sympathetic account. LLOYD (1973) ch. 6 and LLOYD (1975) discuss Erasistratus and Herophilus and Alexandrian dissection. On A.'s claim that the human uterus is bipartite see HA 510b8 and OWEN (1866) vol. 3, pp. 676–708; on the number of human ribs, see HA 583b15, and on why he may have been mistaken see LEWES (1864) pp. 155–70, OGLE (1882) *n.* PA I, 5. None of the common domestic mammals that A. might have seen has eight pairs of ribs. On kidneys of domestic animals see OWEN (1866) vol. 3, pp. 604–9, SISSON (1914) pp. 564–70. A. describes a human foetus at HA 583b14. His excellent, if partly faulty, cardiovascular anatomy is given at HA III, 2–4; cf. HA 496a4 and PA III, 4. He refers to his predecessors: Syennesis at HA 511b24, Diogenes at HA 511b31 and Polybus at HA 512b12. For A.'s relationship to the Hippocratics see OSER-GROTE (2004). A.'s claim to authority in his dissections: HA 513a13, cf. PA 668a22; HA 496a8; PA 668b26. When in PA he refers to 'Dissections' or 'Anatomies' I assume he's referring to books, but he may just mean general studies, LENNOX (2001a) pp. 179, 257, 265. There is an enormous literature on the veracity of A.'s account of the cardiovascular system, in particular why he thought mammals have three-chambered hearts; some of the more important discussions are: HUXLEY (1879), OGLE (1882) pp. 193–6, THOMPSON (1910), *n. HA* 513a30, LONES (1912) pp. 136–47, HARRIS (1973) pp. 121–76, COSANS (1998), KULLMANN (2007) pp. 522–51. A. refers to capillaries at *HA* 513b21, *HA* 514a23 and *PA* 668b1.

XXV

How good is the descriptive zoology? A. describes the urogenitary anatomy of live-bearing tetrapods at *HA* 506b26. BOJANUS (1819–21) illustrates the classic bean shape and modular structure of the kidneys of a tortoise. A. tells of the sea urchins at the *euripos Pyrrhaiōn* at *HA* 544a20. The edible sea urchin is *Paracentrotus lividus*. A. claims that the edible sea urchin may be recognized by the seaweed and other debris that it carries in its spines (*HA* 530b16); in the Aegean, only *P. lividus* does this, though why it does so is something of a mystery, CROOK *et al.* (1999). Even today the natives of Lesbos hunt only decorated urchins – though the inedible, undecorated *Arbacia lixula* is more common. A. describes the structure that would come to be known as 'Aristotle's Lantern' at *HA* 531a3, cf. *HA* 530b24; LENNOX (1984) argues that the part we now refer to as 'Aristotle's Lantern' is only part of what he intended by the simile, but VOULTSIADOU and CHINTRIROGLOU (2008) have clarified the whole matter with a picture of an ancient lantern. For the woodpeckers nesting in olive groves see *HA* 614b11 and, more generally on woodpeckers, Glossary II. CUVIER (1841) vol. I, p. 132 commends A.'s zoology; LEWES (1864) pp. 154–6, BOURGEY (1955) and LLOYD (1987) p. 53 collect similar passages from zoologists past. HALDANE (1955) and BODSON (1983) among others have called for a systematic examination of the quality of A.'s empirical work in the light of modern biology – it still needs to be done.

XXVI

The caring catfish. A. describes parental care in his catfish at *HA* 621a21, its development at *HA* 568a20 and its anatomy at *HA* 490a4, *HA* 505a17 and *HA* 506b8. CUVIER and VALENCIENNES (1828–49) vol. 14, bk 17, ch. 1, pp. 350–1 identified the *glanis* as *S. glanis*; AGASSIZ (1857) proposed the name *S. aristotelis* but did not formally describe it; GARMAN (1890) did. AGASSIZ, GARMAN, HOUGHTON (1873) and GILL (1906), GILL (1907) all repeat the story, but none of them appears actually to have seen male *S. aristotelis* build a nest and guard its eggs;

however I. Leonardos, University of Ioannina (pers. comm. 2010) confirms A.'s facts and adds that the juvenile fish are slow growing. I thank him for this information. At HA 607b18 A. describes parental care in another fish, the *phykis*, and claims that it is the only marine species to do this. Its identity is uncertain, but A. is wrong to suppose that there is only one kind of nest-building marine fish, since several wrasses, gobies and blennies in Aegean waters build nests and guard their young. He speaks of the characters of animals at HA 608a1.

XXVII

The hectocotylus. A. speaks of the *nautílos* at HA 525a19 and HA 622b8. OWEN (1855) pp. 630–1 gives the early history of the discovery of the hectocotylus. A. describes the tentacles of the male octopus at HA 524a4 and HA 541b8 and the octopus' breeding habits at HA 544a8 and GA 720b32. LEWES (1864) pp. 197–201, sour as ever, pours scorn on the idea that A. had seen the hectocotylus, but Lewes was wrong, for STEENSTRUP (1857) and FISCHER (1894) demonstrated what A. saw. THOMPSON (1910) illustrates A.'s passage with an elaborate hectocotylus belonging to a species he could not have seen; the real thing, in *Octopus vulgaris*, is much more subtle.

XXVIII

The reproduction of sharks. A. describes selachean reproductive anatomy at HA VI, 10–11; cf. HA 511a3 and GA III, 3. The famous description of the smooth dogfish's placentation is at HA 565b4; cf. GA 754b28. See MÜLLER (1842), COLE (1944), THOMPSON (1947) pp. 39–42 and BODSON (1983) for the history of A.'s smooth dogfish. On the *batrakhos*, its identity and reproduction see HA 505b4, HA 564b18, HA 570b29, GA 749a23, GA 754a26, GA 754b35, GA 755a8, , GA 749a24. THOMPSON (1940) p. 47 gives his summary of A.'s accomplishments.

XXIX

Natures. Schiller on nature is quoted by THOMPSON (1940) p. 39, but the source is the essay *On Simple and Sentimental Poetry*, 1884. Alcaeus' verse is translated by BARNSTONE (1972) pp. 56–8. The Homeric quote is from *Odyssey* X, 302–3 [trans. MURRAY (1919)]. Democritus refers to natures at DK 68B33 [trans. BARNES (1987)]. LLOYD (1991) ch. 18 discusses the social context of the 'invention of nature' in ancient Greece. A. defines natures at *Metaph* IV, 4 and *Phys* II, 1; see LEAR (1988) pp. 16–17 for an introduction to Aristotelian natures. A. asserts the self-evident quality of natures at *Phys* 193a3.

XXX

The materialists. Plato is said to have wanted Democritus' books burnt at *DL* IX, 38–40. A. by contrast wrote a book about D. which evidently gave a synopsis of the latter's physical theory and its implications for biology: *FR* F208R3. A. repeatedly attacks the materialists, e.g. at *Phys* II, 4–8, *Metaph* I, 3–4, *DA* I, 2–3 and *PA* 640b5. At the heart of his assault is the notion of 'spontaneous', which is the single word that I use for two of A.'s: *automaton* and *tychē*. Both words refer to events or phenomena that appear to be the product of a purposeful agent but aren't. They differ from each other in that *tychē* (often translated as 'luck') could be, but isn't, due to human intelligence while *automaton* (often translated as 'spontaneous', 'the automatic', 'the fortuitous') could be, but isn't, due to *any* purposeful agent, e.g. the desires of some animal. So *automaton* is the more inclusive term. Both words are sometimes translated as 'chance', but that suggests the outcome of a probabilistic process such as flipping a coin which is not what A. has in mind here. I use 'spontaneous' for both since: (i) I do not treat human agency; (ii) A. doesn't consistently distinguish them either; (iii) 'spontaneous' seems to capture the idea of a determinate but undesigned outcome. Finally, it should be noted that A. also uses *automaton* in a different way when describing spontaneously generated animals; see Chs LXXVI–LXXVIII.

Empedocles & selection. E.'s theory of mixing is given at *DK* 31B8 and is quoted by A. at *Metaph* 1015a1. E.'s zoogony can be reconstructed from the following fragments: tissue formation, *DK* 31B96, *DK* 31B98; body parts, *DK* 31B57; random combination, *DK* 31B59. Much about the theory is opaque, in particular whether Love and Strife are to be understood as intrinsic properties of elements, external physical forces or divine powers, or all three. Simplicius analyses E.'s account in his *Physics* 371.33–372.11 [trans. LONG and SEDLEY (1987)]. Although a clear statement of the principle of selection, E. did not imagine continuous evolution. A. also attributes embryonic selection to E. in his critique of preformationism (see ch. LXVII), but it's uncertain whether or not E. did, in fact, believe this – though he did seem to recognize that monsters still occur in our time; see SEDLEY (2007) pp. 31–74. CAMPBELL (2000) claims to detect evolution by natural selection in the Hippocratic *On Ancient Medicine* 3.25. However, although this text clearly discusses selection (by diet) it's unclear whether the more robust individuals transmit their tougher constitutions – that is, evolve. The case for selection in Epicurus/Lucretius is much more convincing; see CAMPBELL (2000) and SEDLEY (2007) pp. 150–5. In *Phys* II, 8 A. expresses himself more abstractly than I have here, but it's clear that he has organic development in mind. LLOYD (1970) ch. 4 and SEDLEY (2007) chs II, V discuss the Pre-Socratic materialists in general; for the texts and commentary see BARNES (1982) chs XV–XX. For Democritus' atomism see BARNES (1982) p. 377. For A.'s critique of the materialists see NUSSBAUM (1978) pp. 59–99, WATERLOW (1982) ch. II and JOHNSON (2005) chs 4, 5.

XXXI

The origin of teleological explanation. A. commends Anaxagoras at *Metaph* 984b15, cf. *DA* 405a20, and then criticizes him, *Metaph* 985a19. Socrates–Plato criticizes Anaxagoras, *Phaedo* 98B–99C; see JOHNSON (2005) pp. 112–15. On the eighteenth-century origin of the term 'teleology' see JOHNSON (2005) p. 30. PALEY (1809/2006) p. 24 sings the praises of the eyelid, so does Socrates according to *Mem* I, 4.6 [trans. DAKYNS (1890)]; for A. on eyelids see *PA* II, 13. For Socrates as the origin of the Argument from Design see JOHNSON (2005) pp. 115–17 and SEDLEY (2007) pp. 78–92. For Plato on the good and the divine *Tim* 29A, *Tim* 30A and *Rep* 530A. For P. on human craftsmen see *Gorgias* 503D–504. P.'s zoology is given in *Tim* 72D–73, *Tim* 74E–75C. For P. on the digestive tract see *Tim* 73A, and on the transformation of fingernails into claws, *Tim* 76D–E. P.'s dislike of materialism is evident at *Laws* 889A–890D. LENNOX (2001b) ch. 13 discusses P.'s unnatural teleology. LLOYD (1991) ch. 14 gives a less jaundiced view of P.'s science than I do.

XXXII

Teleology. 'We all say x is *for the sake of*': *PA* 641b25; see GOTTHELF (2012) pp. 2–5 for other uses of the phrase or its grammatical relations. There is a huge literature on A.'s system of teleological explanation; here is a selection of important recent monographs and collections of essays weighted towards biology: KULLMANN (1979), GOTTHELF and LENNOX (1987), LENNOX (2001b), QUARANTOTTO (2005), JOHNSON (2005), LEUNISSEN (2010a), GOTTHELF (2012). A. speaks of automatic puppets and living things, *MA* 701b2; see Ch. LIX. A. compares artefacts and living things, *Phys* II, 8, *PA* I, 1 and *Metaph* VII, 7, but argues against an intelligent craftsman at *Phys* 199a8 and *Phys* 199b30. A. denies Plato's teleology at *Metaph* 988a7, but P. does use 'for the sake of which' when speaking of generation at *Philebus* 54C; see JOHNSON (2005) pp. 118–27. A. on the workings of the digestive tract, *PA* 675b23; for a parallel argument with respect to reproductive morphology, see *GA* 717a21. A. on the purpose of the body, *PA* 645b15.

XXXIII

Forms. Plato outlines his Intelligible Living Creature and its relationship to subordinate Forms at *Tim* 30C–31A; CORNFORD (1997) pp. 39–42. A. critiques Platonic Forms at *Metaph* I, 9. A. speaks of many *eidē* of birds and fishes at *HA* 486b224. THOMPSON (1910) *n.* 490b16 was one of the first scholars to point out that A. uses *eidos* in several different ways, not all of which are consistent with translating it as 'species'. BALME (1962a) and PELLEGRIN (1986) later developed this line of interpretation further, wielding it against the idea that A. is engaged in a taxonomic project. A. uses the term *atomon eidos* at *PA* 643a13, *Metaph* 1034a5, *DA* 415b6 and *HA* 486a16. Even here there is a debate about whether *atomon eidos* refers to individuals or species or both – A. is by no means clear. Some, e.g. BALME (1987d), HENRY (2006a), HENRY (2006b), have argued that he means individuals, but I find GELBER (2010)'s argument that he usually does mean species, i.e. that two individuals can share the same indivisible form, convincing. This interpretation has consequences for reading A.'s theory of inheritance, for I am thereby compelled to invoke an additional, sub-specific level of heritable variation, which I call 'informal' variation; see Chs LXX and LXXIII. A. explains what forms are by means of the carpenter analogy at *PA* 641a6 and the syllable theory of form at *Metaph* VII, 17. DELBRÜCK (1971) argued for the interpretation of form as information and many have followed him, e.g. FURTH (1988) pp. 11–120, KULLMANN (1998) p. 294 and HENRY (2006a), HENRY (2006b); but see DEPEW (2008) for a different view.

XXXIV

The four ways of explaining. A. frequently states his four basic causal explanations, e.g. *GA* 715a4, from which the quote comes, and *Phys* II, 3 and *PA* 642a2. PECK (1943) pp. xxxviii–xliv, LEUNISSEN (2010a) and LEUNISSEN (2010b) give general discussions of A.'s system of causal explanation. LEAR (1988) pp. 29–31 explains how A.'s 'causes' differ from Hume's. The influence of A.'s division of causal explanations on the history of biology is one of the great themes of RUSSELL (1916), a classic. HUXLEY (1942), MAYR (1961) and TINBERGEN (1963) give the different kinds of causal explanations in modern biology; Mayr cites A. explicitly, Tinbergen does not; see also DEWSBURY (1999).The main difference between their list of causal explanations and A.'s is that A.'s does not have an evolutionary dimension where theirs do. Among those who have seen A. as a mere synthesizer or a species of Platonist are POPPER (1945/1962) vol. 2, ch. 11 and SEDLEY (2007) pp. 167–204 – but the tradition is an ancient one and underlay the entire Neoplatonist project.

XXXV

The Bird Hall & how to carve up nature. For a history of the Natural History Museum and its exhibits see STEARN (1981). LENNOX (2001b) ch. 2 discusses A.'s options in arranging his data in much the same spirit as I do here.

XXXVI

A. as a natural historian. Theodore of Gaza's preface to his edition of A.'s zoology, GAZA (1476), is given by PERFETTI (2000) p. 16, who also discusses Pliny's influence on him. BEULLENS and GOTTHELF (2007) discuss the dating and structure of Theodore's *HA*. Pliny on the elephant is from *Plin* VIII, 1, 13, 32 [trans. RACKHAM *et al.* (1938–62)]. The view that A. was not doing natural history is a commonplace among A. scholars. FRENCH (1994) disagreed, but his view of A. as a natural historian was inconsistent by his own criteria.

XXXVII

A. as a taxonomist. CUVIER (1841)'s encomium to A.'s skills as a taxonomist is quoted by PELLEGRIN (1986) p. 11. Modern Greek fish names are given in KOUTSOGIANNOPOULOS (2010) – essential for those interested in Greek fishes, but so far available only in Greek. A. speaks of the several breeds of dogs at *HA* 574a16, the *hippos* crab at *HA* 525b7 and the *kyanos* at *HA* 617a23. He writes a great deal about cephalopods, notably their anatomy in *HA* IV, 1 (see this text Ch. XXIII) but also in many other passages besides. On the paper nautilus see *HA* 622b8; cf. *HA* 525a19 and Ch. XXVII. The mystery cephalopod 'that lives in its shell like a snail' is described – barely – at *HA* 525a26; see SCHARFENBERG (2001). A. talks about the other kinds of crabs at *HA* 525b6; cf. *PA* 683b26. DIAMOND (1966) describes the ability of New Guinea highlanders to discriminate bird species; ATRAN (1993) discusses folk taxonomies in general; he also has a useful chapter on A.'s systematics.

The greatest kinds. A. gives major statements on greatest kinds: blooded animal, *HA* II; bloodless animals, *HA* 490b7, *HA* 523a31 and other passages in *HA* IV. He says that names such as *ornithes* and *ikthyes* are vernacular at *PA* 644b5; cf. *PA* 643b9 Some of his new technical names (e.g. *malakostraka*) are actually 'name-like expressions', i.e. shorthand descriptions that substitute for a noun: cf. *APo* 93b29–32; PECK (1965) pp. lxvii, 31 and LENNOX (2001a) p. 155. It's likely that the use of such names began at the Academy; Speusippus, it seems, used *malakostraka* and was interested in definition; see WILSON (1997). A.'s hierarchy of *genê* is very incomplete, but he does not allow any given subordinate kind to exist in more than one position in the hierarchy; cf. *Top* IV, 2. Many *genē* are, at best, classified into the *enhaima* or *anhaima*, e.g. humans come under no *genos* other than the *enhaima*, *HA* 490b18. The following passages support A.'s commitment to hierarchical classification: on blooded animals, *HA* 505b26; bloodless ones, *HA* 523a31, *HA* 523b1; soft-shells, *PA* 683b26, cf. *HA* 490b7. He says that in a classification each animal should appear only once at *PA* 642b30 and *PA* 643a8. Borges (1942) 'El idioma analítico de John Wilkins' [The analytical language of John Wilkins] in BORGES (2000) p. 231, tells of the (apocryphal) Chinese encyclopaedia. A.'s orthogonal classification of polities is given at *Pol* III, 7. This classification is the result of the method given in *Pol* IV, 3. Indeed, there A. explicitly compares classifications of states to classifications of animals and recommends that we take all the varieties of organs – of states or animals – and array them orthogonally: 'There *must* be as many forms of government as there are arranging the offices.' But this is exactly what he does not do when classifying animals, for that procedure would necessarily lead to empty classes. For example, suppose you were to classify animals by two kinds of features: oral organs (teeth *v.* beaks) and dermal organs (hair *v.* feathers). An orthogonal classification would yield four classes of animals: (i) toothed-hairy, (ii) beaked-hairy, (iii) toothed-feathered, (iv) beaked-feathered. Of these, (i) is the mammals, (iv) is the birds, while (ii) and (iii) do not exist, allowing that the 'duck-billed' platypus does not, in fact, have a beak. This shows that orthogonal classifications are inefficient for they do not reflect the real covariance structure among biological entities. I think A. originally considered them, recognized their absurdity and abandoned them, perhaps when he started to do biology. In fact he doesn't even follow his recommendation for an orthogonal classification in *Pol* since later he subdivides the *genē* given at *Pol* III, 7 further so that his full classification of political constitutions is also, in fact, nested.

The meanings of genos. *Metaph* V, 28; see PELLEGRIN (1986) ch. 2.

The method of division. Plato defines kings at *States* 257–68e. He says, *States* 266D, that it is amusing that the king has been 'running a race with the man who is of all men best trained for living an easy life'. In his divisional scheme the sister-group of 'herders of featherless bipeds' – kings – are 'herders of feathered bipeds' – goose-herds, evidently an undemanding job. In *Metaph* VII, 12 and *APo* II, 5, 13, 14 A. follows P.'s method of division, but introduces some technical modifications; his critique becomes more extensive in *PA* I, 2–3; see BALME (1987b), LENNOX (2001a) pp. 152–472. Some scholars claim that the objective of A.'s division was definition rather than classification, but in *PA* I, 4 it's clear that he's interested in identifying kinds and works with them; see LENNOX (2001a) pp. 167–9 and *n.* this text Ch. XLI. P. says that 'we shouldn't cut across the joints . . .' at *Phaedrus* 265E.

XXXVIII

Taxonomic methodology. For a list of *diaphorai*, or differentiating characters, see HA I, 1. On 'the more and the less' as distinguishing regular kinds within greatest kinds of birds see PA 692b3, cf. HA II, 12–13, and, in general, HA 486b13, HA 497b4 and PA 644a13; LENNOX (2001b) ch. 7. A. gives his animal geometry at IA 4, PA 665a10 and HA 494a20. Man stands alone: HA 490b18, HA 505b31. A. discusses cuttlefish and gastropod geometry at HA 523b22, PA IV, 9, IA 706a34; see also this text Chs XCI and XCVII; for his plant geometry see IA 706b5, LBV 467b2, *Phys* 199a26 and PA 686b35. A.'s theory of analogues is given at HA 486b18, HA 497b11 and PA 644a22. Some see A.'s usage as being close to OWEN (1843)'s definition of *analogy* as 'a part or organ in one animal which has the same function as a part or organ in another animal'. LENNOX (2001a) p. 168 points out correctly that A. often doesn't make the functional similarity explicit, though he certainly does sometimes (e.g. hearts and heart-analogues). On analogues see LLOYD (1996) ch. 7, LENNOX (2001b) ch. 7 and PELLEGRIN (1986) pp. 88–94 who claims that *analogon* does not serve a classificatory function, but I find his arguments unconvincing. For the cephalopod 'brain' see PA 652b24, HA 494b28 and HA 524b4; LENNOX (2001a) pp. 209–10. RUSSELL (1916) p. 7 and BALME and GOTTHELF (1992) p. 120 agree that A. has an implicit concept of homology; however, it should be noted that parts that are the same 'without qualification' has multiple meanings, LENNOX (2001b) ch. 7. A. compares the skeletons of snakes and snake and egg-laying tetrapod at HA 516b20 and PA 655a20, says that snakes are like footless lizards at HA 508a8 and PA 676a25 and speaks of seals at HA 498a32, PA 657a22 and PA 697b5.

XXXIX

Polythetic classification. A. considers how to divide up some land animals at HA 490b19. This passage occurs in the middle of a discussion on the greatest kinds so seems to be aimed at delineating them. He says that snakes are a *genos* at HA 490b23 and HA 505b5. He points to the need to consider many features simultaneously when dividing at PA 643b9. For polythetic classification and its history see BECKNER (1959), and MAYR (1982) pp. 194–5. MAYR (1982) p. 192, LENNOX (2001a) pp. 165–6, 343 and LENNOX (2001b) ch. 7 agree that A. used polythetic classification. On the ostrich see Ch. XXI. A. discusses three kinds of apes at HA 502a34 and PA 689b31 (a possible fourth is mentioned elsewhere). The *pithekos* is a dualizer not because of convergent evolution but because it falls between two divergent kinds (tetrapods and humans). This is a consequence of A.'s refusal to place humans where they should go, among the *zōotoka tetrapoda*; he has torn a natural kind apart and done so for no reason other than a belief in the specialness of humans (see Ch. XCVII). On dualizers see LLOYD (1983) ch. I, 4 and LLOYD (1996) ch. 3.

XL

Dolphins. Herod I, 24 tells the tale of Arion. For accounts of the dolphin in antiquity, and riders in particular, see THOMPSON (1947) pp. 54–5. A. speaks of paedophilic dolphins at *HA* 631a8 and the features of cetaceans in general at *JSVM* 476b12, *HA* 589a33, *PA* 655a15, *PA* 669a8 and *PA* 697a15. Pliny talks nonsense about dolphins at *Plin* IX, 7–10.

XLI

What is the project of HA? MEYER (1855), BALME (1987b) (a revised version of a 1961 paper) and PELLEGRIN (1986) successively attacked the view that A. constructs a classification or even wants to, and this has become, to varying degrees, dogma. But A. clearly does construct a classification, and uses it, even if it is very incomplete and not his primary objective; see LLOYD (1991) ch. I, LENNOX (2001a) p. 169 and GOTTHELF (2012) ch. 12. A. says why classifications are useful at *PA* 644a34. The order in which he will arrange his information in *HA* is given at *HA* 487a10. My synopsis of *HA* is based on the order of the books given by BALME (1991) rather than D'Arcy Thompson and earlier authors who used the order imposed by Theodore of Gaza; see Balme's Introduction for a discussion of authenticity, plan and order of the component books of *HA*. Balme also makes the case that, among the zoological works, *HA* was not written first. Indeed, it's likely that they were all updated and more or less integrated with each other over A.'s lifetime, and perhaps afterwards by his successors, so that it is now very difficult to discern the order of composition. A describes the ruminant stomach at *HA* 507a32. The view that *HA* provides the basic material for a demonstrative science is a commonplace among scholars of A.'s zoology.

XLII

The need for demonstration. A. alludes to the purpose of *HA* as material for demonstration at *HA* 491a12, cf. *PA* 639a13, *PA* 640a1 and *GA* 742b24; see LEUNISSEN (2010a) ch. 3.1.

XLIII

Demonstration & the syllogistic. The following works, arranged in order of increasing difficulty, discuss A.'s logic and theory of demonstration: BARNES (1996) chs 7–8, ACKRILL (1981) chs 6–7, ROSS (1995) ch. II, ANAGNOSTOPOULOS (2009c), BYRNE (1997) and BARNES (1993), the last of which really requires a grasp of formal logic. What it takes to have scientific knowledge of something: *APo* 71b9. The conditions of demonstration are specified as follows: the premises must be true and immediate, *APo* 71b9; must concern universals, *APo* 71a8, *APo* 73b25, cf. *DA* 417b21, *Metaph* 1036a2, *Metaph* 1039a24, *Metaph* 1086b32; we must have better knowledge of them than the conclusions, *APo* 72a25. For the story of how the stickleback lost its pelvic girdle, see SHAPIRO *et al.* (2004) and CHAN *et al.* (2010). *Gasterosteus aculeatus* ranges into Greece, but there is no Aristotelian fish that can obviously be identified with it. A. used 'definition' in several different ways, see *APo* II esp. *APo* 94a11; here I mean 'conclusion of the demonstration of what something is' [trans. BARNES (1993)]. For the role that such causal definitions play in his science see Ch. XLVII. Teleological demonstrations, e.g. *PA* 640a1, are discussed by LLOYD (1996) ch. 1, LEUNISSEN (2010a). A. discusses the need for primary definitions at *APo* II, 19; see GOTTHELF (2012) ch. 7 for other examples. BYRNE (1997) pp. 207–11 discusses A. *v.* the sophists.

XLIV

Problems with, and critiques of, A.'s theory of demonstration. A. is certainly not oblivious to the problems of (i) falsely inferring causes from associations, (ii) falsely inferring the direction of causation, and (iii) multiple causation. In *APo* I, 13 he distinguishes between the 'fact', by which he seems to mean just the association proved by the syllogism, and the 'reasoned fact' by which he seems to mean the association + some other information that will convince us that there is indeed a causal connection, and which way the causal connection runs. In short he seems to argue, reasonably, that some other source of information, external to the syllogism, will show that there is a causal connection and what it is, but his discussion is not very clear; see LENNOX (2001b) ch. 2. The question of why A.'s works aren't arrayed in syllogistic form has led to much discussion. BARNES (1996) pp. 36–9 lays out one solution, but Kosman, quoted in GOTTHELF (2012) ch. 7, says that the issue is a red herring for nowhere does A. say that science should be *presented* in this way. I think he doesn't present his science syllogistically because he can't. True, *APo* I, 30 claims that demonstrations *can* involve relations that are 'for the most part', but this seems to violate his universality requirement; BARNES (1993) p. 192 and

HANKINSON (1995) discuss the difficulties. A.'s ad hoc explanation for why camels don't have horns is given at *PA* 674a30; LENNOX (2001a) pp. 280–1. Demonstration blends into dialectic at *EN* 1145b2 [trans. modified from NUSSBAUM (1982)]. This leads us to the controversy over the degree to which A.'s official theory of demonstration is found in his biology. Some scholars think that the biology is richly informed by the official theory; others are more ambivalent and point to a diversity of methods of demonstration. Major discussions can be found in BOLTON (1987), LLOYD (1996) ch. 1, LENNOX (2001b) chs 1, 2, LEUNISSEN (2007), LEUNISSEN (2010a), LEUNISSEN (2010b) and GOTTHELF (2012) chs 7–9. A. discusses how to cope with multiple causes – divide and explain – at *APo* II, 13–18; LENNOX (2001b) ch. 1. He talks about dividing and explaining when prescribing for ocular ailments at *APo* 97b25 [trans. BARNES (1993)]; for its modern equivalent in cancer research see HARBOUR *et al.* (2010).

XLV

The functional beauty of birds. On the bird winds see *Meteor* 362a24. A. describes birds and their habits in *HA* VII, 3 and the *tyrannos* at *HA* 592b23. He speaks of the more and the less of bird features at *PA* 692b4 and the relationship between bird diversity and *bios* at *PA* 662a34, *PA* 674b18, *PA* 692b20, *PA* 693a11, *PA* 694a15, *PA* 694b12; cf. *GA* 749a35. See WILSON (1999) for guilds and functional groups in modern ecology. DARWIN (1845) p. 380 gives the famous passage about the birds of the Galapagos. A. speaks of how nature makes instruments to fit the function at *PA* 694b12.

XLVI

Teleology in the zoology. A. speaks of the primacy of final causes at *PA* 639b13 and *PA* 646a25; see LEUNISSEN (2010a) ch. 7.1, and the reasons that organisms reproduce at *GC* 338b1, *DA* 415a25 and *GA* 731b31. Strictly this argument applies only to (i) sublunary organisms (i.e. it excludes heavenly organisms); (ii) organisms that reproduce (i.e. it excludes spontaneous generators). Here, and elsewhere (Ch. XCVI), I argue that the beneficiaries of reproduction are forms; see LENNOX (2001b) ch. 6 for a somewhat different view.

XLVII

Explaining the elephant. A. discusses the elephant's trunk at *HA* 497b26, *HA* 536b20 and *HA* 630b26; he explains it at *PA* 658b34 and *PA* 661a26. There is some inconsistency between *HA* and *PA* concerning the lifestyle of elephants. In *PA*, the aquatic habitats of the elephant are strongly stressed; in *HA*, however, the elephant is not named among the amphibious animals, and although it clearly lives near rivers, it does not live in them, and is a poor swimmer; see LENNOX (2001a) p. 234, KULLMANN (2007) pp. 469–73 and, especially, GOTTHELF (2012) ch. 8 on A.'s analysis of the elephant. JOHNSON (1980) describes snorkelling elephants, but these days you can also see them on YouTube. At *PA* 659a25 A. argues that the elephant's legs are 'unsuitable for bending' but elsewhere – *HA* 498a8, *IA* 709a10, *IA* 712a11 – he suggests that they can bend, though the latter passages are, admittedly, unclear. Ctesias is often blamed for telling A. about the elephant's inflexibility, but no existing text corroborates this, BIGWOOD (1993). For the elephant's leg in history see TENNANT (1867) pp. 32–42; for the modern kinematics of the elephant's legs see REN *et al.* (2008), and for the aquatic ancestry of the elephant see GAETH *et al.* (1999) and WEST *et al.* (2003).

Association between form and lifestyle. At *HA* 487a10 A. gives a quite extensive list of ways in which the lifestyles of animals may differ from each other; however, he uses few of them in teleological explanations in *PA*. Perhaps this is because, as LENNOX (2010) notes, the list of differentiae in lifestyles in *HA* I, 1 is confounded by a list of activities. See also *Pol* 1256a18 for use of lifestyle, with an emphasis on diet. A. speaks of the adaptations of the frogfish and the torpedo ray at *HA* 620b10. Additional cases where A. explains diversity in forms in terms of lifestyle are: fish mouths and diet, *PA* 662a7, *PA* 662a31 and *PA* 696b24; insect wings and mobility and damage, *HA* 490a13, *HA* 532a19 and *PA* 682b12; land *v.* water animals, *PA* 668b35.

Conditional necessity. At *PA* 642a4 A. distinguishes two basic causes: the cause for the sake of which and the cause from necessity. He then goes on to distinguish two forms of necessity. One is 'conditional necessity' by which he means the features that a part must have if it is to function properly, the other is 'material necessity' by which he means the features of a part (or animal) that arise directly from the properties of the matter of which it is composed; cf. *PA* 639b24 and *PA* 645b15. In practice, these kinds of necessity are difficult to disentangle, and A. often doesn't indicate which he's talking about; see COOPER (1987) and LEUNISSEN (2010a) ch. 3.

XLVIII

The power of conditional necessity. Emphasizing the importance of *genē* to the explanations given in *PA* goes, it will be apparent, against the trend of those who believe that A.'s classification is of no importance, or does not exist; see GOTTHELF (2012) ch. 9 for an argument similar to the one I present here. A. says that the following parts or activities are part of the 'definition of substantial being' (GA 778a34) in the following *genē*: flight in birds, PA 669b10, PA 697b1 and PA 693b10; fishes, swimming, PA 695b17; birds, lungs, PA 669b10; fish, blooded, PA 695b17; birds, blooded, PA 693b2–13; blood or its absence, PA 678a26; animals, sensation, PA 653b19. GOTTHELF (2012) ch. 7 and LEUNISSEN (2010a) ch. 3.2 show the import-ance of such arguments for A. He just says that birds have beaks because 'nature has constituted them this way', PA 659b5, and that it is an 'odd and distinctive feature' of birds, PA 692b15. On the consequences of beaks for the alimentary tracts of birds see HA 508b25ff. and PA 674b22. OGLE (1882) p. 241, OWEN (1866) vol. 2, pp. 156–86 and ZISWILER and FARNER (1972) describe the diversity of avian alimentary tracts.

XLIX

Material necessity. BALME (1987d) discusses role of material necessity in A.'s explana-tory scheme. A. outlines the uniform parts, their composition and their functions, at HA III, 2–20 and PA II, 1–9; see LONES (1912) pp. 107–17 for a survey of A.'s knowledge of them. A. says that uniform parts are for the sake of non-uniform parts at PA 646b11; cf. PA 653b30 and PA 654b26. For their physiological relation-ships see Ch. CVII. A. mentions the sea urchin of the deep at HA 530a32 and explains its spines at GA 783a20; see THOMPSON (1947) p. 72 for its identification. He asserts that sea urchins in general are cold at PA 680a25. The Hippocratics and the medical writer Discorides appear to have used sea-urchin spines as a diuretic, PLATT (1910) *n.* GA 783a20. GUIDETTI and MORI (2005) analyse the functional properties of sea-urchin spines; MOUREAUX and DUBOIS (2012) demonstrate their plasticity. A. refers to the sea urchin of the deep as a distinct kind (*genos*) which would seem to imply heritable differences with respect to other sea urchins, but his explanation of the features he discusses is given purely in terms of environmentally determined features so cannot be associated with a differ-ence in *eídos* or inherited form. He uses *genos* in a similarly casual way elsewhere in his zoology (see Ch. LXXXII on bees).

L

The interaction of conditional and material necessity. A. describes the functional prop-erties of the snake's vertebral column at PA 692a1, how rays move at PA 655a23, the structure of the oesophagus at PA 664a32 and the penis at PA 689a20. Such examples are very close to the axe metaphor he gives when explaining conditional necessity in *PA* I, 1; see LENNOX (2001b) ch. 8 for other examples. A. describes the purpose of the epiglottis at *PA* 664b20; for a modern account see EKBERG and SIGURJONSSON (1982). A. discusses spleen and its purpose at *HA* 506a13, *PA* 666a25, *PA* III, 7; see LENNOX (2001a) p. 270 and OGLE (1882) pp. 207–8; the latter assesses his fairly accurate comparative data. The spleen is an example of 'indirect' or 'secondary' teleology, LENNOX (2001a) pp. 248–9, LEUNISSEN (2010a) ch. 4.3. MEBIUS and KRAAL (2005) review the modern view of spleen function. A. discusses gall bladders and bile at *HA* 506a20 and *PA* IV, 2. LENNOX (2001a) pp. 288–90 insists that the Greek *cholē* does not differentiate between 'gall bladder' and 'bile' and so just chooses to translate always as 'bile', but A.'s descriptions of the distribution of *cholē* in different animals seem to make more sense if we allow that sometimes he's talking about gall bladders and at other times about bile. OGLE (1882) p. 218 reviews the comparative distribution of gall bladders and concludes, again, that A.'s comparative anatomy is mostly sound. A. concludes that bile is useless at *PA* 677a16.

LI

The teleology of household economics. A. alludes to Aesop's fable about Momus at *PA* 663a34; the original can be found in Babrius' *Fables*, 59. A. argues the need for auxiliary teleological principles at *IA* 704b11, cf. *IA* 708a9, *IA* 711a18, but he lists only a few of them; FARQUHARSON (1912) *n.* 704b12 identifies many more. A.'s major statement on household economics is in *Pol* I, 2–9. He states and applies a series of economic principles in his zoology at the following places: (i) nature is 'like a good householder' at *GA* 744b12; see LEUNISSEN (2010a) ch. 3.2 who speaks of 'luxury parts'. (ii) 'Nature does nothing in vain' as applied to: fish eyelids, *PA* 658a8; tooth morphology, *PA* 661b23; mouth function, *PA* 691b25; fish don't have legs, *PA* 695b16; fish don't have lungs, *JSVM* 476a13; teeth, *GA* 745a32; males, *GA* 741b4; see LENNOX (2001a) pp. 231, 244, LENNOX (2001b) ch. 9. (iii) 'What nature takes from one place it gives to another' as applied to: selachian cartilage, *PA* 655a27, cf. *PA* 696b5; distribution of body hair, *PA* 658a31; absence of bladder in feathered and scaled animals, *PA* 671a12; teats in lions, *PA* 688b1; no tails in humans, *PA* 689a20; wings *v.* spurs, *PA* 694a8; spurs *v.* claws, *PA* 694a26; bird tails and legs, *PA* 694b18; the reason ducks have short legs, *IA* 714a14; the frogfish's funny shape, *PA* 695b12; life history, see ch. LXXXIII; see LENNOX (2001a) pp. 218–19 and LEROI (2010). (iv) 'Nature is not parsimonious . . .': *Pol* 1252b1 and *PA* 683a22. (v) Multifunctional parts: e.g. *PA* 655b6, see TIPTON (2002) and KULLMANN (2007) p. 444. A. discusses the function and formation of horns at *PA* 655b2, *PA* 661b26, but mostly *PA* III, 2; see OGLE (1882) pp. 186–91, LENNOX (2001a) pp. 246–50 and KULLMANN (2007) pp. 499–514. A. does mention some aggressive behaviour in animals when mating 571b1, but does not refer to stags using their antlers in male–male combat.

LII

The soul of the cuttlefish. A. describes cuttlefish spawning at *HA* 550b6, embryology at *HA* 550a10, and cephalopod copulation at *HA* 541b1, cont. *HA* 541b13. THOMPSON (1928) describes ancient and modern ways of catching cuttlefish.

LIII

Definitions of life. The various definitions can be found in SCHRÖDINGER (1944/1967) ch. 6, LOEB (1906) p. 1 and SPENCER (1864) vol. 1, p. 74; LEWES (1864) pp. 228–31 gives earlier definitions and a commentary on A.'s definition. A. gives his own definition at *DA* 412a14 [trans. modified from HETT (1936)].

LIV

Early conceptions of the soul. Patroclus' fate is described in *Iliad* XVI. A. calls the butterfly *psychē* at *HA* 551a14; see DAVIES and KATHIRITHAMY (1986) pp. 99–108. Plato's conception of the soul and argument for its immortality can be found at *Phaedo* 78B–95D, *Phaedrus* 245C–257B and *Rep* 609C–611C; see LORENZ (Summer 2009) for an account of early theories of the soul. A.'s early conception of the soul is given in the fragments of *Eudemus FR* F37R3–F39R3 and *Protrepticus FR* F55R3, F59R3, F60R3, F61R3. It is generally thought that A.'s conception of the soul changed radically over the course of his life, e.g. LAWSON-TANCRED (1986) pp. 51–2, but BOS (2003) gives a contrary view, on whom, however, see KING (2007). A. says knowledge of the soul is very important in *DA* 402a1 and considers his predecessor's views in *DA* I. He defines the soul as a first actuality of the body at *DA* 412b4 [trans. HETT (1936)], cf. *DA* 412a19, *DA* 412b4, *DA* 414a15, and speaks of seeds as potentially ensouled bodies at *DA* 412b27; see KING (2001) pp. 41–8. A.'s doctrine that the soul is an enmattered form is a special case of his theory of 'hylomorphism' – the idea that a substance (*ousia*) can be thought of as a compound of matter and form. He applies this theory to the soul at *DA* 412b6. This position is sometimes thought to conflict with this general hylomorphic theory which holds that form and matter are contingently related, ACKRILL (1972/1973). A. makes the soul responsible for change at *DA* I, 3; *DA* 415b21. Here I translate A.'s *kinēsis* (pl. *kinēseis*) as 'process' – by which I mean any time-dependent set of states – but it is more commonly translated as 'movement'. A. makes the soul goal directed at *DA*

II, 4 and speaks of it as an entity at DA 412b10, cf. DA 415b8, PA 640b34 and *Meteor* 390b31; he states its relationship to his 'causes' at DA 415b8. A. describes the eyes of moles at HA 491b28, HA 533a1 and DA 425a10.

LV

Spiritual interpretations of the soul. The 'ghost in the machine' is due to RYLE (1949) ch. 1. LAWSON-TANCRED (1986) p. 24 seems to view A.'s theory of soul through the lens of Cartesian mind–body dualism, but FREDE (1992), among others, show that A.'s theory is not Descartes'. A. speaks of the mysterious active intellect at DA 408b19 and DA III, 5. A. argues that souls are not agents at DA 408b11 and DA 408b25; more generally, DA I, 4. See the collections of essays in NUSSBAUM and RORTY (1992) and DURRANT (1993) for A.'s conception of soul, mental states and their relevance to the modern theory of mind. KANT (1793) ❡ 75, Ak. v, p. 400 despairs at explaining teleological processes; see GRENE and DEPEW (2004) ch. 4 on Kant's biology. As LENOIR (1982) points out, not all teleologists were overt vitalists; some were 'telomechanists'; however, a fascination with tele-ology often tips over into vitalism. DRIESCH (1914) gives a self-serving history of vitalism; CONKLIN (1929) p. 30 and SCHRÖDINGER (1944/1967) react to Driesch's vitalism; SANDER (1993a) and SANDER (1993b) give a sympathetic account of it; see also KULLMANN (1998) pp. 308–10. DRIESCH (1914) p. 1 and NEEDHAM (1934) pp. 30 ff. give an explicitly vitalist interpretation of A.'s biol-ogy; nowadays few scholars subscribe to this but FREUDENTHAL (1995)'s account of A.'s theory of *pneuma* often seems vitalist; see KING (2001) n. p. 141. Among the scholars who agree that A. is neither a vitalist nor a Democritean materialist are NUSSBAUM (1978), COOPER (1987), BALME (1987c), GOTTHELF (2012) ch. 1, KING (2001) ch. 3, KULLMANN (1998) ch. IV, QUARANTOTTO (2010). My own label of A. as an 'informed materialist' is merely a restatement of his hylomor-phism. A. identifies soul with form at DA 412b6 and DA 414b20 and the moving principle of life at DA 415b21.

LVI

The capacities of the soul. For the hierarchical capacities of the soul see DA 414a2. For the capacities of the nutritive soul see DA 415a22, DA 416b3 and DA 432b7. Living things defined by possession of a nutritive soul at DA 416b20; it's said to be found in all living things at DA 414a29 and DA 434a22. The nutritive soul is the first to appear in ontogeny, GA 735a12; see also ch. LXV. The soul holds living things together, DA 411b5 and DA 415a6; see QUARANTOTTO (2010). A. discusses metabolism at DA 416a33 and compares growth to the flow of a river at GC 321b24; cf. GC 322a22. This is similar to modern growth models, e.g. BERTALANFFY (1968) p. 180; A. also distinguishes between nutrition used for somatic maintenance and nutrition used for growth, e.g. GA 744b33, see n. PECK (1943) p. 232. A. speaks of chemical transformation at DA 416a21; see Ch. LVII. His account of digestion and assimilation in blooded animals is given at PA III. For an example of a modern energy budget see WARE (1982).

LVII

The chemistry of uniform parts. The proportions of elements specify, indeed begin to define, the uniform parts, PA 642a18 and Metaph 993a17. For Empedocles on the chemical constituents of bone see DK 31B96 and FURTH (1987) pp. 30–3. SOLMSEN (1960) p. 375 and KING (2001) p. 168, n. 12, are sceptical about expressing A.'s compounds in terms of actual ratios, but in addition to the passages cited above the idea of a numerical ratio is implicit in many others where he discusses the composition of various uniform parts, e.g. PA II, 4 for blood; PA 653a20 for brain; PA 654a29 for insect exoskeletons and GA 743a14 for nails. A. sometimes refers to uniform parts as also being composed of a 'hot substance', e.g. GA 743a14; he may mean *pneuma* – see Chs LIX and LXV. A. berates Empedocles for his theory of mixtures at GC 334a27; for the difference between mixtures and compounds see BOGAARD (1979). A.'s general theory of compounds is given at Meteor IV, 8, GC I, 10 and GC II, 7–8. I speak here of the uniform parts as varying in elemental proportions, but A. often couches his discussion of the composition of uniform parts in terms of contrary elemental *powers* (hot/cold, dry/wet) present and actually says these powers are more fundamental, e.g. PA 646a12. These powers are not isomorphic with the elements since each element is a combination of them, see Ch. LXXX, but, in fact, he often segues between talking about elements and their powers, e.g. GA 743a14; see WATERLOW (1982) pp. 83–6, SORABJI (1988) p. 70, KING (2001) pp. 74–80 and SCALITAS (2009).

The meaning of 'hot' and 'cold'. A. discusses the various meanings of 'hot' and 'cold' at PA II, 2. When reading him on heat there are at least three possible sources of confusion. (i) He does not distinguish very clearly between the roles of heat in 'cooking' and 'burning', i.e. between endothermic and exothermic reactions, but see PA 648b35. (ii) When he says that something is 'hot' he does not necessarily mean that it has a high temperature compared to its surroundings, but often means that it is easily altered by the application of heat – in other words, that it burns, melts or cooks easily, cf. *PA* 648b16, i.e. he is talking about something akin to its relative thermodynamic stability. Fat is 'hot' in this sense (though it may also have a high temperature). (iii) Finally, there's the question of 'vital heat'. FREUDENTHAL (1995) argues that it is a very exotic kind of 'informed heat' rather than regular heat that happens to be in a living thing and he ties this to *pneuma*, on which see Ch. LIX. Although vital heat is not the same as conventional fire, this is unnecessarily vitalistic and we may doubt that *pneuma* is really so important for it does not appear in A.'s adult nutritional physiology in *JSVM*, but only in his embryology and sensory physiology where it seems to be a vehicle of the soul that permits action at a distance; see KING (2001) for a discussion.

The role of heat in the workings of the nutritive soul. Animals are said to have an internal source of heat at *JSVM* 469b8; cf. *PA* 682a24. A. compares fire to a river at *JSVM* 470a3 [trans. HETT (1936)], but says that vital heat is not conventional fire, *GA* 736b33, even though he often speaks of an internal 'fire'. On the sufficiency of heat to effect transformation, see e.g. *Meteor* 390b2. For concoction and transformation see *Meteor* IV, 2–3, *DA* 416b28. When A. explains how heat produces various uniform parts he can be very confusing, e.g. *GA* 743a5. This is because he says that some uniform parts are formed by heating and others by cooling, and sometimes (e.g. flesh) are formed by both. The solution seems to be that the *blood* is heated, thereby separating it into hotter and colder components; the colder components then congeal into flesh or bone or other solid uniform parts, cf. *Meteor* IV, 7–8. Fire is said not to be the main cause of nutrition and growth at *DA* 416a9. A. emphasizes the necessity of regulating the internal fire at *JSVM* 469b10 and *JSVM* 474b10.

LVIII

The seat of the soul. A. speaks of vivisecting tortoises at *JSVM* 468b9 and *JSVM* 479a3; chameleons at *HA* 503b23, cf. *PA* 692a20; insects and plants at *DA* 411b19, *JSVM* 468a23, *JSVM* 471b20, *JSVM* 479a3 and *PA* 682a2; see LLOYD (1991) ch. 10. A. talks of the heart as the seat of the soul at *JSVM* 1, 3. Concoction and the internal fire in the heart are described at *JSVM* 469b10, its boiling action at *JSVM* 479b28. He says the heart is the citadel of the body at *PA* 670a25 and that it has supreme control at *JSVM* 469a5. His general account of the heart is given in *PA* III, 4; see KING (2001) pp. 64–73 on A.'s cardiocentrism. He asserts that only organs with blood (in blooded animals) are viscera at *PA* 665a28. His contrast between centralized and distributed souls is developed at *JSVM* 468b9, cf. *PA* 682a2, *PA* 682b30, *PA* 666a13. COSANS (1998) 'vivisected' a terrapin.

LIX

The structure of the sensitive soul. The CIOM model is from GREGORIC and CORCILIUS (2013) who would not, however, call the entire system the sensitive soul. This difference in interpretation arises from the tension, pervasive in A.'s writings, between the cardiocentric and hylomorphic accounts of the soul. Perception as transmission of form: *DA* 435a4. Empedocles' and Plato's theory of vision are given at *DA* II, 7 and *Sens* 2; A. has other anatomical arguments against it too, but it's hard to interpret them since his ocular anatomy is so hazy; see LLOYD (1991) ch. 10. A.'s theory of light and vision are given at *DA* II, 7 and *DA* 434b24. The precise nature of the change that occurs in the eyeball is controversial. Some scholars argue that it is a material change, others deny this. I am inclined to believe that it is a material change since it's hard to see how a non-material change could effect further physical changes; and this model is consistent with the plainly material changes that occur with touch-perception; see JOHANSEN (1997) for a discussion. A. identifies the heart as the sensorium at *PA* 657a28 and *JSVM* 467b27 and argues against the brain at *JSVM* 469a10, *JSVM* 469a20 and *PA* 656a15. His account of the communication between the sense organs and heart is given at *Sens* 2; see LLOYD (1991) ch. 10 and FRAMPTON (1991). GREGORIC and CORCILIUS (2013) p. 63 discuss the homeostatic role of the sensitive soul; see *DA* 431a8. Desire has to drink at *MA* 701a32; see NUSSBAUM (1978) Essay 5 and CASTON (2009) on *phantasia*. A. alludes to the higher cognitive processes involved in the perception of smells at *DA* 424b16. Pleasant and painful desires are discussed at *MA* 701b35.

Pneuma. It's hard to know what, exactly, A. thinks *pneuma* is since the whole theory seems to be rather poorly worked out. The problem is that A. first says *pneuma* is just 'hot air', *GA* 736a1, and then, just a few dozen lines later, says it's something more 'divine' than the basic terrestrial elements, indeed it appears to be analogous to the element of which stars are made, *aithēr*, *GA* 736b33, cf. *DC* I, 3. Between these mundane and exotic options scholars have found much to dispute, see PECK (1943) Appendix B, BALME and GOTTHELF (1992) pp. 158–65, FREUDENTHAL (1995) ch. 3 and KING (2001) ch. 4. For the role of *pneuma* in animal locomotion see *MA* 10. FRAMPTON (1991) and GREGORIC and CORCILIUS (2013) give slightly different accounts of the distribution of *pneuma* in the body and hence the extent of the connectivity problem; see also NUSSBAUM (1978) Essay 3. The communication between the heart and locomotor appendages, and the metaphor of the automatic puppets is given at *MA* 701b2 [trans. Nussbaum, 1978], but I have omitted references to 'little carts', another mechanical simile. PREUS (1975) p. 291 and LOECK (1991) discuss what A. might have meant by these devices. For the Greeks and muscles see OSBORNE (2011) pp. 39–40. Mechanical amplification in the rudder and city analogies can be found at MA 701b27 and MA 702a21. The whole CIOM model is put together with a diagram at MA 703b27 [trans. Nussbaum, 1978]. A. discusses the mental faculties of human beings in DA III, 3–4; I do not consider them further.

LX

The cybernetic soul. A.'s account of thermoregulation is mostly given in the book traditionally known as *de Respiratione*; following KING (2001) pp. 38–40 I include it in *JSVM*. A. discusses the need for cooling at *JSVM* 5 and the heart–lung cycle at *JSVM* 480a16; see KING (2001) pp. 127–9. He explains respiration in insects at *JSVM* 471b20, *JSVM* 474b25 and *JSVM* 475a29 and fishes at *JSVM* 480b19. The cybernetic interpretation of A.'s theory of the soul is originally due to NUSSBAUM (1978) pp. 70–4 and adopted to varying degrees by FREDE (1992), WHITING (1992), KING (2001), SHIELDS (2008), QUARANTOTTO (2010), MILLER and MILLER (2010), among others. For the history of homeostasis, cybernetics and systems biology see BERNARD (1878), CANNON (1932), ROSENBLUETH *et al.* (1943), WIENER (1948) – who, on p. 19, gives the etymology of governor/*kybernētēs*/cybernetics – ADOLPH (1961) and COOPER (2008). For a history of feedback control devices see MAYR (1971) and, more generally on Greek technology, BERRYMAN (2009). The relationship between teleology and goal-seeking behaviour is discussed by AYALA (1968) and RUSE (1989). 'Many of the characteristics of organismic systems . . .' is from BERTALANFFY (1968) p. 141. For the general properties of systems see SIMON (1996). 'Components come and go . . .' is from PALSSON (2006) p. 13. A. uses the steersman metaphor in another context at DA 413a8 and DA 416b26. He speaks of methodological reductionism at *Pol* 1252a17. Souls hold living things together at DA 410b10, DA 411b6 and DA 415a6; see QUARANTOTTO (2010) for further references and discussion.

LXI

The end of development. A. attacks Empedocles on the vertebral column at PA 639a20. He describes a spontaneously aborted human foetus at HA 583b14; at least some of the information in surrounding passages is Hippocratic; this may be too.

LXII

Mating behaviour. Most of A.'s information on mating in blooded animals is in HA VI, 18–37. Animals are excited by desire at HA 571b9. A. describes mating calls at

HA 536a11, pigeon courtship at HA 560b25, the wantonness of mares and cats at HA 572a9 and HA 540a9 respectively, and the reluctance of hinds at HA 540a4, cf. HA 578b5. He describes male–male conflict at HA 571b11. Males are initially defined at GA 716a14. A.'s initial definition of the sexes is anatomical and functional; later, at GA 765b13, he amplifies it with a physiological one; see MAYHEW (2004) and NIELSEN (2008). For the copulatory techniques of blooded animals see HA V, 2–6, GA I, 4; for how hedgehogs mate see GA 717b26, and for how fishes do it, GA 756a32.

Reproductive fluids. A. describes the origin of *sperma* at PA 651b15 and GA 725a21. Although I usually translate *sperma* as 'seed' – which could be either male or female reproductive residues – it's clear that sometimes A. uses it in the more restricted sense of 'semen', i.e. male residue, and I translate accordingly. A. discusses the formation of the menses at GA 738a10ff. and elsewhere; describes vaginal discharges at HA VI, 18–19, HA 582a34 and GA 738a5; see PREUS (1975), pp. 54–7, n. pp. 286–7. He conflates the menstrual and oestral discharges at GA 728b12. He discusses exceptions to his menstrual fluid model at GA 727b12 and GA 739a26. A. claims that wind eggs and fish roe are the avian and piscine equivalent of menses at GA 750b3. For a modern view on the distribution and function of menstruation see STRASSMANN (1996).

LXIII

Anatomy of generative organs. A. describes the external genitals of blooded animals at HA 500a33, HA III, 1, HA V, 5, HA 566a2 and GA I, 3–8. He describes the cloaca of the ovipara at GA 719b29. For anatid penises see BRENNAN *et al.* (2007); for penis construction in general see KELLEY (2002). A. explains the function of the testes at GA I, 4–7 and GA 787b20. He explains the absence of testes and penises in fish and snakes and other differences in male reproductive anatomy at GA I, 4–7. Here A. also tackles the question of why, if the business of animals is to reproduce, they would want to limit their sperm production at all. For a modern explanation of the looping vas deferens see WILLIAMS (1996) pp. 141–3. The anatomy of male generative organs in blooded animals is described at HA 510a13 and of females at HA 510b7 and GA I, 3, 8–17. Here A. also explains why the uterus is so variously arranged in different kinds.

LXIV

Female sexual desire. A. discusses sexual desire in girls and women at HA 581b12 and GA 773b25, the role of female pleasure during sex, its relationship to conception, the production of menstrual fluids and the production of vaginal lubrication at HA 583a11, GA 727b7, GA 728a31 and GA 739a29. He names the glans at HA 493a25. HA X is usually excluded from HA because it's devoted to causal explanation; it is even sometimes thought not to be Aristotelian at all; see BALME (1991), Introduction, p. 26 and NIELSEN (2008). The accounts of the mechanics of reproduction in HA X and GA are similar, but differ in two ways. In HA X A. argues that intercourse brings down female seed (= menstrual fluid) to a region in front of the uterus where it mingles with the male seed, but at GA 739b16 he denies it. Second, in HA X, A. argues that the female orgasm is needed to suck the mixture of seeds back up into the uterus, but in GA this is apparently not necessary. For a comparison of the two accounts see BALME (1991), n. pp. 487–9. For modern views on the function of the female orgasm, if any, see JUDSON (2005) *contra* LLOYD (2006). Montaigne's spurious quote is from his *Essays* III, 5. 783.

LXV

Fertilization. For *GA*'s subject as the moving cause of life see GA 715a12. A.'s set of sexual dichotomies between the male and female contribution to reproduction are known as his theory of 'reproductive hylomorphism', HENRY (2006b). Here are some typical passages in which he claims that males supply form and females matter: GA 729a9, GA 730a27, GA 732a1, GA 737a29, GA 738b9 and GA 740b20. The theory apparently conflicts with many different aspects of his mechanistic accounts, and I consider some of these conflicts in more detail below. See HENRY (2006b) for an entrée to the literature on how, or indeed if, these conflicts can be resolved.

Wind eggs. A. repeatedly returns to the subject of wind eggs. For wind eggs in birds in general see HA 539a31, HA 560a5, GA 730a32, GA 737a30, GA 741a16 and GA III, 1; for wind eggs in partridges see HA 560b10, GA 751a14 and once more at HA 541a27, which, however, appears to be an interpolation. I thank Chris McDaniel of Mississippi State University, and Tommaso Pizzari of the University of Oxford and Nick Willcox of Pheasants UK, for telling me about wind eggs. A. discusses possibly parthenogenetic fish at HA 538a18, HA 539a27, HA 567a26, GA 741a32, GA 757b22 and GA 760a8; see CAVOLINI (1787) and SMITH (1965) for hermaphroditism in the *Serranidae*. Interestingly, A. not only misses the dual gonads of these fishes, he also claims that functional hermaphrodites can't exist, GA 727a25.

The transmission of soul. A. speaks of the menses' potential for soul at GA 736a31, and how the semen is the animal potentially at GA 726b15 [trans. PECK (1943)]. The term translated here as 'potential' is, once again, *dynamis*; A. discusses the potential/ actual distinction extensively at GA II, 1; see PECK (1943) pp. xiix–lv. A. applies the carpenter analogy to the action of semen at GA 730b6. He gives zoological arguments against the physical transmission of seminal matter at GA 729a34 and GA 736a24; cf. GA 721a13. Besides these passages he describes grasshopper copulation at HA 555b18; see DAVIES and KATHIRITHAMY (1986) p. 81.

Pneuma *in reproduction. Pneuma* as found in semen, GA 736b33; its action in fertilization, GA 737a7, GA 741b5. A. alludes to the homonymy of *Aphros*/Aphrodite at GA 736a19. Semen as foam is an early idea and appears in the Hippocratic corpus, *Littré* VII, *On Generation*, 1; see LONIE (1981), and in a fragment of Diogenes of Apollonia, DK 64B6. See COLES (1995) for a discussion of fifth-century models of reproduction.

LXVI

Descriptive embryology. On Hippocratic embryology see *Littré* VII, *On Generation*, 29; LONIE (1981) and NEEDHAM (1934) p. 17. A. describes the embryogenesis of the chicken at HA 561a7, cf. GA II, 4–6 and GA III, 1–2; THOMPSON (1910) *n. HA* 561a7 explains what A. is seeing and PECK (1943) p. 396 illustrates the various membranes. A. describes teleost embryology at HA 564b24 – see OPPENHEIMER (1936) – and mammalian embryology at GA 745b23 and GA 771b15. He thinks that mice and bats and hares also have cotyledonary 'uteruses', HA 511a28, but their placentas are now classified as discoidal. He discusses insect ontogeny at HA 550b22, GA 732a25 and GA 758a30; see DAVIES and KATHIRITHAMY (1986) p. 102; and compares viviparous and oviparous embryos at GA 753b31. He describes the relative perfection of embryos at GA 732a25, cf. HA 489b7, and in GA II, 1 A. argues that within the blooded creatures at least, the perfection of the offspring is associated with how much heat and moisture the parent has (cold/dry being least perfect and hot/moist being most perfect). This will become part of an arrangement of the animals into grades, a kind of *scala naturae*, that is orthogonal to his classification system, see Chs LXXXVII and XCVII. A.'s anticipation of von Baer's first law, BAER (1828), can be found at GA 736b2; see NEEDHAM (1934) p. 31 and PECK (1943) *n.* p. 166. For the embryological hourglass see KALINKA *et al.* (2010).

LXVII

Developmental mechanics. A. compares the effect of semen on the menses to the action of rennet and fig juice on milk at GA 737a11 and GA 739b21 [trans. PLATT (1910)]; cf. HA 516a4, GA 729a11, GA 771b23 and GA 772a22. A. also compares embryonic growth to the growth of yeast at GA 775a17; see PREUS (1975) pp. 56 and 77. NEEDHAM (1934) p. 34 draws attention to the fact that he is talking about enzymes and traces the fate of the cheese-making metaphor in, for example, the Book of Job. A. says that the heart develops first at HA 561b10, PA III, 4, JSVM 468b28, GA 734a11, GA 735a23, GA 738b15, GA 740b2, GA 741b15 and GA 742a16. He speaks of the yolk as the supply of nutrient in GA III, 2 and of the blood vessels as roots at GA 739b33. The pottery metaphor is from GA 743a10 and the furrow is from GA 746a18.

Epigenesis v. preformationism. A. argues against the Pre-Socratic preformationists at GA I, 17. PREUS (1975) p. 285 suggests that certain passages in Aeschylus' and Euripides' tragedies as well as Plato's *Symposium* are preformationist in the broad sense, but their embryology is sufficiently sketchy that you can read any theory you want into them. A more convincing case can be made for Anaxagoras DK 59B10 [trans. BARNES (1982)] and Empedocles; see BARNES (1982) pp. 332, 436–42. His own, epigenetic account is given in two metaphor-rich passages in which he compares the embryo to a painter, GA 743b20, and then to a net, GA 734a11. He asserts the homogeneity of semen at GA 724b21. The origin of each organ or uniform part in raw maternal material is given at GA 734a25. His *automaton*-causality is given at GA 734b9, cf. GA 741b8; see Ch. CIX for the role of these puppets in locomotion. *Automaton*-causality in embryogenesis appears to conflict with A.'s reproductive hylomorphism insofar as it gives a substantial formative role to the mother. PECK (1943) p. xiii simply accepts that maternal matter is 'informed to a high degree', but BALME (1987c) pp. 281–2, cf. BALME (1987d) p. 292, resolves the conflict by arguing that the *automaton* refers to movements in the semen and not the embryo. The *kordylos* is described at HA 589b22; cf. HA 490a4, JSVM 476a5 and PA 695b24. Thompson (1910) and PECK (1965) suggest this animal is a larval newt; OGLE (1882) p. 248 that it is a tadpole. He says that 'it is strange and yet, as it appears to me, indisputably true, that A. was perfectly ignorant of the fact that tadpoles are the larval forms of frogs and newts'. See also KULLMANN (2007) pp. 741–2 on the mysterious *kordylos*.

LXVIII

Embryology after A. The classic history of embryology is NEEDHAM (1934) who assesses the Renaissance 'macroiconographers' as well as Harvey's Aristotelianism, p. 118, on which see LENNOX (2006) too. Traditionally all theories that postulate that the embryo or its parts exist in the unfertilized material of its parents, be it the sperm or eggs, have been labelled 'preformationist', NEEDHAM (1934), and this is the sense in which I use the term; but see BOWLER (1971) and PYLE (2006) for more subtle distinctions among the various theoretical strands. NEEDHAM (1934) pp. 29–30 suggested that A.'s account of *automaton*-causality is an anomaly in what is otherwise a more or less vitalistic account of embryogenesis, but it is, in fact, at the heart of his account of embryogenesis – as the *kordylos* and sex determination show (Ch. LXXIII); also PECK (1943) p. 577. See PINTO-CORREIA (1997) and COBB (2006) for the elucidation of the role of semen and MAYR (1982) ch. 15 for the work of the mostly German microscopists of the nineteenth century.

LXIX

Variation under domestication. A. writes about sheep husbandry at HA 573b18, HA 596a13 and elsewhere; see THOMPSON (1932) on leader rams. He describes morphological variation in sheep at HA 496b25, HA 522b23, HA 596b4, in particular the Syrian sheep and the humped cattle at HA 606a13. Darwin's passage on the same subject is compounded from DARWIN (1837–8/2002–) 233e and DARWIN (1838–9/2002–) 12e.

LXX

Intra-specific/informal variation. For Darwin on pigeons see DARWIN (1859) ch. 1; for justification of the term 'informal variation' see *n.* Chs XXXIII and LXXIII. On hoofed pigs in A., HA 499b12, GA 774b15; in DARWIN (1868) vol. 1, p. 75 who cites A. For A. on domesticated *v.* wild animals see HA 488a30 and PA 643b5. A. mentions Ethiopians often, e.g. HA 517a18, HA 586a4, GA 722a10, GA 736a10, GA 782b35 and *Metaph* X, 9, but never says they're a distinct *genos*. In the *Politics*, e.g. *Pol* VII, 7, A. occasionally speaks of different *genē* of men, distinguishing Greeks from various non-Greeks. This seems to be a casual use of *genos* since elsewhere he is clear that the difference among men is due to a difference in environment not form. He will, occasionally, use *genos* in this more casual way, e.g. the *genē* of bees which are clearly reproductively linked and the sea urchin of the deep which differs from other sea urchins in its material aspects only. On Greeks *v.* barbarians: e.g. *Pol* 1252b5; see HANNAFORD (1996) pp. 43–57; SIMPSON (1998) p. 19 and this text Ch. XCIX.

The only domesticated breeds that A. distinguishes as 'kinds' are dog breeds, *HA* 574a16 and *HA* 608a27, which he seems to think are as different from each other as are wolves and foxes (cf. Theophrastus *CP* IV, 11.3); he accordingly treats crosses between them as hybrids: *HA* 607a1, *HA* 608a31, *GA* 738b27 and *GA* 746a29. A. makes his interest in informal (intra-specific) variation explicit at *LBV* 465a1, where he uses *eidos* in the sense of 'species'. For more about essentialism see Chs XXXVI–XXXVIII. For A.'s environmental determinism see *HA* 605b22, and specifically with respect to large reptiles in Egypt, *LBV* 466b21; small mammals in Egypt, *HA* 606a22; bees and wasps, *GA* 786a35; hair, *GA* 782a19; sheep fleece colour, *HA* 518b15. In *GA* V most of the variety that he considers cannot be teleologically or formally explained, but is the consequence of material necessity; see Gotthelf and Leunissen in GOTTHELF (2012) ch. 5. For Plato on selective breeding in animals and humans see *Rep* 459A, *Rep* 546A and POPPER (1945/1962) vol. I, pp. 51–4, 81–4 n. pp. 227–8, 242–6; for A. on the regulation of marriage see *Pol* VII, 16.

LXXI

Theophrastus on nature v. nurture. T. discusses early- v. late-sprouting wheat and other plants at *CP* I, 10.1–2 and *CP* IV, 11.1–7; the differences in environmental sensitivities between plants and animals at *CP* IV, 11.9; the effects of environmental factors on plant growth at *CP* II, 1–6, *CP* II, 13.1–5 and *HP* II, 2.7–12; and the waters at Pyrrha at *CP* II, 6.4. T. on nature v. nurture see *CP* IV, 11.7 [trans. EINARSON and LINK (1976–90)]. There is a fascinating cross-reference between T. and A. where each compares the influence of the soil on a plant to an animal mother's influence on her offspring – T.: *CP* I, 9.3 and *CP* II, 13.3, and A.: *GA* 738b28; see *n.* this text Ch. XCIV.

LXXII

A.'s model of inheritance. For the absence of an heredity theory of hair, eye, skin colour and hair type, see *GA* V. Much of my account of A.'s genetics is modelled on HENRY (2006a)'s insightful analysis of *GA* IV. Nevertheless, my interpretation differs from his in several ways: see below. A. discusses the inheritance of deformities at *HA* 585b29, *GA* 724a3 and teratology at *GA* IV, 3. The basic phenomena of inheritance are given at *GA* 767b1. His attack on pangenesis is in *GA* I, 17–18, specifically as applied to the children of deformed people at *GA* 724a4, cf. *GA* 721b28 [trans. PECK (1943)], as applied to plants, *GA* 722a13. MORSINK (1982) pp. 46–7 argues that A.'s target is the Hippocratic author of *On Generation* rather than Democritus – see *Littré* VII, *On Generation*, 3, 8, 11 for crippling and LONIE (1981). Morsink is surely right to suppose that A.'s opponent is the Hippocratic author but at GA 769a7 A. discusses two flavours of the theory, one of which may be Democritus' who may have held some such theory, DK 68B32, DK 68A141 and DK 68A143. DARWIN (1868) vol. II, ch. 27 gives his theory of pangenesis; PECK (1943), MORSINK (1982), HENRY (2006a) among others have applied C.D.'s term to A.'s theory. C.D. acknowledges ancient Ppangenesis in DARWIN (1875) 2nd edition, vol. II, p. 370, footnote. See MORSINK (1982) ch. III for an analysis of A.'s argument against pangenesis; HENRY (2006a) notes the plant example.

LXXIII

Dual-inheritance theory. This term is a small novelty of mine; it emerges from a solution to a problem in A.'s theory of inheritance. In A.'s standard theory of reproductive hylomorphism males supply the form and females the matter, see HENRY (2006b), but GA IV allows that maternal matter (the menses) can also encode hereditary information. One solution to this apparent conflict is to allow that when A. talks about indivisible forms he means individuals not species. This is the solution that HENRY (2006a, b), among others, adopt – and it implies that both parents transmit form. I, however, think that the weight of evidence favours the idea that form picks out the essential features of kinds and that only fathers supply it (see *n*. Chs XXXIII and LXX). If this is so, then we need another term for variation within an *atomon eidos*, hence 'informal variation'. Since such informal variation can come from both mothers and fathers and is also encoded in movements in the seed, we have then a dual-inheritance system: one (paternal) that encodes essential, functional features; the other (bi-parental) that encodes non-essential features (snub noses, sex, etc.), both of which depend on seminal movements and are susceptible to mutations. GA 767b24 speaks of several levels of inheritance.

Sex determination. A. critiques existing sex-determination theories at GA IV, 1. His own theory is framed in terms of hot/cold, GA 766b8. It is important to remember that for A. 'hot' does not merely denote the presence of heat (thermal energy) and 'cold' its absence, rather 'hot' and 'cold' are opposing qualities that are more like forces – hence the language of conflict and conquest. The idea of a proportion in sperm and menses (a *logos* or *symmetria*) occurs at GA 767a16, cf. GA 723a29; later, A. will restate the hot/cold theory in terms of actual and potential movements and tie it to a theory of general inheritance in GA IV, 3. A. speaks of environmental sex determination at GA 767a28 and refers to the parts (heart) as principles at GA 766a28. PLATT (1910) *n*. GA 716b5 points out the distinction between primary and secondary sex determination; PECK (1943) *n.* GA 776a30 points out that A. often seems ambivalent about whether the sexual parts are 'principles' or not, but clarifies his position at GA 766a31 and identifies the heart, PECK (1943) *n.* GA 766b8. A. discusses castration and eunuchs at GA 716b4, GA 766a26. He does not explain how castration might affect the heart. Perhaps he didn't recognize just how direct his analogy was, for post-natal castration affects only some secondary sexual characteristics such as balding and voice pitch, but not the genitalia. See LEROI (2003) ch. 7 for an account of Jost's experiments and sex determination.

LXXIV

A general theory of inheritance. A. explains his model of inheritance at GA IV, 3 and sex-associated features at GA 768a24. The woman of Elis features at HA 586a4 and GA 722a8. A. argues that the Hippocratic theory cannot explain ancestral similarities of this sort at GA 769a24; see HENRY (2006a). Failure of the semen's heat as the cause of atavism, GA 768a9. *Littré* VII, *On Generation*, 8, shows that the Hippocratic theory is a blending theory since the author states: 'If from any part of the father's body a *greater* quantity of seed is derived than from the corresponding part in the mother's body the child will, in that part, bear a *closer* resemblance to its father; and vice versa' [trans. LONIE (1981), modified, italics mine]. Thus any trait has a continuous, rather than a discrete distribution, and depends on proportionate contribution; see also GA 769a7 where A. reports, much less precisely, of this (or a related) theory that if 'the same amount comes from each of the two [parents], then, they say, the offspring formed resembles neither'. That probably means that the offspring is a blend of the two parents, but, admittedly, it could also mean that it's something completely different. Monsters are said not to be hybrids at GA 769b11. A. gives his reversion theory of monstrosity at GA 767b1. For early modern theories of genetics see GLASS (1947) on Maupertuis, DARWIN (1868) vol. 2 pp. 399–401 and ch. XIII and MAYR (1982) ch. 14 for the dismal record of early theories of genetics. PA 642a29 tells how Democritus was brought to the theory of substantial definition by the facts.

LXXV

Shellfish of the lagoon. The biology of the *ostrakoderma*, HA IV, 4–7; the *porphyra* (murex), HA 528b36, HA 546b18 and PA 679b2; see THOMPSON (1910) *n.* HA 547a3 on the royal purple industry. On oyster gonads see GA 763b5; cf. HA 607b2.

LXXVI

Spontaneous generators. Some animals are generated from animals, *HA* 539a21. Cockles, clams, razorfish, scallops, oysters, fan mussel, ascidians, limpets, barnacles, murex, other snails, hermit crabs are said to be spontaneous generators at *HA* V, 15; sea anemones and sponges at *HA* V, 16; fish lice at *HA* 557a21; worms at *HA* 551a8; cockchafers, scarabs, flies, horseflies, pseudoscorpions, clothes moths at *HA* V, 19 and fish fry and Cnidian mullet at *HA* VI, 15–16. Oysters are provided as evidence for spontaneous generation at *GA* 763a26; cf. for mullets and eels *HA* 569a10 and *HA* 570a3. The recipe for an oyster is given at *GA* 762a19, *GA* 763a25; cf. *HA* 569a10. A. discusses eel reproduction at *HA* 538a3, *HA* 570a3 and *GA* 762b27. The *gēs entera* appears at *HA* 570a15, *GA* 762b22; see PLATT (1910) and PECK (1943) *n. GA* 762b22; THOMPSON (1947) p. 59 for varying ideas as to what it might be. THOMPSON (1910) *n. HA* 538a12, BERTIN (1956) and PROMAN and REYNOLDS (2000) discuss eel-head shape. On not removing the foundations of a science without replacing them, *DC* 299a5.

LXXVII

The fate of A.'s theory of spontaneous generation. A.'s theory of spontaneous generation and early modern science are discussed by FARLEY (1977), RUESTOW (1984) and ROGER (1997). Oyster's gonads and larvae first observed by Brach in 1690; Leeuwenhoek independently described them in the following letters: 151 (1695), 157 (1695), 170 (1696) in LEEUWENHOEK (1931–99). Sea-urchin pluteus larvae were identified by Müller in 1846, barnacle nauplius larvae by Thompson in 1835, ascidian tadpole larvae by Kowalevsky in 1866. See WINSOR (1969) and WINSOR (1976) for accounts of their significance. Leeuwenhoek discusses his observations on eels and contemporary theories of eel reproduction in the following letters: 33 (1677), 15 (1691), 123 (1693), 169 (1696) in LEEUWENHOEK (1931–99). Leeuwenhoek initially observed putative eel progeny in the intestines of eels, but later identified them as parasites; he remained convinced, however, that he had identified the womb and progeny of the eel. For the discovery of the eel's gonads see BERTIN (1956).

LXXVIII

Flies. Flies copulate and produce larvae, *HA* 539b10, cf. *HA* 542a6, *GA* 721a8; flies are produced from larvae, *HA* 552a20; flies are spontaneously generated, *HA* 552a20 and *GA* 721a8. The same confusion applies to fleas and lice: e.g. *HA* 556b21. A. also considers what would happen if maggots were to reproduce. He says they

NOTES TO PAGES 232–239

can't since, if they did, they would necessarily produce a third kind of animal – some sort of 'nondescript' – whose progeny, in turn, would be yet another kind of animal, and so on to infinity. They would engender an endlessly mutating lineage of living things, and that cannot be, for as he says, 'nature flies from infinity', HA 539b7 and GA 715b14.

Spontaneous generation recipe v. sexual reproduction. For the comparison see GA 762b1; specificity of the spontaneous generation recipe, GA 762a25. Many scholars have noted the tension between A.'s theory of spontaneous generation and his metaphysical commitments, though they agree neither on the exact nature of the problem nor on its solution, see PECK (1943) pp. 583–5, BALME (1962b), LLOYD (1996), ch. 5; LENNOX (2001b), ch. 10; GOTTHELF (2012) ch. 6; ZWIER (in prep.). *Why believe in spontaneous generation.* ZWIER (in prep.) argues that A. is investigating how spontaneous putative spontaneous generators actually are. My solution differs from hers only in the relative emphasis placed on the influence of A.'s predecessors on his thought and the degree to which 'spontaneous' generation and 'spontaneous' events *sensu Physics* II are intended in the same way. Following BALME (1962b) and LLOYD (1996) ch. 5, I think they're being used in quite different ways. Theophrastus discusses spontaneous generation at CP I, 5.1–4; cf. CP I, 1.2, HP III, 1.3–6 and among the *physiologoi* HP III, 1.4. On origin-of-life theories and spontanous generation, see Prob X, 13; cf. GA 762b28. On traditional beliefs about spontaneous generation in cicadas see CAMPBELL (2003) p. 72. A.'s empiricism is evident in his discussion of spontaneous generation in mullets, HA 569a23, and muricids, HA V, 15 and GA 762a34. He gives the life cycle of the cicada at HA 556a25.

LXXIX

Life cycles. On the need for life cycles see Ch. XCVI and KING (2010). A. describes the natural history of the tuna at HA 537a19, HA 543b32, HA 543a9, HA 543a12, HA 571a8, HA 597a23, HA 598a18, HA 598a27, HA 599b9, HA 602a26, HA 607b28 and HA 610b4. He speaks of the regulation of monthly menstrual cycles in women at HA 582a34 and how most animals mate in spring in HA 542a20. A. talks of the *alkyōn* at HA 542b1, cf. HA 616a14; see PECK (1970) *n.* pp. 368–72 and ARNOTT (2007) for its identity and mythological associations. Most of A.'s information on the seasonal habits of animals, other than reproductive, is in HA VII, 12–30. He gives fish spawning times at HA VI, 17, cf. HA V, 9–11 and elsewhere, speaks of the hibernation of bees at HA 599a21 and bears at HA 600b28, the migration of cranes at HA 597a4, cf. HA 597b30, and the reasons why fish migrate at HA 598a30.

Animals adjust their habits to the season at *HA* 596b20 and have certain thermal tolerances at *HA* 597a14. He discusses the relationship of life cycles to celestial cycles at *GA* 778a5 and *GA* IV, 10, *GC* 336b16 and *LBV* 465b26.

LXXX

Theory of elemental movement and transformation. On the natural movements of the elements see *Phys* 225a28, *Phys* 255b14 and *DC* 297a30. My account rests on claims in *Physics* VIII and *DM* that the elements are not, strictly speaking, self-movers. It follows COHEN (1996) ch. II and FALCON (2005) p. 11; see WATERLOW (1982) pp. 167–8 and GILL (1989) p. 238 for different accounts. On the transmutation of the elements see *GC* II, 1–5. On seasons and elemental transformation see *GC* 336a13, *GC* 336b16, *GC* 337a4 and *GC* 338b1; FALCON (2005) p. 11. LEUNISSEN (2010a) ch. 5.2–3 discusses the teleological connection between the theory of elemental formation and the celestial movements. The following passages in the *Meteorology* contain A.'s theory of winds and rains: *Meteor* I, 9, *Meteor* II, 4–6; but the wind has a life cycle in *GA* 778a2; on rivers *Meteor* 347a2 and geological cycles *Meteor* I, 14; see WILSON (2013). Many scholars have discussed *Phys* II, 8 198b16ff. on the winter rain; see JOHNSON (2005) ch. 5.5 and WILSON (2013) ch. 5. Wilson rightly weighs the ambiguities of this passage against the complete lack of teleological explanation in the *Meteorology*. See WILSON (2013) ch. 5 for a rich discussion of the use of biological metaphors in the *Meteorology*. He also offers the intriguing suggestion that meteorological phenomena should be viewed as dualizing between elements and spontaneous generators; and spontaneous generators as dualizing between meteorological phenomena and sexually reproducing animals.

LXXXI

Figs. A. on figs, *HA* 557b25; T. on figs, *HP* II, 8.1–3, *CP* II, 9.5–15. T. on seasonal flowers, *HP* VI, 8.1–5; on flower structures, *HP* I, 12. The quotes are respectively from *HA* 557b25 and *GA* 715b21; cf. *GA* 755b10. T. on the date palm, *HP* II, 6.6, *HP* II, 8.4, *CP* II, 9.15; AMIGUES (1988–2006) vol. I, p. xxiii, discusses the source of T.'s information about date palms; cf. *Herod* I, 193. See LLOYD (1983) ch. III, 2 on T.'s background. A. on plant sexes, *GA* 715b16 and *GA* 731a21; T. on plant sexes, *CP* II, 10; NEGBI (1995) discusses Theophrastus' concept of male and female though gives him more assurance in distinguishing plant sexes than I think is his due. For the identity of fig-related insects see DAVIES and KATHIRITHAMY (1986) pp. 81–2, 92 and figs on Lesbos CANDARGY (1899) p. 29. I thank Charles

Godfray, University of Oxford, for suggesting the identification of the *kentrines*, and Filios Akriotis and Theodora Petanidou, both of the University of the Aegean, Mytilene, for telling me respectively about fig varietal names and fig culture; I also thank Dimitrios Karidis, an Erresos fig farmer, for further information about the last at Erresos. For the history of the study of caprification see Gasparrini quoted in LELONG (1891). See KJELLBERG *et al.* (1987) and WEIBLEN (2002) for fig wasp life cycles.

LXXXII

Bees. For the origin of honey see *HA* VIII, 40 and Theophrastus *HP* VI, 11.2–4; SHARPLES (1995) pp. 208–10 for the missing Theophrastan work on honey. A. discusses the generation of bees at *GA* III, 10; MAYHEW (2004) ch. 2 defends A. from sexism on bees. A. speaks of his uncertainties about bees at *GA* 760b27; cf. *DC* 287b28 for a similar look to the future possible resolution of explanatory difficulties. MADERSPACHER (2007) gives a brief history of the elucidation of bee life cycles.

LXXXIII

Life history. The swallow winds, *HP* VII, 15. For the migratory and nesting habits of swallows see *HA* VII, 16 and *HA* VIII, 8. Eye regeneration in swallows, *HA* 508b4, *HA* 563a15, *GA* 774b31; and in chicks, DEL RIO-TSONIS and TSONIS (2003). On the altricial cubs of bears see *HA* 579a20 and PECK (1970) pp. 376–8.

Life-history patterns. A.'s life-history data on mammals and birds are mostly in *GA* IV, 4–10. Important passages telling of particular associations are: *GA* 771a17ff. (litter size and body size); *GA* 773b5 (adult body size and neonate body size); *GA* 774b5 (neonate perfection, litter size and gestation time); *GA* 774b30 (neonate perfection, gestation time); *GA* 777a32 (gestation time, longevity, neonate size). All this material is interwoven with explanations of abnormalities. Besides these passages, for the predicted longevity from gestation time of deer see *HA* 578b23; and *LBV* 466 b7 (longevity and fecundity); see *n.* Ch. LXXXV; and on birds see *GA* 749a35 (also below). SUNDEVALL (1835) coined the modern terms altricial and precocial; see STARCK and RICKLEFS (1998) for some of this history.

Explaining life-history associations. A. argues that the negative association between body size and fecundity is causal at *GA* 771b8, and that the positive association between gestation time and longevity is *not* causal at *GA* 777a35. On confounding variables in the comparative method see LEROI *et al.* (1994). For weakness after sex see *GA*

725b6. On the infertility of fat people see GA 725b32 and PA 651b12. For the effect of castration on longevity and growth see HA 575a31, HA 578a33 and HA 631b19; also LEROI (2010) and Chs LXXXV and XCVII. A. discusses the Adrianic fowl at HA 558b16, GA 749b25; Aldrovandi does too, LIND (1963) pp. 27–9. For the connection between feet, wings and way of life in birds see Ch. XLV; for their connection, in turn, to life history see GA 749a30, cf. GA 771a17. I emphasize the allocative aspect of his argument, but A. also argues that some raptors acquire less nutrition than other birds. See Appendix V for further discussion of the way that life-history features covary in mammals.

LXXXIV

Fish life history. T. lists the summer flowers at HP VI, 8.1–5. For A.'s observations on fish life history see HA VI, 10–17 and GA III, 3–6. That it is the function of (egg-laying) fishes and plants to be prolific is asserted at GA 718b8 and that this is, in the case of egg-laying fishes, due to high embryonic mortality is explained at GA 755a30, cf. HA 570b30. The features of egg-laying fishes that permit high fecundity are: (i) reverse sexual dimorphism, GA 720a16; (ii) small eggs, GA 755a30; (iii) external 'perfection' (fertilization? – see below) to avoid uterine space constraints, GA 718b8å, cf. GA 755a26; (iv) rapid growth of the embryos, GA 755a26; (v) parental care in the glanis and its explanation, HA 568b15. A. describes brooding in the belonê at HA 567b22, HA 571a2, GA 755a30. On the contrast between the fecundity of viviparous selacheans and oviparous scaly fishes see HA 570b29.

Perfect v. imperfect eggs. When speaking of the relative perfection of birds and mammal progeny, it's clear that A. means something like altricial v. precocial. When he speaks of perfect and imperfect eggs (e.g. GA 718b8, GA 732b1, GA 754a22 and GA 755a11), he means something related but rather different. Once again, his technical vocabulary is seriously underdetermined. Among the reproductive products of fishes, A. thinks that those that are live born (those of most selacheans) are the most perfect, then eggs with a hard shell-like case (other elasmobranchs, e.g. the rays and skates) are less perfect; soft eggs (e.g. most scaly fishes) are the least perfect of all. The distinction lies in how much development the reproductive product (i.e. the thing that emerges from the mother) has to undergo before it becomes a functional creature (little, some, lots). This difference in egg morphology is, in fact, closely associated with fertilization mode: elasmobranchs have internal fertilization while most bony fishes have external fertilization, and it's probable that A. recognizes this, but it's hard to be sure since he's very vague about how fish copulate. Even so, it's likely that he views fertilization itself as a 'perfecting' of the female matter; and the stage at which this 'perfecting' occurs (early, internal v. late, external) partially determines how perfect the progeny are at birth. For the expansion of the jelly coat in fish eggs that occurs at fertilization see COWARD *et al.* (2002).

Modern life-history theory. For an introduction to life-history theory see ROFF (2002); for a typical paper on fish life history see WINEMILLER and ROSE (1993).

LXXXV

Greek life expectancy. Life expectancy at birth on Ikaria appears to be within the national range for Greece (C. Tsimabos pers. comm.), but a closer look at survival of the oldest old suggests that Ikarian women, at least, have a significant survival advantage at late age relative to Greece as a whole (M. Poulain pers. comm.).

The length and shortness of life. See KING (2001) for the authoritative account of A.'s theory of ageing. A. says we must investigate why some animals are long lived and others short lived at LBV 464b19. He tells of mayflies at HA 552b18 and how winged insects die at the end of summer at HA 553a12. He summarizes the comparative biology of lifespan at LBV 466a1 and says that the old are cold and dry at LBV 466a21. Is there a single explanation for death? JSVM 478b22. Is there one cause for the diversity of longevity? LBV 464b19. His account of the relative heat and moisture of different animals is given at LBV 5, 6. For the role of fat in promoting life see LBV 466a24, cf. PA 651b1; see FREUDENTHAL (1995) ch. IV. On the longevity cost of reproduction, LBV 466b7, HA 576b2 and GA 750a20; see LEROI (2010). For the same idea of a senescence cost of reproduction in modern evolutionary biology, see WILLIAMS (1966), ROSE (1991), LEROI (2001), ROFF (2002). On regeneration in plants see LBV 467a7, snakes and lizards HA 508b4, and *Hydra* BOSCH (2009). On death by failure of cooling systems, see JSVM 470b10; on the seizing up of cooling organs, JSVM 479a8 and JSVM 479a31; on the (false) etymology of earth/age, GA 783b7. On the vulnerability of old animals to variation in the environment, see JSVM 474b30, JSVM 478a15 and JSVM 479a16. On the role of the soul and ageing, see DA 415b25, DA 434a22 and throughout JSVM. Death is said to be in the nature of living things at JSVM 464b29, PA 644b23 and GA 731b24. For modern mechanistic theories of ageing see FINCH (2007) and, more recently, GEMS and PARTRIDGE (2013). For thermoregulation and ageing in humans see SOMEREN (2007). For evolutionary theories of ageing see WEISMANN (1889); the modern theory is due to MEDAWAR (1951/1981) and WILLIAMS (1957); see LEROI (2003) ch. IX for a popular review.

For the general theory of the destruction and regeneration of natural objects: *LBV* 2, 3 and *DC* 288b15 and Ch. LXXX.

LXXXVI

The story of Daphnis and Chloe. I used the Loeb edition of Longus' *Daphnis & Chloe*, translated by Jeffrey Henderson, 2009. The idyllic scene is described in I, 9–10; see MASON (1979), GREEN (1982) and GREEN (1989) ch. 3.

LXXXVII

Sponges and other plant–animal dualizers. On sponges, *HA* 487b10, *HA* 548a32, *HA* 548b8, *HA* 588b21, *PA* 681a10. VOULTSIADOU (2007) discusses the cultural role of sponges in antiquity. For the other plant–animal dualizers, sea anemones, sea squirts, etc., see *HA* 487b10, *HA* 547b12, *HA* 548a22, *PA* 681a10, *PA* 683b18. It is hard to say whether A. thinks that some or all of these creatures are animals, plants or something in between. For example, he says of sponges that they are 'plant-like' or even 'plant-like in every respect'. I take it that A. thinks that they are, on balance, animals since they all seem to have at least one capacity of the sensitive soul (loco-motion or sensation or appetite); that would be consistent with his polythetic approach to classification. But perhaps the most compelling reason to believe that A. thinks they are animals is that he discusses them in *HA* – where he might have discussed them elsewhere, e.g. in his lost *On Plants* (*Peri phytōn*). Similarly, in his *Enquiries into Plants* (*HP* IV, 6.10), T. touches on sponges but then says 'they're of a different character' – presumably animal, to be dealt with elsewhere. The pseudo-Aristotelian text *de Plantis*, which is thought to be a commentary by Nicholas of Damascus on A.'s *Peri phytōn*, also struggles with this question since it asserts that animals have sensation, plants do not, shellfish have sensation but are at once both animals and plants: *DP*, 1; DROSSART LULOFS (1957). See LLOYD (1983) ch. 1, 4, LLOYD (1996) ch. 3 and LENNOX (2001a) p. 301 for how A. deals with these crea-tures. For T. on corals and other sea 'plants' see *St* 38 and *HP* IV, 6 where the precious coral is the 'sea palm'; *HP* IV, 7.2 on the growths in the Gulf of Heroes (Aqaba). See THOMPSON (1947) p. 250 for scepticism on sponge contraction; I thank Sally Leys, University of Alberta, for telling me about movement in sponges; see also NICKEL (2004).

'Nature proceeds from the inanimate . . .': *HA* 588a1, cf. *PA* 681a10, *Meteor* IV, 12 and *GA* 731a25. It may seem that there is a tension between this claim and A.'s belief that the world is composed of discrete kinds each possessed of its own inherited form

and teleologically defined essence; however, by 'continuity' A. did not mean that the continuum of kinds is infinitely divisible, nor that kinds overlap so that the boundaries of one cannot be distinguished from the other, but only that they form a graduated series that progresses in small but discrete steps; see GRANGER (1985) contra LOVEJOY (1936).

LXXXVIII

Aristotelian themes in Darwin and vice versa. For the history of Natura non facit saltum and its use by Darwin see FISHBURN (2004). 'Ever since Darwin' was the title of Stephen Jay Gould's 1977 anthology of essays from Natural History magazine, so named for the cliché used as the starting point of so many papers by evolutionary biologists. On the meanings of genos see Metaph V, 28. Some scholars (LENNOX (2001b) ch. 6 and PELLEGRIN (1986) ch. 2) have emphasized that in the zoology A. uses genos to designate a group of organisms that are related by descent. That's not unreasonable so long as it is recognized that, in the case of a megista genos (e.g. birds) with subordinate genē (sparrows, cranes) we recognize that A. is not saying that sparrows and cranes are related to each other by descent since that would imply a common ancestor, i.e. evolution. His use of genos in the sense of common descent (definitions 1 and 2 in Metaph V, 28 which are very close) can, then, apply only to genē that are atoma eidē, i.e. that actually interbreed (e.g. humans). In general, he must be using genos in the third sense given in Metaph V, 28 which is purely classificatory and does not imply anything about ancestry. For a discussion of A.'s. anticipations of evolutionary themes (without being an evolutionist) see KULLMANN (2008).

LXXXIX

Darwin on A. See LENNOX (2001b) ch. 5 and GOTTHELF (2012) ch. 15 on A.'s indirect influence on Darwin. 'Read A. . . .': DARWIN (1838/2002–) p. 267. STOTT (2012) tells the story of how A. infiltrated himself into the Origin as an evolutionary precursor. Translations of The Parts of Animals: today LENNOX (2001a) is the authoritative English translation of PA. His superb commentary focuses on philosophical and theoretical aspects; the zoological facts still have to come from OGLE (1882), in this case p. 240, n. 36: 'the camel, the cats, and many rodents including the hare are retromingent'. The German translation and commentary of KULLMANN (2007) are excellent on both philosophical and zoological matters.

XC

A., *Linnaeus and the* scala naturae. On the origin of Linnaean names see HELLER and
PENHALLURICK (2007). In neither of the two passages – HA 588b30 and PA
681a10 – that are generally held to show the Ladder of Nature or *scala naturae* is the
idea very explicit, LENNOX (2001a) pp. 300–1; however, in other passages, particularly in GA, it is clear that A. has a strong sense of how animals should be arranged
in a continuum of increasing perfection. See, for example, GA 733a32 on the relative perfection of progeny, and GA 733a1 on the link between parental and parental
perfection. The physiology of relative perfection is given in the following passages:
hot animals have lungs, PA 669b1; tend to upright, PA 686b26; tend to be larger,
GA 732a17; tend to live longer than cold animals (Ch. LXXXV). A.'s theory – PA
648a2 and PA II, 4 – of how the constitution of blood influences intelligence and
temperament is in the same spirit, but more complicated. In brief, there are three
properties of blood that influence intelligence and temperament: heat, thickness
and purity. Although correlated, these properties vary, to some degree, independently in both blooded and bloodless animals, which allows A. to explain the
various behaviours of various animals (bulls, bees and so on). Those with hot, thin
and pure blood are best – for they are both courageous and clever. Humans have
the thinnest and purest blood of all animals; see LLOYD (1983) ch. I, 3.

The naturalists and the scala naturae. For the history of the *scala naturae* in Western
thought see LOVEJOY (1936) ch. 2 who, p. 79, quotes Albert Magnus. *Systema natu-
rae*, 1st edition, LINNAEUS (1735); 13th edition, LINNAEUS and GMELIN (1788–93).
For a history of zoophytology see JOHNSTON (1838) pp. 407–37; ELLIS (1765).
Cuvier classification, first given in 1812, is best known in the version given by
CUVIER and LATREILLE (1817).

XCI

The great cuttlefish debate. For accounts of the debate see RUSSELL (1916) chs 3, 5, 6,
APPEL (1987), GUYADER (2004) and STOTT (2012). For A.'s analysis of cephalopod geometry see PA IV, 9 and Chs XXIII, XXXVIII and XCVII. For Geoffroy
and Cuvier on A. see GUYADER (2004) pp. 143, 155, 181. The terms 'homology'
and 'analogy' have a complicated history but they were first distinguished by OWEN
(1843), pp. 374, 378 and OWEN (1868); their meaning, however, has continued to
evolve, HALL (2003). On Cuvier's method see CUVIER (1834) vol. I, pp. 97,
179–89; and why natural history should not have its Newton, CUVIER (1834) vol.
I, p. 96. 'The form of the tooth . . .': CUVIER (1834) vol. I, p. 181; 'Natural history

has a rational ...': CUVIER and LATREILLE (1817) vol. I, p. 6, trans. OUTRAM
(1986). For Geoffroy on the vertebrate breastbone and the *loi de balancement* see
GUYADER (2004).

XCII

The evolution of concepts. For Cuvier's relationship to other thinkers see RUSSELL
(1916) ch. 3, OUTRAM (1986), RUDWICK (1997), GRENE and DEPEW (2004)
ch. 5, REISS (2009) pp. 103–13. Cuvier's Conditions of Existence appear in
DARWIN (1859) p. 206 and PALEY (1809/2006) ch. 15. For the same idea in
modern genetics see LEROI *et al.* (2003) on cancer in *Xiphophorus* hybrids and
PHILLIPS (2008) on epistasis. Geoffroy's *loi de balancement* appears as the correlation of growth in DARWIN (1859) p. 143 and as pleiotropies in LEROI (2001).
A. speaks of recurring ideas at DC 270b16, *Meteor* 339b28, *Metaph* 1074b1 and *Pol*
1329b25.

XCIII

A.'s anti-evolutionism. On the Pre-Socratic zoogonies and transformism see
CAMPBELL (2000), LLOYD (2006) ch. 11 and SEDLEY (2007). Plato's transformism is evident from *Tim* 91D–92C; see SEDLEY (2007) ch. 4. A. considers
the idea that all animals might be 'earth-born' at GA 762b23. With most
commentators, I argue that A. is committed to the fixity of forms. BALME and
GOTTHELF (1992) pp. 97–8, BALME (1987d) and GRANGER (1987) argue that
he is not, but not convincingly; see LENNOX (2001b) ch. 6. Eternity of kinds/
forms: DA 415a25, GA 731b31, *Metaph* VII, 8–9, GC II, 10–11. A. discusses the
deleterious effects of congenital deformities at GA 771a12 and GA 772b35.
HENRY (2006a) suggests that for A. the fit to the environment of a given kind is
maintained by selection against mutations that fall outside some range so that
any animal that bears them 'will no longer be adapted to that environment and
so possessing that feature will be detrimental to its ability to survive and reproduce'. That kind of selection, which is very similar to Empedoclean selection, is
known as stabilizing or purifying selection, but I don't think that A. invokes it.
He only says that unconditionally unfit creatures (those lacking essential organs)
die and he never ties this to the maintenance of forms, be it with respect to a
particular environment or not.

Origin of new kinds/species by hybridization. On Linnaeus' hybridism see MÜLLER-WILLE
and OREL (2007). A. discusses hybrids at *Metaph* 1033b33, GA 738b32, GA 746a29,

HA 566a27, *HA* 606b25, *HA* 608a32; cf. *Mirab* 60. The question of whether A. thinks that new animal kinds can arise by hybridization is a tricky one. Scholars such as HENRY (2006b) who would not accept the dual-inheritance system that I do here, argue that both mother and father contribute their forms to the embryo. If so, then hybrids that are a stable mix of the parental forms would be possible. But there isn't much evidence that A. thought this. In fact, at GA 738b28, where A. discusses dog × fox crosses, he claims that the hybrids will revert to the female's form. This is incompatible with either exclusively paternal forms or biparental forms since it gives an unexplained priority to the mother's form or matter. In fact, I think it's an un-Aristotelian interpolation, probably by Theophrastus – the language of soil and seed suggest a botanist's meddling with the texts; see *CP* I, 9.3, *CP* II, 13.3 and this text *n*. Ch. LXXI. Geoffroy's teratological transformism, APPEL (1987) pp. 128, 130–42; GUYADER (2004).

Terato-transformism. On the relationship between the monstrous and natural in A. see GA 770b15 and GA 769b27. The following animals are said to be naturally deformed: seals, *HA* 498a33 and *PA* 657a22, and moles, *HA* 491b28, *HA* 533a1 and *DA* 425a10; for other examples and discussions of what exactly A. meant by 'deformed' or 'warped' see LLOYD (1983) ch. I, 4, GRANGER (1987) and WITT (2013). On how tetrapods came to walk on all fours see *PA* 686a32; cf. *PA* 686b21, *Tim* 91D–92C.

Evolutionism. There is a story, told by Ernst Mayr, David Hull and Arthur Cain, that Aristotle's 'essentialism', via Linnaeus, held up the theory of evolution for 2,000 years. To refute this would require a detailed analysis of what these scholars thought Aristotle and Linnaeus said, and this book is not the place to do it. I shall do so in a future paper.

XCIV

Fossils. For Darwin's predecessors see MAYR (1982) and STOTT (2012). Some scholars have argued that A. simply did not have evidence for evolution, e.g. BALME (1987d), BALME and GOTTHELF (1992) pp. 97–8; LENNOX (2001b) ch. 6. See *Strab* I, 3.19 for a theory of Lesbos' geological history. For Xenophanes' fossil fish see PEASE (1942); for Xanthus, Eratosthenes and Strato on fossils see *Strab* I, 3.3–4; Theophrastus on fossil ivory, *St* 37 and MAYOR (2000). The papers collected in DERMITZAKIS (1999) give an overview of Lesbos' vertebrate palaeontology. SOLOUNIAS and MAYOR (2004) describe the elephants of Samos and their remains. *Herod* II, 75 speaks of the winged serpents of Arabia; see RADNER (2007).

For Theophrastus on petrified reeds see *HP* IV, 7.3, and *fossiles, Meteor* 378a20. There are also a number of references to petrification in pseudo-Aristotelian works, e.g. *Prob* XXIV, 11, *Mirab* 52 and *Mirab* 95.

XCV

Theophrastus' transformism. For assimilation to the country in wheat see *CP* IV, 11.5–9; new natures in plants, *CP* IV, 11.7 [trans. EINARSON and LINK (1976–90)]; degeneration of seed (reversion to wild), *CP* I, 9.1–3 and *HP* II, 2.4–6. On darnel see *CP* II, 16.3, *CP* IV, 4.5–5.5, *HP* II, 4.1 and *HP* VIII, 8.3, where T. recognizes that darnel might just be a weed. See THOMAS *et al.* (2011) for the evolution and cultural significance of darnel. T. is often said, on the basis of some fragments, to be more resistant to teleology than A. but there is no doubt that his biology is underpinned by a thoroughgoing teleology even if it is less of an ostensible concern for him; see LENNOX (2001b) ch. 12.

XCVI

Aristotelian and evolutionary explanations compared. 'Nothing in biology makes sense except . . .', DOBZHANSKY (1973). 'Nature does that which, among the possibilities', *IA* 704b11; cf. *GA* 788b20. For optimality thinking in A. see LEROI (2010); in evolutionary biology and its formal link to the theory of natural selection see GRAFEN (2007). Individuals are the beneficiaries of Darwinian adaptations, see DARWIN (1859) p. 186 and RUSE (1980); for a discussion on how to distinguish levels of selection from levels of adaptation see GARDENER and GRAFEN (2009). Living things participate in the eternal and the divine at *DA* 415a25, *DA* 415a22, *GA* 731b18, *GC* II, 10–11; cf. the dubiously Aristotelian *MM* 1187a30. When A. discusses the ultimate purpose of life, he generally frames it in terms of the purpose of the soul, i.e. the physiological system that controls nutrition, growth and reproduction, as well as other functions; see Ch. LIV. I claim that the features of Aristotelian creatures are ultimately for the sake of forms/kinds. Some scholars, e.g. BALME and GOTTHELF (1992) pp. 96–7 and LENNOX (2001b) ch. 6 and pers. comm., deny this and argue that eternal survival of kinds is just a secondary consequence of the desire of individuals to reproduce. However, *DA* 415b2 points out that there are two senses in which we can speak of 'that for the sake of which'. The first is 'that for the purpose of which', the second is 'that for the benefit of which'. He then clearly goes on to identify the 'that for the benefit of which' as the form/kind; see also *DA* 416b22. It is also worth pointing out that, when speaking of a particular adaptation, A. usually does not specify whether it (say, horns) is 'good for the individual' or 'good for the species'. He doesn't need to – it is good for both. Sometimes, however, he does explicitly say that some feature is good for the species, as when discussing the features of fish life history, GA 755a30; Ch. LXXXV. In this respect, his teleology *is* different from Darwin's insofar as, for Darwin, individuals are beneficiaries of adaptation; for a Neo-Darwinian, genes are; see also Ch. CI. On reductionism in Aristotelian explanations see GOTTHELF (2012) ch. 3. A. claims that it is better to exist than not exist at GA 731b30.

XCVII

Honour teleology. For the definition of the body axes of animals and plants see Ch. XXXVIII. For the relative value of the poles see IA 5. SOLMSEN (1955), LENNOX (2001a) p. 275 and SEDLEY (2007) p. 172 discuss Platonic values in A.'s biology. Much of A.'s discussion about the genesis, position and structure of the heart is in PA III, 4–5, specifically PA 665b20; see LENNOX (2001a) pp. 254–65. On the symmetry of the liver and spleen: PA 666a25 and PA 669b13ff. Honour teleology also partly explains the existence of the diaphragm separating the lower, and less valuable, digestive organs from those in the thoracic cavity, in particular the heart, the centre of cognition, see PA 672b17. For a general discussion on honour teleology see GOTTHELF (2012) ch. 2.

Humans v. animals. A. discusses the differentiation of body axes between humans and animals at HA 494a27 and differences in character at HA 588a19, HA 608a10 and HA 608b4. He claims that females are disabled, deviant, deformed or monstrous at GA 728a17 (females are infertile males), GA 737a22 (females are deformed males), GA 767b6 (females are the result of deviations in development from the kind, and in a way monstrous) and GA 775a15 (females are natural deformities). He explains the reason for having separate sexes at GA 732a1 and notes that females are needed for the perpetuation of form at GA 767b8; cf. GA 731b34, *Metaph* X, 9. The production of the sexes is, however, an 'accidental' feature – *Metaph* X, 9 – thus due to the informal system of inheritance. Eunuchs are said to be feminized at GA 716b5; cf. GA 766a26. MAYHEW (2004), HENRY (2007) and NIELSEN (2008) discuss whether A.'s theory of sex determination is sexist. The 'scale of perfection' interpretation given here is due to WITT (1998).

Explaining human uniqueness. For the disproportionate production of seed in humans and its explanation, see HA 521a25, HA 572b30, HA 582b28, GA 728b14 and GA 776b26. I thank Tim Birkhead, University of Sheffield, for information about the relative volume of semen production in mammals. Humans and horses are said to have sex during pregnancy at HA 585a4. On the lechery of bald men see GA 783b27; cf. GA 774a34. On why women and eunuchs don't go bald see HA 583b33, GA 728b15, GA 784a4–7 and LEROI (2010). A. discusses the peculiarities of human physiology at HA 521a2 and PA 669b1; the relationship between nudity and the use of hands as weapons at PA 687a22 and between human upright posture and divinity at PA 686a25, cf. PA 656a7; see LLOYD (1983) ch. I, 3, LENNOX (2001) pp. 317–18 and KULLMANN (2007) p. 690.

XCVIII

Political animals. A speaks of political activity among animals at Pol 1253a7, HA 488a10, cf. HA 589a3; see KULLMANN (1991) and DEPEW (1995). Crane sociality: HA 488a7, HA 614b18.

The behaviour of bees. On single flower visitation and the waggle dance see HA 624b5; on the latter see also HALDANE (1955). On comb construction see HA 623b26; drone expulsion see HA 626a10; division of labour see HA 625b18 and HA 627a20; queen specialization see GA 760a11. Xenophon, Econ VII, tells how the queen rules the hive. A. is much less clear on the subject. At HA 488a10 he says that bees 'live under a ruler', but he gives no further specifics. He describes bee regicide at HA 625a17 and HA 625b15.

The missing Habits of Animals. Most of A.'s ecological information is in HA VII and VIII (Balme's numbering), the former focusing on their food and habitats, the latter on their habits and characters. Since the latter, which includes bee habits, is sometimes suspected of being un-Aristotelian, one explanation for the absence of a causal analysis of animal habits is simply that it's not A.'s data. Most modern scholars, however, accept these books as being largely authentic.

XCIX

State formation. A. describes the origin of the household and state at Pol I, 1–2. He explains the purpose of the family household at Pol 1252b9 and argues for specialization at Pol 1252b1; cf. Ch. LI. The initial end of the state is said to be self-sufficiency, Pol 1253a1. The Cyclopes are said to be lawless at Pol 1252b35. See KULLMANN

(1998) ch. V for the relationship between A.'s political thought and that of his successors.

Natural slavery. A. gives his theory of natural slavery at *Pol* 1254a9ff. See *Pol* 1254b16, cf. *Pol* 1260a1, for the mental ability of natural slaves. Various commentators, e.g. HEATH (2008), have tried to refine exactly what mental capacity A. thought natural slaves lack. Setting the question of ownership aside: at *Pol* 1260a35, A. says that a free craftsman, working for a master, is in a restricted sense in a condition of slavery. A. muses on automatic lyres at *Pol* 1253b30. He suggests that barbarians make natural slaves at *Pol* 1256b20, cf. *Pol* 1252b5, *Pol* 1255a28 and *Pol* 1285a19; see HEATH (2008).

C

The cyborg state. Plato famously describes the Greeks clustering around the pond of the Mediterranean at *Phaedo* 109B. A. speaks of the state as having organs at *Pol* IV, 4 and compares its organs of central control to a soul at *Pol* 1254a28, *Pol* 1254a34, *DA* 410b10; cf. *Pol* 1253a20. The constitution is envisioned as a river at *Pol* 1276a35. The state is said to be a creation of nature at *Pol* I, 2, *Pol* 1263a1, but is in fact a natural–artificial hybrid, *Pol* 1265a29; see KULLMANN (1991) and LEUNISSEN (2013). Without the rule of law humans are the worst of animals, *Pol* 1253a29. For the classification of sciences see *Metaph* XI, 7. A. discusses the best state in *Pol* IV, 11 and *Pol* VII. For professional and property bars to citizenship see *Pol* 1328b35, *Pol* 1329a20. BURKHARDT (1872/1999) ch. 5 gives a particularly harsh and detailed assessment of fourth-century Athenian democracy, but A. also criticizes it in the *Athenian Constitution*; cf. *Pol* V, 5. A. classifies the organs of the state according to the distribution of their organs, *Pol* IV, 4. He describes the material causes for different kinds of states and constitutions at *Pol* 1321a5, *Pol* 1318b10; cf. *Pol* 1326a5 and the characters of Europeans, Asians and Greeks at *Pol* VII, 7. He speaks of the revolution and destruction of states at *Pol* V, 1, and inveighs against Plato's marital communism at *Pol* II, 1–3 – even if his account of the scheme given in *Rep* is a bit of a caricature. He talks of natural order and living well at *PA* 656a5.

CI

Ecology and Metaphysics λ. The war between the eagle and the dragon snake is mentioned at *HA* 609a4; see WITTKOWER (1939) and RODRÍGUEZ PÉREZ (2011) for the origin and spread of this symbolic motif. Other references to the dragon snake are: *HA* 602b25 and *HA* 612a33. 'We must consider also': *Metaph* XII (λ)

1075a16 [trans. SEDLEY (1991)]. The three monographs and one paper that tackle this passage are JOHNSON (2005) ch. 9, SEDLEY (2007) ch. V, LEUNISSEN (2010a) and BODNÁR (2005), but many others have too. The position on global teleology that I have adopted here is close to that of NUSSBAUM (1978) pp. 93–9, BODNÁR (2005) and MATTHEN (2009). I thank István Bodnár for guidance here. See SCHMIDTT (1965) for the Renaissance trope of A. as a cuttlefish. For the origin of the term 'ecology' see HAECKEL (1866) vol. II, pp. 286–8 and STAUFFER (1957). The *pinnophylax* is said to inhabit the *pinna* at HA 547b16. For the biology of Pandora see SWIRE and LEROI (2010). A.'s discussion at DA I, 3, cf. DC II, 3, shows that A. thinks that there is no world soul.

Ecological relations. The shark's face, PA 696b25, cf. HA 591b25, has also been frequently discussed in terms of global teleology. LENNOX (2001a) pp. 341–2 considers the options for explaining this odd passage away, but admits it's hard to do; see him for earlier references. A. speaks of the extraordinary fecundity of fish at HA 567a34 and mice at HA 580b10; of incontinent animals at EN 1149b30 and of the state of war that exists when food runs out, HA 608b19; cf. HA 610a12. At HA 610b2 he suggests that hostile fish will shoal together when there's an abundance of food. Herodotus alludes to the balance of nature at *Hist.* III, 108–9 [RAWLINSON et al. (1858–60/1997)]; see EGERTON (1968), EGERTON (2001a) and EGERTON (2001b) for a cool look at ancient zoology and this idea. A. makes no use of this passage even though it crops up when H. is talking about his winged serpents, a passage that A. clearly knows. Besides *Metaph* XII (λ) 1075a16, *Pol* 1256b7 is the other major passage appealed to by supporters of global teleology. SEDLEY (1991), SEDLEY (2007) ch. 5 gives the strongest anthropocentric interpretation of this passage, but see JOHNSON (2005) ch. 9 for a rebuttal. A. speaks of prudent fishes at EN 1141a20. JOHNSON (2005) ch. 8 effectively uses this passage to demolish an anthropocentric teleology, but fails to wonder how fishes can be prudent at all. An alternative reading, it's true, is that fishes must be prudent for the sake of some direct, physiological benefit – and in the shark's-face passage he mentions such benefits. But that does not seem to be what he's getting at here, since he adds: 'For the one observing each thing in relation to itself is prudent, and such things are entrusted to this one' – trans. JOHNSON (2005). This is rather cryptic, but I suggest that it can be read as meaning that each kind is prudent about particular things (a man is prudent with money, a shark is prudent with sardines), indeed, that each kind has those things in trust, i.e. its nature binds it not to destroy the things it needs. QUARANTOTTO (2010) discusses wholes and their properties. See PIMM (1991) for a critique of the 'balance of nature' in modern ecology. '[Natural selection] does not plan for the future' is from DAWKINS (1986) p. 5.

CII

The eternity of the cosmos. On the Pre-Socratics and the origin of the cosmos, DC 297b14. A. argues against the origin of change, *Phys* (VIII) 250b7, and gives a proof of eternal change, *Phys* 251a8; see GRAHAM (1999) pp. 41–4 for an analysis of this argument. In DC I, 10–13 A. gives another set of arguments, some of which are related to that given in *Phys*, others of which are linguistic. A. argues the need for a continuous source of change, *Phys* VIII, 5; see GRAHAM (1999) pp. 93–4 and BODNÁR (Spring 2012). A. has no theory of inertia, see BALME (1939).

Astronomy. On studying the stars, PA 644b22, DC 286a5, DC 291b24 and DC 292a14; see FALCON (2005) p. 99. A. defers to the experts in mathematical astronomy at *Metaph* 1073b10 and *Metaph* 1074a16; see LLOYD (1996) ch. 8 on A.'s relationship to the mathematical astronomers and a rather severe analysis of his own astronomical efforts. The geometrical model of the cosmos is sketched at *Metaph* XII (λ) 8; see LLOYD (1996) ch. 8. On Eudoxus see DL VIII, 86–91 and JAEGER (1948) ch. 1. A. gives estimates of the size of the Earth at DC 298b15. On the science of nature and bodies see DC 268a1; FALCON (2005) ch. 2. Saving the appearances *v.* explanation: *Phys* 193b22; LLOYD (1991) ch. 11, LEUNISSEN (2010a) ch. 5. A. asserts, on the basis of ancient astronomical records, that the cosmos is unvarying, DC 270b13, DC 292a7, cf. *Metaph* 342b9, LLOYD (1996) ch. 8. On the first element, *aithēr*, DC I, 2–3; FALCON (2005) p. 115 suggests that the traditional identification of the 'first element' with *aithēr* arose in later antiquity; for the reception and properties of *aithēr*, see FALCON (2005) ch. 3. The virtues of circular motion, DC I, 2; cf. *Phys* VIII, 9. A. discusses the final causes of circular movement of the celestial bodies: DC II, 3, DC II, 12; see LEUNISSEN (2010a) ch. 5.2.

The celestial bodies are alive. A. explains why the stars do not have locomotor append-ages at DC II, 8 – LEUNISSEN (2010a) ch. 5.4 – and compares them to ships in a stream at DC 291a11. A. claims that the stars (or spheres) are alive, DC 292a18 [trans. I. Bodnár], cf. DC 285a29; see GUTHRIE (1981) p. 256 text and note. On the properties of celestial life see DC 279a20, and on the celestial hierarchy, DC II, 12. On the motions of the sun and moon, DC II, 3. Here A. does not use the term 'for the sake of' and so it could be that the motions of the sun and the moon merely keep the sublunary elemental cycle going out of material necessity. At GC 336b1 he says that if coming to be and passing away (cycling of the elements) is to be contin-uous there *must* be some body (the sun) that moves with secondary motions. At GC 336b32 (see Ch. LXXX) he goes so far as to forget himself and talk of God who *arranged* the motions of the sun and the moon precisely to secure the greatest possible coherence to existence. LEUNISSEN (2010a) ch. 5.2, a stern opponent of global teleology when reading *Metaph* XII (λ), 10, concedes that conditional necessity and, by implication, global teleology are at work in *DC* II, 3: 'Here, the use of the teleological principle allows A. to draw an organic picture of the cosmological system . . .' There is an additional teleological argument for the relative perfection of the celestial bodies' motions in *DC* II, 12. 'A man and the sun . . .': *Phys* 194b13, cf. *Metaph* XII (λ), 10; FALCON (2005) p. 9. A. argues against the materialist explanation (chance) for order in the cosmos, *Phys* 196a26. For Democritus on infinity see SEDLEY (2007) p. 138 who alludes to modern infinite-universe cosmological theory. Cosmological Selection Theory: see REES (1999) on fine-tuning, TEGMARK (2007) on multiverses in general and GARDNER and CONLON (2013) on CST and the Price equation.

CIII

Approaching God. The hallmarks of life reprised, *DA* 412a14. See Ch. LIII. A. justifies living celestial beings on religious grounds, *DC* 270b5; more generally for the religious motivation of A.'s cosmology see *DC* 270b5, cf. *DC* 278b14, *DC* 283b26; NUSSBAUM (1978) pp. 134ff. and FALCON (2005) p. 112. A. engages in religious archaeology at *Metaph* 1074b1. On the distinction between first and second philosophy see 1026a27, *Metaph* 1026a27 and GRENE (1998). For a specific denial that animals are truly self-movers see *Phys* 252b16, *Phys* 259b1, *MA* 2–5. GUTHRIE (1939) Introduction, GUTHRIE (1981) ch. 8 and SORABJI (1988) ch. 13 discuss the evidence for at least two, intertwined theories of cosmology, theology and physical motion in A.'s works. The problem is that UMs seem redundant as moving causes if the stars are already rotating because they are made of *aithêr*. Even so, it might be possible to construct a unified account of the UMs and celestial spheres made of *aithêr* if we allow that *aithêr*, like *pneuma*, is just part of a chain of moving causes; Bodnár (pers. comm.) points out to me that the UMs appear in the early, lost dialogue *de Philosophia*. A. lays out the argument for the unmoved movers at *Metaph* 1073a23, *Phys* VIII, 8–10. At *Metaph* 1073a1 he says there are fifty-five of them, but actually this is just one of several totals A. gives; another is forty-nine. He seems to be working, a bit ineptly, with several different models, LLOYD (1996) ch. 8. Considering A.'s mature theory of motion I have omitted most of that dark book, *Physics* VIII; see BODNÁR (Spring 2012) for a crisp account of the theory, GRAHAM (1999) for a textual commentary and WATERLOW (1982) for a full analysis not much easier than A.'s text. A. does not have laws of motion: see DELBRÜCK (1971), NUSSBAUM (1978) pp. 130, 305ff. A. discusses how the unmoved movers move the things they move: *Metaph* 1072a26, *Phys* VIII, 10. There is an apparent conflict between the claim that there are many UMs, *Metaph* 1074a14, and his usual focus on just one (e.g. throughout *Phys* VIII); GUTHRIE (1981) pp. 267–79, drawing on work by Philip Merlan, reconciles these passages by appealing to a hierarchy. A. lays out the nature of the ultimate UM at *Metaph* 1072b13ff. [trans. ROSS (1915)]; how God thinks at *Metaph* 1074b33. A. describes the best sort of life at *EN* X, 7. He quotes Anaxagoras at *FR* B18–19 (*Protrepticus*) and *EE* 1216a10.

CIV

The Lyceum and its texts. For a picture of life at the Lyceum see JAEGER (1948) chs 12, 13. The first of the opposed quotes is from *DC* 276a18; the second from *GA* 745b23. ANAGNOSTOPOULOS (2009b) gives an entry into the modern literature on A.'s development.

CV

Last days. The accusation against A. is related in *DL* V, 6–8 including the problematic hymn itself. A.'s will is given at *DL* V, 12–16; JAEGER (1948) p. 325 speaks of the Delphian honours. A.'s sayings and letters: 'I will not allow the Athenians . . .', *FR* F666R3; on regrets over revoked honours, *FR* F667R3; 'The more alone . . .', F668R3; JAEGER (1948) pp. 320–1. T.'s will is given at *DL* V, 51–7. *Strab* XIII, 1.54–5 describes the fate of the library; see BARNES (1995a), ANAGNOSTOPOULOS (2009b) for evaluations of the story. LENNOX (2001b) ch. 5 discusses the disappearance of biology. I thank William S. Morison, Grand Valley State University, for telling me about the archaeology of the Lyceum; see LYGOURI-TOLIA (2002) for the original excavation report.

CVI

Modern assessments of A. MEDAWAR and MEDAWAR (1985) pp. 26–7 has often been quoted by Aristotelian scholars as an example of stark insensibility.

CVII

The fate of A. in early modern times. For the Paris condemnations see GAUKROGER (2007) ch. 2 and GARBER (2000). For Thomist Aristotelianism see GAUKROGER (2007) ch. 2. BALME and GOTTHELF (2002) pp. 6–35 discuss the manuscript tradition of *HA.* GAUKROGER (2007) ch. 3 discusses fifteenth-century countercurrents to Aristotelian scholasticism. Galileo's debate is from his *Dialogues on the Two Chief World Systems,* Day 2, 1632.

CVIII

The fate of the biology. Albert Magnus' quotes are from his *de Miner.*, lib. II, tr. ii, I; *de Veg.*, lib. VI, tr. ii, i. On Pomponazzi see PERFETTI (2000) ch. I, I, GAUKROGER (2001) p. 92 and GAUKROGER (2007) ch. 3.

CIX

Francis Bacon. 'And herein I cannot a little marvel': *Advancement of Learning* (1605) bk. 2, cf. *Cogitata et visa* (1607); see GAUKROGER (2001) pp. 10ff. on scientific discourse. For Bacon on teleology see *Advancement of Learning* bk. 2; on forms *Novum organum* (1620) ch. 63 and JARDINE (1974) ch. 5. For Bacon on artificial science see GAUKROGER (2001) p. 39. Glanvill is quoted by MEDAWAR (1984) p. 95 for which see a general discussion of experiment and critique of A.'s method. GRENE and DEPEW (2004) ch. 2 and GAUKROGER (2007) ch. 9 discuss Descartes' *bête machine* as given in his *Discourse on Method* (1637), V. Steno (1666) is quoted by GRENE and DEPEW (2004) p. 63.

Vitalism. See Chs LV and LXVIII; CRICK (1967) inveighs against vitalism; SCHRÖDINGER (1954/1996) ignores A.

CX

Experiment. Classical philosophers often refer to 'experiment' in the more general sense when talking about A.'s empirical investigations. LENNOX (Fall 2011), for example, refers to A.'s studies of chicken embryogenesis as an 'experiment'. It's not: it's just a really nice observational study. HANKINSON (1995) refers to the wax vessel as an experiment – again, it's not. LLOYD (1991) ch. 4 summarizes ancient Greek experiment, and views on it, but also doesn't clearly distinguish between true experiments and various observations. See also LLOYD (1987) for the relationship of empirical data to theory in A. BUTTERFIELD (1957) ch. 5 tells the complicated story of Galileo and the cannonball.

Hero. FARRINGTON (1944–9) vol. 2, ch. I is vastly enthusiastic over the *Pneumatics* and, following Diels, credits Strato for it, but see LLOYD (1973) ch. 7 and BERRYMAN (2009) ch. 5.

CXI

Styles of science. For the distinction between the two styles of science see KELL and OLIVER (2004). A. says we should theorize even when we have few facts, *DC* 292a14ff (on stars); cf. *GA* 760b28–32 (on bees). See also *DC* 293a25–31 for the relationship between proof and theory.

CXII

A.'s natures and his critics. See HENRY (2008) for a discussion of Aristotelian organismal natures and its critics. LEAR (1988) pp. 23–4, remarkably, admits and defends
virtus dormitiva arguments in A. BERRYMAN (2007), BERRYMAN (2009) and
JOHNSON (in press) both give valuable, if rather different, discussions of the
meaning of mechanistic and whether or not A.'s theories can be judged as such. I
hope to discuss Thompson's On Growth and Form, and its relationship to antiquity, in
a future paper. It is now quite commonplace for scholars of A.'s biology to speak of
mechanism, e.g. KULLMANN (1998) p. 292 and HENRY (2006a) on the mechanism of inheritance and GREGORIC and CORCILIUS (2013) on the mechanism of
the animal motion. SHIELDS (2008) pins down what ousia means to A. with respect
to artefacts and organisms. The invitation to biomedical science is from LBV
480b20. ANAGNOSTOPOULOS (2009a) discusses A.'s interest in medicine. 'This
our science': THOMPSON (1913) p. 30.

CXIII

A. and Darwin. The Tuco-Tuco appears in DARWIN (1845) where he suggests that it
is en route to becoming a blind fossorial animal such as a Proteus, mole or Aspalax –
but cautiously attributes the thought to Lamarck. In DARWIN (1859), he draws the
parallel with moles again, no longer attributes the evolutionary thought to Lamarck
and suggests that natural selection for eye loss combined with the effects of disuse
(for Darwin remains, in part, Lamarckian) might be responsible for the loss of eyes
in burrowing animals. BORGHI (2002) investigates eye reduction in various fossorial mammals, and shows that Ctenomys has slightly smaller eyes than a squirrel,
but much larger than any other fossorial mammals, and that they protect their eyes
when burrowing by closing them. 'Lack of experience is a cause': GC 316a5; see
LENNOX (2011).

CXIV

What A. teaches us. All men are born . . .: Deutsche Requiem (1949) in BORGES (1999)
p. 233. Samuel Taylor Coleridge said it first, Table Talk, 2 July 1830.

BIBLIOGRAPHY

Ackrill, J. L. 1972/1973. Aristotle's definitions of '*psuche*'. *Proceedings of the Aristotelian Society*, New Series 73:119–33.

Ackrill, J. L. 1981. *Aristotle the philosopher*. Oxford University Press, Oxford.

Adolph, E. F. 1961. Early concepts of physiological regulations. *Physiological Reviews* 41:737–70.

Agassiz, L. 1857. Quarterly meeting report. *Proceedings of the American Academy of Arts and Sciences* 3:325–84.

Amigues, S. 1988–2006. *Théophraste: recherches sur les plantes (5 vols)*. Les Belles Lettres, Paris.

Amigues, S. 2012. *Théophraste: les causes des phénomènes végétaux – Tome 1: livres I et II*. Les Belles Lettres, Paris.

Anagnostopoulos, G. 2009a. Aristotle's life *in* G. Anagnostopoulos, ed. *A companion to Aristotle*. e-book edn. Blackwell, Oxford.

Anagnostopoulos, G. 2009b. Aristotle's works and the development of his thought *in* G. Anagnostopoulos, ed. e-book edn. *A companion to Aristotle*. Blackwell, Oxford.

Anagnostopoulos, G. 2009c. Aristotle's methods *in* G. Anagnostopoulos, ed. *A companion to Aristotle*. e-book edn. Blackwell, Oxford.

Andrews, P. 1952. Aristotle, Politics IV. II. 1296a38–40. *Classical Review*, New Series 2:141–4.

Appel, T. A. 1987. *The Cuvier–Geoffroy debate: French biology in the decades before Darwin*. Oxford University Press, New York, NY.

Arnott, W. G. 2007. *Birds in the ancient world from A to Z*. Routledge, London.

Atran, S. 1993. *Cognitive foundations of natural history: towards an anthropology of science*. Cambridge University Press, Cambridge.

Ayala, F. 1968. Biology as an autonomous science. *American Scientist* 56:207–21.

Baer, K. E. von. 1828. *Über die Entwicklungsgeschichte der Thiere*. Bornträger, Königsberg.

Balme, D. M. 1939. Greek science and mechanism I. Aristotle on nature and chance. *Classical Quarterly* 33:129–38.

Balme, D. M. 1962a. *Genos* and *eidos* in Aristotle's biology. *Classical Quarterly*, New Series 12:81–98.

Balme, D. M. 1962b. Development of biology in Aristotle and Theophrastus: theory of spontaneous generation. *Phronesis* 7:91–104.

Balme, D. M. 1987a. The place of biology in Aristotle's philosophy. pp. 9–20 *in* A. Gotthelf and J. G. Lennox, eds. *Philosophical issues in Aristotle's biology*. Cambridge University Press, Cambridge.

Balme, D. M. 1987b. Aristotle's use of division and differentiae. pp. 69–89 in A. Gotthelf and J. G. Lennox, eds. *Philosophical issues in Aristotle's biology*. Cambridge University Press, Cambridge.

Balme, D. M. 1987c. Teleology and necessity. pp. 275–85 in A. Gotthelf and J. G. Lennox, eds. *Philosophical issues in Aristotle's biology*. Cambridge University Press, Cambridge.

Balme, D. M. 1987d. Aristotle's biology was not essentialist. pp. 291–312 in A. Gotthelf and J. G. Lennox, eds. *Philosophical issues in Aristotle's biology*. Cambridge University Press, Cambridge.

Balme, D. M. 1991. *History of animals: books VII–X*. Harvard University Press, Cambridge, MA.

Balme, D. M. and A. Gotthelf. 1992. *Aristotle's De partibus animalium I and De generatione animalium I (with passages from II. 1–3)*. Clarendon Press, Oxford.

Balme, D. M. and A. Gotthelf. 2002. *Historia animalium*. Cambridge University Press, Cambridge.

Barnes, J. 1982. *The Presocratic philosophers*. Routledge, London.

Barnes, J. 1987. *Early Greek philosophy*. Penguin Books, Harmondsworth.

Barnes, J. 1993. *Aristotle's Posterior Analytics*. Clarendon Press, Oxford.

Barnes, J. 1995a. Life and work. pp. 1–26 in J. Barnes, ed. *The Cambridge companion to Aristotle*. Cambridge University Press, Cambridge.

Barnes, J. 1995b. Metaphysics. pp. 66–108 in J. Barnes, ed. *The Cambridge companion to Aristotle*. Cambridge University Press, Cambridge.

Barnes, J. 1996. *Aristotle*. Oxford University Press, Oxford.

Barnstone, W. 1972. *Greek lyric poetry*. Schocken, New York, NY.

Bazos, I. and A. Yannitsaros. 2000. The history of botanical investigations in Lesvos island (East Aegean, Greece). *Biologia Gallo-Hellenica, Supplementum* 26:55–68.

Beckner, M. 1959. *The biological way of thought*. Columbia University Press, New York, NY.

Bernard, C. 1878. *Leçons sur les phénomènes de la vie communs aux animaux et aux végétaux*. Baillière, Paris.

Berryman, S. 2007. Teleology without tears: Aristotle and the role of mechanistic conceptions of organisms. *Canadian Journal of Philosophy* 37:357–70.

Berryman, S. 2009. *The mechanical hypothesis in ancient Greek natural philosophy*. Cambridge University Press, Cambridge.

Bertalanffy, L. von. 1968. *General system theory: foundations, development, applications*. Penguin Books, Harmondsworth.

Bertin, L. 1956. *Eels: a biological study*. Cleaver-Hume Press, London.

Beullens, P. and A. Gotthelf. 2007. Theodore Gaza's translation of Aristotle's *De Animalibus*: content, influence, and date. *Greek, Roman and Byzantine Studies* 47:469–513.

Biel, B. 2002. Contributions to the flora of the Aegean islands of Lesvos and Limnos, Greece. *Willdenowia* 32:209–19.

Bielby, J. *et al.* 2007. The fast-slow continuum in mammalian life history: an empirical reevaluation. *American Naturalist* 169:748–57.

Bigwood, J. M. 1993. Aristotle and the elephant again. *American Journal of Philology* 114:537–55.

Bodnár, I. 2005. Teleology across natures. *Rhizai* 2:9–29.

Bodnár, I. Spring 2012. Aristotle's natural philosophy *in* E. N. Zalta, ed. *The Stanford encyclopedia of philosophy*. (http://plato.stanford.edu/entries/aristotle-natphil/)

Bodson, L. 1983. Aristotle's statement on the reproduction of sharks. *Journal of the History of Biology* 16:391–407.

Bogaard, P. A. 1979. Aristotle's explanation of compound bodies. *Isis* 70:11–29.

Bojanus, L. V. 1819–21. *Anatome testudinis Europaeae*. Josephi Zawadzki, Vilnius.

Bolton, R. 1987. Definition and scientific method in Aristotle's *Posterior Analytics* and *Generation of Animals*. pp. 120–66 *in* A. Gotthelf and J. G. Lennox, eds. *Philosophical issues in Aristotle's biology*. Cambridge University Press, Cambridge.

Borges, J. L. 1999. *Collected fictions*. Penguin, Harmondsworth.

Borges, J. L. 2000. *Selected non-fictions*. ed. E. Weinberger. Penguin Books, New York, NY.

Borghi, C. 2002. Eye reduction in subterranean mammals and eye protective behaviour in *Ctenomys*. *Journal of Neotropical Mammology* 9:123–34.

Bos, A. P. 2003. *The soul and its instrumental body: a reinterpretation of Aristotle's philosophy of living nature*. E. J. Brill, Leiden.

Bosch, T. C. G. 2009. Hydra and the evolution of stem cells. *Bioessays* 31:478–86.

Bourgey, L. 1955. *Observation et expérience chez Aristote*. Paris.

Bowler, P. J. 1971. Preformation and pre-existence in the seventeenth century: a brief analysis. *Journal of the History of Biology-X* 4:221–44.

Brennan, P. L. R. *et al*. 2007. Coevolution of male and female genital morphology in waterfowl. *PLoS ONE* 2:e418.

Brown, T. S. 1949. Callisthenes and Alexander. *American Journal of Philology* 70.

Burkhardt, J. 1872/1999. *The Greeks and Greek civilization*. St Martin's Griffin, New York, NY.

Burnyeat, M. F. 2005. *eikōs mythos*. *Rhizai* 2:143–65.

Butterfield, H. 1957. *The origins of modern science, 1300–1800*. The Free Press, New York, NY.

Byrne, P. H. 1997. *Analysis and science in Aristotle*. State University of New York Press, Albany, NY.

Campbell, G. 2000. Zoogony and evolution in *Timaeus*, the Presocratics, Lucretius and Darwin. pp. 145–80 *in* M. R. Wright, ed. *Reason and necessity: essays on Plato's Timaeus*. Classical Press of Wales, Swansea.

Campbell, G. 2003. *Lucretius on creation and evolution: a commentary on De Rerum Natura book five, lines 772–1104*. Oxford University Press, Oxford.

Candargy, C. A. 1899. *La végétation de l'île de Lesbos*. A. Diggelmann, Uster, Zurich.

Cannon, W. B. 1932. *The wisdom of the body*. W. W. Norton, New York, NY.

Caston, V. 2009. *Phantasia* and thought *in* G. Anagnostopoulos, ed. e-book edn. *A companion to Aristotle*. Blackwell, Oxford.

Cavolini, F. 1787. *Memoria sulla generazione dei pesci e dei granchi*. Naples.

Chan, Y. F. *et al*. 2010. Adaptive evolution of pelvic reduction in sticklebacks by recurrent deletion of a PitxI enhancer. *Science* 327:302–5.

Clarke, J. T. *et al*. 1882. *Report on the investigations at Assos, 1881*. A. Williams, Boston.

Cobb, M. 2006. *The egg and sperm race: the seventeenth-century scientists who unravelled the secrets of sex, life and growth.* Pocket Books, London.

Cohen, S. M. 1996. *Aristotle on nature and incomplete substance.* Cambridge University Press, Cambridge.

Cole, F. J. 1944. *A history of comparative anatomy, from Aristotle to the eighteenth century.* Macmillan, London.

Coles, A. 1995. Biomedical models of reproduction in the fifth century BC and Aristotle's *Generation of animals. Phronesis* 40:48–88.

Conklin, E. G. 1929. Problems of development. *American Naturalist* 63:5–36.

Cooper, J. 1987. Hypothetical necessity and natural teleology in pp. 243–74 *in* A. Gotthelf and J. G. Lennox, eds. *Philosophical issues in Aristotle's biology.* Cambridge University Press, Cambridge.

Cooper, S. J. 2008. From Claude Bernard to Walter Cannon: emergence of the concept of homeostasis. *Appetite* 51:419–27.

Cornford, F. M. 1997. *Plato's cosmology: the Timaeus of Plato.* Hackett, Indianapolis.

Cosans, C. E. 1998. Aristotle's anatomical philosophy of nature. *Biology and Philosophy* 13:311–39.

Coward, K. *et al.* 2002. Gamete physiology, fertilization and egg activation in teleost fish. *Reviews in Fish Biology and Fisheries* 12:33–58.

Cresswell, R. and J. G. Schneider. 1862. *Aristotle's History of animals in ten books.* H. G. Bohn, London.

Crick, F. 1967. *Of molecules and men.* University of Washington Press, Seattle.

Crook, A. C. *et al.* 1999. Comparative study of the covering reaction of the purple sea urchin, *Paracentrotus lividus,* under laboratory and field conditions. *Journal of the Marine Biological Association of the UK* 79:1117–21.

Cuvier, G. 1834. *Recherches sur les ossemens fossiles, où l'on rétablit les caractères de plusieurs animaux dont les révolutions du globe ont détruit les espèces.* Editions d'Ocagne, Paris.

Cuvier, G. 1841. *Histoire des sciences naturelles depuis leur origine jusqu'à nos jours.* Fortin Masson, Paris.

Cuvier, G. and P. A. Latreille. 1817. *Le règne animal distribué d'après son organisation, pour servir de base à l'histoire naturelle des animaux et d'introduction à l'anatomie comparée.* Deterville, Paris.

Cuvier, G. and A. Valenciennes. 1828–49. *Histoire naturelle des poissons.* F. G. Levrault, Paris.

Dakyns, H. G. 1890. *The works of Xenophon.* Macmillan, London.

Darwin, C. R. 1837–8/2002–. Transmutation Notebook, B *in* J. van Wyhe, ed. *The complete work of Charles Darwin online (http://darwin-online.org.uk/),* Cambridge.

Darwin, C. R. 1838/2002–. Transmutation Notebook, C *in* J. van Wyhe, ed. *The complete work of Charles Darwin online (http://darwin-online.org.uk/),* Cambridge.

Darwin, C. R. 1838–9/2002–. Transmutation Notebook, E *in* J. van Wyhe, ed. *The complete work of Charles Darwin online (http://darwin-online.org.uk/),* Cambridge.

Darwin, C. R. 1845. *Journal of researches into the natural history and geology of the countries visited during the voyage of H.M.S. Beagle round the world.* 2nd edn. John Murray, London.

Darwin, C. R. 1859. *On the origin of species by means of natural selection, or, the preservation of favoured races in the struggle for life.* 1st edn. John Murray, London.

Darwin, C. R. 1868. *The variation of animals and plants under domestication.* 1st edn. John Murray, London.

Darwin, C. R. 1875. *The variation of animals and plants under domestication.* 2nd edn. John Murray, London.

Davidson, J. N. 1998. *Courtesans & fishcakes: the consuming passions of classical Athens.* St Martin's Press, New York, NY.

Davies, M. and J. Kathirithamy. 1986. *Greek insects.* Duckworth, London.

Dawkins, R. 1986. *The blind watchmaker.* W. W. Norton, New York, NY.

Del Rio-Tsonis, K. and P. A. Tsonis. 2003. Eye regeneration at the molecular age. *Developmental Dynamics* 226:211–24.

Delbrück, M. 1971. Aristotle-totle-totle. pp. 50–5 *in* J. Monod and E. Borek, eds. *Of microbes and life: festschrift for André Lwoff.* Columbia University Press, New York, NY.

Depew, D. J. 1995. Humans and other political animals in Aristotle's *History of Animals. Phronesis* 40:156–81.

Depew, D. J. 2008. Consequence etiology and biological teleology in Aristotle and Darwin. *Studies in History and Philosophy of Science Part C: Studies in History and Philosophy of Biological and Biomedical Sciences* 39:379–90.

Dermitzakis, M. D., ed. 1999. *Natural history collection of Vrisa-Lesvos Island.* National and Kapodistrian University of Athens, Athens.

Derrickson, E. M. 1992. Comparative reproductive strategies of altricial and precocial Eutherian mammals. *Functional Ecology* 6:57–65.

Dewsbury, D. A. 1999. The proximate and the ultimate: past, present, and future. *Behavioural Processes* 46:189–99.

Diamond, J. M. 1966. Zoological classification system of a primitive people. *Science* 151:1102–4.

Dobzhansky, T. 1973. Nothing in biology makes sense except in the light of evolution. *American Biology Teacher* 35:125–9.

Driesch, H. 1914. *The history and theory of vitalism.* Macmillan, London.

Drossart Lulofs, H. J. 1957. Aristotle's *Peri phyton. Journal of Hellenic Studies* 57:75–80.

Dudley, S. 2009. *A birdwatching guide to Lesvos.* Subbuteo Natural History Books, Shrewsbury.

Düring, I. 1957. *Aristotle in the ancient biographical tradition.* Almqvist & Wiksell, Göteborg.

Durrant, M. 1993. *Aristotle's De anima in focus.* Routledge, London.

Egerton, F. N. 1968. Ancient sources for animal demography. *Isis* 59:175–89.

Egerton, F. N. 2001a. A history of the ecological sciences: early Greek origins. *Bulletin of the Ecological Society of America* 82:93–7.

Egerton, F. N. 2001b. A history of the ecological sciences: Aristotle and Theophrastos. *Bulletin of the Ecological Society of America* 82:149–52.

Einarson, B. and G. K. K. Link. 1976–90. *Theophrastus: De causis plantarum.* 2 vols. Harvard University Press, Cambridge, MA.

474 THE LAGOON

Ekberg, O. and S. V. Sigurjonsson. 1982. Movement of the epiglottis during de-glutition – a cineradiographic study. *Gastrointestinal Radiology* 7:101–7.

Ellis, J. 1765. On the nature and formation of sponges: in a letter from John Ellis, Esquire, F.R.S. to Dr. Solander, F.R.S. *Philosophical Transactions of the Royal Society* 55:280–9.

Estes, R. 1991. *The behavior guide to African mammals: including hoofed mammals, carnivores, primates.* University of California Press, Berkeley, CA.

Falcon, A. 2005. *Aristotle and the science of nature: unity without uniformity.* Cambridge University Press, Cambridge.

Farley, J. 1977. *The spontaneous generation controversy from Descartes to Oparin.* Johns Hopkins University Press, Baltimore, MD.

Farquharson, A. S. L. 1912. *De incessu animalium.* W. D. Ross and J. A. Smith, *The works of Aristotle translated into English.* Clarendon Press, Oxford.

Farrington, B. 1944–9. *Greek science, its meaning for us.* 2 vols. Penguin Books, Harmondsworth.

Finch, C. 2007. *The biology of human longevity: inflammation, nutrition, and aging in the evolution of lifespans.* Academic Press, New York, NY.

Fischer, H. 1894. Note sur le bras hectocotylisé de l'*Octopus vulgaris*, Lamarck. *Journal de Conchyliologie* 42:13–19.

Fishburn. 2004. 'Natura non facit saltum' in Alfred Marshall (and Charles Darwin). *History of Economics Review* 40:59–68.

Frampton, M. F. 1991. Aristotle's cardiocentric model of animal locomotion. *Journal of the History of Biology-X* 24:291–330.

Frede, M. 1992. On Aristotle's conception of soul. pp. 93–107 *in* M. C. Nussbaum and A. Rorty, eds. *Essays on Aristotle's De anima.* Clarendon Press, Oxford.

French, R. K. 1994. *Ancient natural history: histories of nature.* Routledge, London.

Freudenthal, G. 1995. *Aristotle's theory of material substance: heat and pneuma, form and soul.* Clarendon Press, Oxford.

Funk, H. 2012. R. J. Gordon's discovery of the spotted hyena's extraordinary genitalia in 1777. *Journal of the History of Biology* 45:301–28.

Furth, M. 1987. Aristotle's biological universe: an overview. pp. 21–52 *in* A. Gotthelf and J. G. Lennox, eds. *Philosophical issues in Aristotle's biology.* Cambridge University Press, Cambridge.

Furth, M. 1988. *Substance, form, and psyche: an Aristotelean metaphysics.* Cambridge University Press, Cambridge.

Gaeth, A. P. et al. 1999. The developing renal, reproductive, and respiratory systems of the African elephant suggest an aquatic ancestry. *Proceedings of the National Academy of Sciences, USA* 96.

Garber, D. 2000. Defending Aristotle/defending society in early 17th century Paris. pp. 135–60 *in* W. Detel and C. Zittel, eds. *Wissensideale und Wissenskulturen in der frühen Neuzeit/Ideals and cultures of knowledge in early modern Europe.* Akademie Verlag, Berlin.

Gardener, A. and A. Grafen. 2009. Capturing the superorganism: a formal theory of group adaptation. *Journal of Evolutionary Biology* 22:1–13.

Gardner, A. and J. P. Conlon. 2013. Cosmological natural selection and the purpose of the universe. *Complexity* 18:48–56.

Garman, S. 1890. *Silurus (Parasilurus) aristotelis. Bulletin of the Essex Institute* 22:56−9.

Gaukroger, S. 2001. *Francis Bacon and the transformation of early-modern philosophy.* Cambridge University Press, Cambridge.

Gaukroger, S. 2007. *The emergence of a scientific culture: science and the shaping of modernity 1201−1685.* Oxford University Press, Oxford.

Gaza, T. 1476. *De animalibus.* Johannes de Colonia and Johannes Manthen, Venice.

Gelber, J. 2010. Form and inheritance in Aristotle's embryology. *Oxford Studies in Ancient Philosophy* 39:183−212.

Gems, D. and L. Partridge. 2013. Genetics of longevity in model organisms: debates and paradigm shifts. *Annual Review of Physiology* 75:621−44.

Gill, M. J. 1989. *Aristotle on substance: the paradox of unity.* Princeton University Press, Princeton, NJ.

Gill, T. 1906. Parental care in fishes. *Annual report of the Smithsonian Institution, Washington for the year ending June 30,* 1905:403−531.

Gill, T. 1907. The remarkable story of a Greek fish. *Washington University Bulletin* 5:5−15.

Glass, B. 1947. Maupertuis and the beginnings of genetics. *Quarterly Review of Biology* 22:196−210.

Gotthelf, A. 2012. *Teleology, first principles and scientific method in Aristotle's biology.* Oxford University Press, Oxford.

Gotthelf, A. and J. G. Lennox, eds. 1987. *Philosophical issues in Aristotle's biology.* Cambridge University Press, Cambridge.

Grafen, A. 2007. The formal Darwinism project: a mid-term report. *Journal of Evolutionary Biology* 20:1243−54.

Graham, D. W. 1999. *Aristotle: Physics, book VIII: translated with a commentary.* Clarendon Press, Oxford.

Granger, H. 1985. Continuity of kinds. *Phronesis* 30:181−200.

Granger, H. 1987. Deformed kinds and the fixity of species. *Classical Quarterly,* New Series 37:110−16.

Green, P. 1982. Longus, Antiphon, and the Topography of Lesbos. *Journal of Hellenic Studies* 102:210−14.

Green, P. 1989. *Classical bearings: interpreting ancient history.* University of California Press, Berkeley, CA.

Gregoric, P. and K. Corcilius. 2013. Aristotle's model of animal motion. *Phronesis* 58:52−97.

Gregory, A. 2000. *Plato's philosophy of science.* Duckworth, London.

Grene, M. 1998. *A portrait of Aristotle.* Thoemmes Press, Bristol.

Grene, M. and D. J. Depew. 2004. *The philosophy of biology: an episodic history.* Cambridge University Press, Cambridge.

Guidetti, P. and M. Mori. 2005. Morpho-functional defences of Mediterranean sea urchins, *Paracentrotus lividus* and *Arbacia lixula,* against fish predators. *Marine Biology* 797−802.

Guthrie, W. K. C. 1939. *Aristotle on the heavens.* Harvard University Press, Cambridge, MA.

Guthrie, W. K. C. 1981. *Aristotle: an encounter.* Cambridge University Press.

Guyader, H. le. 2004. *Étienne Geoffroy Saint-Hilaire, 1772−1844: a visionary naturalist.* University of Chicago Press, Chicago, IL.

Haeckel, E. 1866. *Generelle Morphologie der Organismen. Allgemeine Grundzüge der organischen Formen-Wissenschaft, mechanisch begründet durch die von Charles Darwin reformirte Descendenz-Theorie.* Reimer, Berlin.

Haldane, J. B. S. 1955. Aristotle's account of bees' 'dances'. *Journal of Hellenic Studies* 75:24–5.

Hall, B. K. 2003. Descent with modification: the unity underlying homology and homoplasy as seen through an analysis of development and evolution. *Biological Reviews* 78:409–33.

Hankinson, J. 1995. Philosophy of science. pp. 109–39 *in* J. Barnes, ed. *The Cambridge companion to Aristotle.* Cambridge University Press, Cambridge.

Hannaford, I. 1996. *Race: the history of an idea in the west.* Johns Hopkins University Press, Baltimore, MD.

Harbour, J. W. *et al.* 2010. Frequent mutation of BAP1 in metastasizing uveal melanomas. *Science* 330:1410–13.

Harris, C. R. S. 1973. *The heart and the vascular system in ancient Greek medicine, from Alcmaeon to Galen.* Clarendon Press, Oxford.

Hawking, S. 1988. *A brief history of time: from big bang to black holes.* Bantam, New York, NY.

Heath, M. 2008. Aristotle on natural slavery. *Phronesis* 53:243–70.

Heller, J. L. and J. M. Penhallurick. 2007. *The index of books and authors cited in the zoological works of Linnaeus.* Ray Society, London.

Henry, D. 2006a. Aristotle on the mechanism of inheritance. *Journal of the History of Biology* 39:425–55.

Henry, D. 2006b. Understanding Aristotle's reproductive hylomorphism. *Apeiron* 39:257–87.

Henry, D. 2007. How sexist is Aristotle's developmental biology? *Phronesis* 52:251–69.

Henry, D. 2008. Organismal natures. *Apeiron* 41:47–74.

Hett, W. S. 1936. *On the soul.* Harvard University Press, Cambridge, MA.

Hicks, R. D. 1925. *Diogenes Laertius: lives of eminent philosophers.* Harvard University Press, Cambridge, MA.

Hort, A. F. 1916. *Theophrastus: Enquiry into Plants.* 3 vols. Harvard University Press, Cambridge, MA.

Houghton, R. W. 1873. On the *silurus* and *glanis* of the ancient Greeks and Romans. *Annals and Magazine of Natural History* 11:199–206.

Houghton, S. 1863. On the form of the cells made by various wasps and by the honey bee; with an appendix on the origin of species. *Annals and Magazine of Natural*

Huxley, J. 1942. *Evolution, the modern synthesis.* Allen & Unwin, London.

Huxley, T. H. 1879. On certain errors respecting the structure of the heart attributed to Aristotle. *Nature* 21:1–5.

Jaeger, W. 1948. *Aristotle: fundamentals of the history of his development.* Clarendon Press, Oxford.

Jardine, L. 1974. *Francis Bacon: discovery and the art of discourse.* Cambridge University Press, Cambridge.

Johansen, T. K. 1997. *Aristotle on the sense organs.* Cambridge University Press, Cambridge.

Johansen, T. K. 2004. *Plato's natural philosophy: a study of the Timaeus-Critias.* Cambridge University Press, Cambridge.

Johnson, D. L. 1980. Problems in the land vertebrate zoogeography of certain islands and the swimming powers of elephants. *Journal of Biogeography* 7:383–98.

Johnson, M. R. 2005. *Aristotle on teleology.* Clarendon Press, Oxford.

Johnson, M. R. in press. Aristotelian mechanistic explanation *in* J. Rocca, ed. *Teleology in the ancient world: the dispensation of nature.* Cambridge University Press, Cambridge.

Johnston, G. 1838. *A history of the British zoophytes.* Lizars, Edinburgh.

Jones, K. E. *et al.* 2009. PanTHERIA: a species-level database of life history, ecology, and geography of extant and recently extinct mammals. *Ecology* 90: 2648

Jones, W. H. S. *et al.* 1923–2012. *Hippocrates.* 11 vols. Harvard University Press, Cambridge, MA.

Judson, O. P. 2005. The case of the female orgasm: bias in the science of evolution. *Nature* 436:916–17.

Kalinka, A. T. *et al.* 2010. Gene expression divergence recapitulates the developmental hourglass model. *Nature* 468:811–14.

Kant, E. 1793. *Kritik der Urteilskraft.* 2nd edn.

Keaveney, A. 1982. *Sulla, the last republican.* Croom Helm, London.

Kell, D. B. and S. G. Oliver. 2004. Here is the evidence, now what is the hypothesis? The complementary roles of inductive and hypothesis-driven science in the post-genomic era. *Bioessays* 26:99–105.

Kelley, D. A. 2002. The functional morphology of penile erection: tissue designs for increasing and maintaining stiffness. *Integrative and Comparative Biology* 42:216–21.

King, R. A. H. 2001. *Aristotle on life and death.* Duckworth, London.

King, R. A. H. 2007. Review of *The soul and its instrumental body: a reinterpretation of Aristotle's philosophy of living nature* by A. P. Bos. *Classical Review,* New Series 57:322–3.

King, R. A. H. 2010. The concept of life and the life-cycle in *De juventute.* pp. 171–87 *in* S. Föllinger, ed. *Was ist 'Leben'? Aristoteles' Anschauungen zur Entstehung und Funktionsweise von Leben.* Franz Steiner Verlag, Stuttgart.

Kirk, G. S. *et al.* 1983. *The presocratic philosophers: a critical history with a selection of texts.* Cambridge University Press, Cambridge.

Kitchell, K. F. 2014. *Animals in the ancient world from A–Z.* Routledge, London.

Kjellberg, F. *et al.* 1987. The stability of the symbiosis between dioecious figs and their pollinators – a study of *Ficus caria* L. and *Blastophaga psenes* L. *Evolution* 41:693–704.

Koutsogiannopoulos, D. 2010. *Ta psara tis Hellas* [Fishes of Greece]. Athens.

Kullmann, W. 1979. *Die Teleologie in der aristotelischen Biologie: Aristoteles als Zoologe, Embryologe und Genetiker.* Winter, Heidelberg.

Kullmann, W. 1991. Man as a political animal *in* D. Keyt and F. D. Mill, eds. *A companion to Aristotle's Politics*. Blackwell, Oxford.

Kullmann, W. 1998. *Aristoteles und die moderne Wissenschaft*. Franz Steiner Verlag, Stuttgart.

Kullmann, W. 2007. *Aristoteles: über die teile der Lebewesen*. Akademie Verlag, Berlin.

Kullmann, W. 2008. Evolutionsbiologie vorstellungen bei Aristoteles. pp. 70–80 *in* K.-M. Hingst and M. Liatisi, eds. *Pragmata: Festscrhift für Klaus Ohler zum 80. Geburtstag*. Gunter Narr Verlag, Tübingen.

Lane-Fox, R. 1973. *Alexander the Great*. Allen Lane, London.

Lawson-Tancred, H. 1986. *De anima (On the soul)*. Penguin Books, Harmondsworth.

Lear, J. 1988. *Aristotle: the desire to understand*. Cambridge University Press, Cambridge.

Lee, H. D. P. 1948. Place-names and dates of Aristotle's biological works. *Classical Quarterly* 42:61–7.

Lee, H. D. P. 1985. The fishes of Lesbos again *in* A. Gotthelf, ed. *Aristotle on nature and living things: philosophical and historical studies presented to David M. Balme on his seventieth birthday*. Mathesis Publications, Pittsburgh, PA.

Leeuwenhoek, A. 1931–99. *Alle de brieven van Antoni van Leeuwenhoek*. Swets & Zeitlinger, Amsterdam.

Lelong, B. M. 1891. *California fig industry with a chapter on fig caprification*. California State Office, Sacramento, CA.

Lennox, J. G. 1984. Aristotle's lantern. *Journal of Hellenic Studies* 103:147–51.

Lennox, J. G. 2001a. *Aristotle on the Parts of Animals I–IV*. Clarendon Press, Oxford.

Lennox, J. G. 2001b. *Aristotle's philosophy of biology: studies in the origin of biology*. Cambridge University Press, Cambridge.

Lennox, J. G. 2006. The comparative study of animal development: from Aristotle to William Harvey's Aristotelianism. pp. 21–46 *in* J. E. H. Smith, ed. *The problem of animal generation in early modern philosophy*. Cambridge University Press, Cambridge.

Lennox , J. G. 2010. Bios, praxis and the unity of life. pp. 239–59 *in* S. Föllinger, ed. *Was ist 'Leben'? Aristoteles' Anschauungen zur Entstehung und Funktionsweise von Leben*. Franz Steiner Verlag, Stuttgart.

Lennox, J. G. 2011. Aristotle on norms of inquiry. *HOPOS: The Journal of the International Society for the History of Philosophy of Science* 1:23–46.

Lennox, J. G. Fall 2011. Aristotle's biology *in* E. N. Zalta, ed. *The Stanford encyclopedia of philosophy*. (http://plato.stanford.edu/entries/aristotle-biology/)

Lenoir, T. 1982. *The strategy of life: teleology and mechanics in nineteenth century German biology*. Reidel, Dordrecht.

Leroi, A. M. 2001. Molecular signals versus the *loi de balancement*. *Trends in Ecology and Evolution* 16:24–9.

Leroi, A. M. 2003. *Mutants: on the forms, varieties and errors of the human body*. HarperCollins, London.

Leroi, A. M. 2010. Function and constraint in Aristotle and evolutionary theory. pp. 261–84 *in* S. Föllinger, ed. *Was ist 'Leben'? Aristoteles' Anschauungen zur Entstehung und Funktionsweise von Leben*. Franz Steiner Verlag, Stuttgart.

Leroi, A. M. *et al.* 1994. What does the comparative method reveal about adaptation? *American Naturalist* 143:381–402.

Leroi, A. M. *et al.* 2003. Cancer selection. *Nature Reviews Cancer* 3:226–31.

Leunissen, M. 2007. The structure of teleological explanations in Aristotle: theory and practice. *Oxford Studies in Ancient Philosophy* 33:145–78.

Leunissen, M. 2010a. *Explanation and teleology in Aristotle's science of nature.* Cambridge University Press, Cambridge.

Leunissen, M. 2010b. Aristotle's syllogistic model of knowledge and the biological sciences: demonstrating natural processes. *Apeiron* 43:31–60.

Leunissen, M. 2013. Biology and teleology in Aristotle's account of the city *in* J. Rocca, ed. *Teleology in the ancient world: the dispensation of nature.* e-book edn. Cambridge University Press, Cambridge.

Lewes, G. H. 1864. *Aristotle: a chapter from the history of science, including analyses of Aristotle's scientific writings.* Smith, Elder, London.

Lind, L. R. 1963. *Aldrovandi on chickens: the ornithology of Ulisse Aldrovandi,* vol. II, book XIV. University of Oklahoma Press, Norman, OK.

Linnaeus, C. 1735. *Systema naturæ, sive regna tria naturæ systematice proposita per classes, ordines, genera, & species.* 1st edn. Joannis Wilhelmi de Groot, Leiden.

Linnaeus, C. and J. F. Gmelin. 1788–93. *Systema naturae per regna tria naturae: secundum classes, ordines, genera, species, cum characteribus, differentiis, synonymis, locis.* 13th edn. Georg Emanuel Beer, Leipzig.

Littré, E. 1839–61. *Hippocrate: oeuvres complètes.* 10 vols. Baillière, Paris.

Lloyd, E. A. 2006. *The case of the female orgasm: bias in the science of evolution.* Harvard University Press, Cambridge, MA.

Lloyd, G. E. R. 1970. *Early Greek science: Thales to Aristotle.* Chatto & Windus, London.

Lloyd, G. E. R. 1973. *Greek science after Aristotle.* Chatto & Windus, London.

Lloyd, G. E. R. 1975. A Note on Erasistratus of Ceos. *Journal of Hellenic Studies* 95:172–5.

Lloyd, G. E. R. 1979. *Magic, reason, and experience: studies in the origin and development of Greek science.* Cambridge University Press, Cambridge.

Lloyd, G. E. R. 1983. *Science, folklore, and ideology: studies in the life sciences in ancient Greece.* Cambridge University Press, Cambridge.

Lloyd, G. E. R. 1987. Empirical research in Aristotle's biology. pp. 53–63 *in* A. Gotthelf and J. G. Lennox, eds. *Philosophical issues in Aristotle's biology.* Cambridge University Press, Cambridge.

Lloyd, G. E. R. 1991. *Methods and problems in Greek science.* Cambridge University Press, Cambridge.

Lloyd, G. E. R. 1996. *Aristotelian explorations.* Cambridge University Press, Cambridge.

Lloyd, G. E. R. 2006. *Principles and practices in ancient Greek and Chinese science.* Ashgate, Aldershot.

Loeb, J. 1906. *The dynamics of living matter.* Columbia University Press, New York, NY.

Loeck, G. 1991. Aristotle's technical simulation and its logic of causal relations. *History and Philosophy of the Life Sciences* 13:3–32.

Lones, T. E. 1912. *Aristotle's researches in natural science.* West, Newman, London.

480   THE LAGOON

Long, A. A. and D. N. Sedley. 1987. *The Hellenistic philosophers: vol. 1. translations of principal sources, with philosophical commentary.* Cambridge University Press, Cambridge.

Lonie, I. M. 1981. *The Hippocratic treatises 'On generation', 'On the nature of the child', 'Diseases IV': a commentary.* Walter de Gruyter, Berlin.

Lorenz, H. Summer 2009. Ancient theories of the soul *in* E. N. Zalta, ed. *The Stanford encyclopedia of philosophy.* (http://plato.stanford.edu/entries/ancient-soul/)

Lovejoy, A. O. 1936. *The great chain of being: a study of the history of an idea.* Harvard University Press, Cambridge, MA.

Lygouri-Tolia, E. 2002. Excavating an ancient palaestra in Athens. pp. 203–12 *in* M. Stamatopoulou and M. Yeroulanou, eds. *Excavating Classical Culture.* Oxford University Press, Oxford.

Lynch, J. 1972. *Aristotle's school: a study of a Greek educational institution.* University of California Press, Berkeley, CA.

Maderspacher, F. 2007. All the queen's men. *Current Biology* 17:R191–R195.

Mason, H. J. 1979. Longus and the topography of Lesbos. *Transactions of the American Philological Association* 59:149–63.

Matthen, M. 2009. Teleology in living things *in* G. Anagnostopoulos, ed. e-book edn. *A companion to Aristotle.* Blackwell, Oxford.

Mayhew, R. 2004. *The female in Aristotle's biology: reason or rationalization.* University of Chicago Press, Chicago, IL.

Mayor, A. 2000. *The first fossil hunters: paleontology in Greek and Roman times.* Princeton University Press, Princeton, NJ.

Mayr, E. 1961. Cause and effect in biology. *Science* 134:1501–6.

Mayr, E. 1982. *The growth of biological thought: diversity, evolution, and inheritance.* Belknap Press, Cambridge, MA.

Mayr, O. 1971. *Origins of feedback control.* MIT Press, Cambridge, MA.

Mebius, R. E. and G. Kraal. 2005. Structure and function of the spleen. *Nature Reviews Immunology* 5:606–16.

Medawar, P. B. 1951/1981. *The uniqueness of the individual.* Dover, New York, NY.

Medawar, P. B. 1984. *Pluto's Republic.* Oxford University Press, Oxford.

Medawar, P. B. and J. S. Medawar. 1985. *Aristotle to zoos: a philosophical dictionary of biology.* Harvard University Press, Cambridge, MA.

Meyer, J. B. 1855. *Aristoteles Tierkunde: ein Beitrag zur Geschichte der Zoologie, Physiologie und alten Philosophie.* Reimer, Berlin.

Millar, J. S. and R. M. Zammuto. 1983. Life histories of mammals: an analysis of life tables. *Ecology* 64:631–5.

Miller, M. G. and A. E. Miller. 2010. Aristotle's dynamic conception of the *psuchē* as being alive. pp. 55–88 *in* S. Föllinger, ed. *Was ist 'Leben'? Aristoteles' Anschauungen zur Entstehung und Funktionsweise von Leben.* Franz Steiner Verlag, Stuttgart.

Morsink, J. 1982. *Aristotle on The generation of animals: a philosophical study.* University of America Press, Washington, DC.

Moureaux, C. and P. Dubois. 2012. Plasticity of biometrical and mechanical properties of *Echinocardium cordatum* spines according to environment. *Marine Biology* 159:471–9.

Müller, J. 1842. *Über den glatten Hai des Aristoteles und über die Verschiedenheiten unter den Haifischen und Rochen in der Entwicklung des Eies*. Königlichen Akademie des Wissenschafte, Berlin.

Müller-Wille, S. and V. Orel. 2007. From Linnaean species to Mendelian factors: elements of hybridism, 1751–1870. *Annals of Science* 64:171–215.

Murray, A. T. 1919. *The Odyssey: books 1–12*. Harvard University Press, Cambridge, MA.

Natali, C. 2013. *Aristotle: his life and school*. Princeton University Press, Princeton, NJ.

Needham, J. 1934. *A history of embryology*. Cambridge University Press, Cambridge.

Negbi, M. 1995. Male and female in Theophrastus' botanical works. *Journal of the History of Biology* 28:317–32.

Nickel, M. 2004. Kinetics and rhythm of body contractions in the sponge *Tethya wilhelma* (Porifera: Demospongiae). *Journal of Experimental Biology* 207:4515–24.

Nielsen, K. M. 2008. The private parts of animals: Aristotle on the teleology of sexual difference. *Phronesis* 53:373–405.

Nussbaum, M. C. 1978. *Aristotle's De motu animalium: text with translation, commentary, and interpretive essays*. Princeton University Press, Princeton, NJ.

Nussbaum, M. C. 1982. Saving Aristotle's appearances. pp. 267–94 *in* M. Schofield and M. Nussbaum, eds. *Language and Logos*. Cambridge University Press, Cambridge.

Nussbaum, M. C. and A. Rorty. 1992. *Essays on Aristotle's De anima*. Clarendon Press, Oxford.

Ogle, W. 1882. *Aristotle on the Parts of Animals*. K. Paul, French, London.

Onuki, A. and H. Somiya. 2004. Two types of sounds and additional spinal nerve innervation to the sonic muscle in John Dory, *Zeus faber* (Zeiformes: Teleostei). *Journal of the Marine Biological Association of the United Kingdom* 84:843–50.

Oppenheimer, J. M. 1936. Historical introduction to the study of teleostean development. *Osiris* 2:124–48.

Osborne, R. 2011. *The history written on the classical Greek body*. Cambridge University Press, Cambridge.

Oser-Grote, C. 2004. *Aristoteles und das Corpus Hippocraticum*. Franz Steiner Verlag, Stuttgart.

Outram, D. 1986. Uncertain legislator: Georges Cuvier's laws of nature in their intellectual context. *Journal of the History of Biology* 19:323–68.

Owen, G. E. L. 1961/1986. *Tithenai ta phainomena in* M. Nussbaum, ed. *Logic, science and dialectic: collected papers in Greek philosophy*. Cornell University Press, Ithaca, NY.

Owen, R. 1843. *Lectures on the comparative anatomy and physiology of the invertebrate animals, delivered at the Royal College of Surgeons, in 1843*. Longman, Brown, London.

Owen, R. 1855. *Lectures on the comparative anatomy and physiology of the invertebrate animals: delivered at the Royal College of Surgeons*. Longman, Brown, Green & Longmans, London.

Owen, R. 1866. *On the anatomy of vertebrates*. Longmans, Green, London.

Owen, R. 1868. *Derivative hypothesis of life and species*. Longmans, Green, London.

Paley, W. 1809/2006. *Natural theology*. Oxford University Press, Oxford.

Palsson, B. Ø. 2006. *Systems biology: properties of reconstructed networks.* Cambridge University Press, Cambridge.

Pease, A. S. 1942. Fossil fishes again. *Isis* 33:689–90.

Peck, A. L. 1943. *The generation of animals.* Harvard University Press, Cambridge, MA.

Peck, A. L. 1965. *Historia animalium: books I–III.* Harvard University Press, Cambridge, MA.

Peck, A. L. 1970. *Historia animalium: books IV–VI.* Harvard University Press, Cambridge, MA.

Pellegrin, P. 1986. *Aristotle's classification of animals: biology and the conceptual unity of the Aristotelian corpus.* University of California Press, Berkeley, CA.

Perfetti, S. 2000. *Aristotle's zoology and its Renaissance commentators, 1521–1601.* Leuven University Press, Leuven.

Phillips, P. C. 2008. Epistasis – the essential role of gene interactions in the structure and evolution of genetic systems. *Nature Reviews Genetics* 9:855–67.

Pimm, S. L. 1991. *The balance of nature? Ecological issues in the conservation of species and communities.* University of Chicago Press, Chicago, IL.

Pinto-Correia, C. 1997. *The ovary of Eve: egg and sperm and preformation.* University of Chicago Press, Chicago, IL.

Platt, A. 1910. *De generatione animalium.* J. A. Smith and W. D. Ross, eds. *The works of Aristotle translated into English.* Vol. 5. Clarendon Press, Oxford.

Popper, K. R. 1945/1962. *The open society and its enemies.* Harper Torchbooks, New York, NY.

Preus, A. 1975. *Science and philosophy in Aristotle's biological works.* G. Olms, New York, NY.

Proman, J. M. and J. D. Reynolds. 2000. Differences in head shape of the European eel, *Anguilla anguilla. Fisheries Management and Ecology* 7:349–54.

Pyle, A. J. 2006. Malebranche on animal generation: pre-existence and the microscope. pp. 194–214 *in* J. E. H. Smith, ed. *The problem of animal generation in early modern philosophy.* Cambridge University Press, Cambridge.

Quarantotto, D. 2005. *Causa finale, sostanza, essenza in Aristotele: saggio sulla struttura dei processi teleologici naturali e sulla funzione del telos.* Bibliopolis, Naples.

Quarantotto, D. 2010. Aristotle on the soul as a principle of unity. pp. 35–53 *in* S. Föllinger, ed. *Was ist 'Leben'? Aristoteles' Anschauungen zur Entstehung und Funktionsweise von Leben.* Franz Steiner Verlag, Stuttgart.

Rackham, H. *et al.* 1938–62. *Pliny: natural history.* 10 vols. Harvard University Press, Cambridge, MA.

Radner, K. 2007. The winged snakes of Arabia and the fossil site of Makhtesh Ramon in the Negev. pp. 353–65 *in* M. Köhbach, S. Procházka, G. J. Selz, and L. Rüdiger, eds. *Festschrift für Hermann Hunger zum 65. Geburtstag.* Institut für Orientalistik, Vienna.

Rawlinson, G. *et al.* 1858–60/1997. *Herodotus: the histories.* Everyman, New York, NY.

Rees, M. 1999. *Just six numbers: the deep forces that shape the universe.* HarperCollins, London.

Reiss, J. 2009. *Retiring Darwin's watchmaker.* University of California Press, Berkeley, CA.

Ren, L. *et al.* 2008. The movements of limb segments and joints during locomotion in African and Asian elephants. *Journal of Experimental Biology* 211:2735–51.

Rodríguez Pérez, D. 2011. Contextualizing symbols: 'the eagle and the snake' in the ancient Greek world. *Boreas: Münstersche Beiträge zur Archäologie* 33:1–18.

Roff, D. A. 2002. *Life history evolution.* Sinauer Associates Sunderland, MA.

Roger, J. 1997. *The life sciences in eighteenth-century French thought.* Stanford University Press, Redwood City, CA.

Romm, J. S. 1989. Aristotle's elephant and the myth of Alexander's scientific patronage. *American Journal of Philology* 110:566–75.

Rose, M. R. 1991. *Evolutionary biology of aging.* Oxford University Press, New York, NY.

Rosenblueth, A. *et al.* 1943. Behavior, purpose and teleology. *Philosophy of Science* 10:8–24.

Ross, W. D. 1915. *Metaphysica.* J. A. Smith and W. D. Ross, *The works of Aristotle translated into English.* Vol. VIII. Clarendon Press, Oxford.

Ross, W. D. 1995. *Aristotle.* Routledge, London.

Rudwick, M. J. S. 1997. *Georges Cuvier, fossil bones, and geological catastrophes: new translations & interpretations of the primary texts.* University of Chicago Press, Chicago, IL.

Ruestow, E. G. 1984. Leeuwenhoek and the campaign against spontaneous generation. *Journal of the History of Biology* 17:225–48.

Ruse, M. 1980. Charles Darwin and group selection. *Annals of Science* 37:615–30.

Ruse, M. 1989. Do organisms exist? *American Zoologist* 29:1061–6.

Russell, E. S. 1916. *Form and function: a contribution to the history of animal morphology.* John Murray, London.

Ryle, G. 1949. *The concept of mind.* Hutchinson, London.

Sander, K. 1993a. Hans Driesch's 'philosophy really *ab ovo*', or, why to be a vitalist. *Roux's Archives of Developmental Biology* 202:1–3.

Sander, K. 1993b. Entelechy and the ontogenetic machine: work and views of Hans Driesch from 1895–1910. *Roux's Archives of Developmental Biology* 202:67–9.

Scalitas, R. 2009. Mixing the elements *in* G. Anagnostopoulos, ed. e-book edn. *A companion to Aristotle.* Blackwell, Oxford.

Scharfenberg, L. N. 2001. *Die Cephalopoden des Aristoteles im Lichte der modernen Biologie.* Wissenschaftlicher Verlag, Trier.

Schaus, G. P. and N. Spencer. 1994. Notes on the Topography of Eresos. *American Journal of Archaeology* 98:411–30.

Schmidtt, C. B. 1965. Aristotle as a cuttlefish: the origin and development of a Renaissance image. *Studies in the Renaissance* 12:60–72.

Schnitzler, A. R. 2011. Past and present distribution of the North African–Asian lion subgroup: a review. *Mammal Review* 41:220–43.

Schrödinger, E. 1944/1967. *What is life?* Cambridge University Press, Cambridge.

Schrödinger, E. 1954/1996. *Nature and the Greeks and science and humanism.* Cambridge University Press, Cambridge.

Sedley, D. 2007. *Creationism and its critics in antiquity.* University of California Press, Berkeley, CA.

Sedley, D. N. 1991. Is Aristotle's teleology anthropocentric? *Phronesis* 36:179–96.

Shapiro, M. D. *et al.* 2004. Genetic and developmental basis of evolutionary pelvic reduction in threespine sticklebacks. *Nature* 428:717–23.

Sharples, R. W. 1995. *Theophrastus of Eresus: sources for his life, writings, thought and influence: commentary.* Vol. 5: *Sources on biology (human physiology, living creatures, botany: texts 328–435).* E. J. Brill, Leiden.

Shields, C. 2008. Substance and life in Aristotle. *Apeiron* 41:129–51.

Simon, H. A. 1996. *The sciences of the artificial.* 3rd edn. MIT Press, Cambridge, MA.

Simpson, R. L. P. 1998. *A philosophical commentary on the Politics of Aristotle.* University of North Carolina Press, Chapel Hill, NC.

Sisson, S. 1914. *The anatomy of the domestic animals.* W. B. Saunders, Philadelphia.

Smith, C. L. 1965. The patterns of sexuality and the classification of serranid fishes. *American Museum Novitates* 2207:1–20.

Solmsen, F. 1955. Antecedents of Aristotle's psychology and scale of beings. *American Journal of Philology* 76:148–64.

Solmsen, F. 1960. *Aristotle's system of the physical world: a comparison with his predecessors.* Cornell University Press, Ithaca, NY.

Solmsen, F. 1978. The fishes of Lesvos and their alleged significance for the development of Aristotle. *Hermes* 106:467–84.

Solounias, N. and A. Mayor. 2004. Ancient references to the fossils from the land of Pythagoras. *Earth Science History* 23:283–96.

Someren, E. J. W. von. 2007. Thermoregulation and aging. *American Journal of Physiology – Regulatory, Integrative and Comparative Physiology* 292:R99–R102.

Sorabji, R. 1988. *Matter, space and motion: theories in antiquity and their sequel.* Duckworth, London.

Spencer, H. 1864. *Principles of biology.* Williams & Norgate, London.

Starck, J. M. and R. E. Ricklefs. 1998. Patterns of development: the altricial–precocial spectrum. pp. 3–30 *in* J. M. Starck and R. E. Ricklefs, eds. *Avian growth and development: evolution within the altricial–precocial spectrum.* Academic Press, New York.

Stauffer, R. C. 1957. Haeckel, Darwin, and ecology. *Quarterly Review of Biology* 32:138–44.

Stearn, W. T. 1981. *The Natural History Museum at South Kensington.* Natural History Museum Publishing, London.

Steenstrup, J. 1857. Hectocotylus formation in *Argonauta* and *Tremoctopus* explained by observations on similar formations in the Cephalopoda in general. *Annals and Magazine of Natural History* 20:81–114.

Stott, R. 2012. *Darwin's ghosts: the secret history of evolution.* London, Bloomsbury.

Strassmann, B. I. 1996. The evolution of endometrial cycles and menstruation. *Quarterly Review of Biology* 71:181–220.

Sundevall, C. J. 1835. *Ornithologiskt System.* Kongliga Svenska Vetenskap Akademie, Stockholm.

Swire, J. and A. M. Leroi. 2010. Planet Cameron: return to Pandora. *Trends in Ecology and Evolution* 25:432–3.

Tegmark, M. 2007. The multiverse hierarchy. pp. 99–125 *in* B. Carr, ed. *Universe or multiverse?* Cambridge University Press, Cambridge.

Tennant, J. E. 1867. *The wild elephant and the method of capturing it in Ceylon.* Longmans, Green, London.

Thomas, H. *et al.* 2011. Evolution, physiology and phytochemistry of the psycho-toxic arable mimic weed Darnel (*Lolium temulentum* L.). *Progress in Botany* 72:72–103.

Thompson, D. 1998. *The people of the sea: a journey in search of the seal legend.* Canongate, Edinburgh.

Thompson, D. W. 1895. *A glossary of Greek birds.* Oxford University Press, London.

Thompson, D. W. 1910. *Historia animalium.* W. D. Ross and J. A. Smith, *The works of Aristotle translated into English.* Vol. 4. Clarendon Press, Oxford.

Thompson, D. W. 1913. *On Aristotle as a biologist with a prooemion on Herbert Spencer.* Clarendon Press, Oxford.

Thompson, D. W. 1928. How to catch cuttlefish. *Classical Review* 42:14–18.

Thompson, D. W. 1932. *Ktilos. Classical Review* 46:53–4.

Thompson, D. W. 1940. *Science and the classics.* Oxford, Oxford University Press.

Thompson, D. W. 1947. *A glossary of Greek fishes.* Oxford University Press, Oxford.

Thompson, R. D. A. 1958. *D'Arcy Wentworth Thompson, the scholar-naturalist, 1860–1948.* Oxford University Press, Oxford.

Tinbergen, N. 1963. On the aims and methods of ethology. *Zeitschrift für Tierpschyologie* 20:410–33.

Tipton, J. A. 2002. Division and combination in Aristotle's biological writings. *Journal of Bioeconomics* 3:51–5.

Tipton, J. A. 2006. Aristotle's study of the animal world: the case of the kobios and phucis. *Perspectives in Biology and Medicine* 49:369–83.

Tóth, L. F. 1964. What the bees know and what they do not know. *Bulletin of the American Mathematical Society* 70:468–81.

Voultsiadou, E. 2007. Sponges: an historical survey of their knowledge in Greek antiquity. *Journal of the Marine Biological Association of the United Kingdom* 87:1757–63.

Voultsiadou, E. and C. Chintriroglou. 2008. Aristotle's lantern in echinoderms: an ancient riddle. *Cahiers de Biologie Marine* 49:299–302.

Voultsiadou, E. and D. Vafidis. 2007. Marine invertebrate diversity in Aristotle's zoology. *Contributions to Zoology* 76:103–20.

Wang, J. L. *et al.* 2000. Anatomical subdivisions of the stomach of the Bactrian camel (*Camelus bactrianus*). *Journal of Morphology* 245:161–7.

Ware, D. M. 1982. Power and evolutionary fitness of teleosts. *Canadian Journal of Fisheries and Aquatic Sciences* 39:3–13.

Warren, J. 2007. *Presocratics: natural philosophers before Socrates.* University of California Press, Berkeley, CA.

Waterlow, S. 1982. *Nature, change, and agency in Aristotle's physics.* Clarendon Press, Oxford.

Watson, M. 1877. On the female generative organs of *Hyaena crocuta. Proceedings of the Zoological Society of London* 369–79.

Weiblen, G. D. 2002. How to be a fig wasp. *Annual Review of Entomology* 47:299–330.

Weismann, A. 1889. *Essays upon heredity and kindred biological problems.* Clarendon Press, Oxford.

West, J. B. *et al.* 2003. Fetal lung development in the elephant reflects the adaptations required for snorkeling in adult life. *Respiratory Physiology and Neurobiology* 138:325–33.

Whewell, W. 1840. *The philosophy of the inductive sciences, founded upon their history.* J. W. Parker, London.

Whiting, J. 1992. Living bodies. pp. 75–91 *in* M. C. Nussbaum and A. Rorty, eds. *Essays on Aristotle's De anima.* Clarendon Press, Oxford.

Wiener, N. 1948. *Cybernetics, or, control and communication in the animal and the machine.* John Wiley, New York, NY.

Wilkins, J. *et al.* 2011. *Archestratus: fragments from the life of luxury.* Prospect Books, Totnes.

Wilkins, J. S. 2009. *Species: a history of the idea.* University of California Press, Berkeley, CA.

Williams, G. C. 1957. Pleiotropy, natural selection and the evolution of senescence. *Evolution* 11:398–411.

Williams, G. C. 1966. Natural selection, the costs of reproduction and a refinement of Lack's principle. *American Naturalist* 100:687–90.

Williams, G. C. 1996. *Plan and purpose in nature.* Trafalgar Square, North Pomfret, VT.

Wilson, J. B. 1999. Guilds, functional types and ecological groups. *Oikos* 86:507–22.

Wilson, M. 1997. Speusippus on knowledge and division. pp. 13–25 *in* W. Kullmann and S. Föllinger, eds. *Aristotelische Biologie: Intentionen, Methoden, Ergebnisse.* Franz Steiner Verlag, Stuttgart.

Wilson, M. 2013. *Structure and method in Aristotle's Meteorologica: a more disorderly nature.* Cambridge University Press, Cambridge.

Winemiller, K. G. and K. A. Rose. 1993. Why do most fish produce so many tiny offspring? *American Naturalist* 142:585–603.

Winsor, M. P. 1969. Barnacle larvae in the nineteenth century: a case study in taxonomic theory. *Journal of the History of Medicine and Allied Sciences* 24:294–309.

Winsor, M. P. 1976. *Starfish, jellyfish and the order of life: issues in nineteeth-century science.* Yale University Press, New Haven, CT.

Witt, C. 1998. Form and normativity in Aristotle: a feminist perspective. pp. 118–37 *in* C. Freeland, ed. *Re-reading the canon: feminist essays on Aristotle.* Penn State University Press, University Park, PA.

Witt, C. 2013. Aristotle on deformed animal kinds. *Oxford Studies in Ancient Philosophy* 43:83–106.

Wittkower, R. 1939. Eagle and serpent: a study in the migration of symbols. *Journal of the Warburg Institute* 2:293–325.

Ziswiler, V. and D. S. Farner. 1972 Digestion and the digestive system. pp. 343–430 *in* D. S. Farner, J. R. King and K. C. Parkes, eds. *Avian Biology.* Academic Press, New York, NY.

Zouros, N. *et al.* 2008. *Guide to the Plaka and Sigri petrified forest parks.* Natural History Museum of the Lesvos Petrified Forest, Ministry of Culture, Lesvos.

Zwier, K. in prep. *Aristotle on Spontaneous Generation.*

ILLUSTRATING
ARISTOTLE

A NY ILLUSTRATIONS THAT Aristotle's zoological works may have contained have been long lost. Rather than plunder the rather thin collections of animal representations in ancient Greek art, little of which is contemporaneous with Aristotle anyway, I have chosen to illustrate his animals by modern – post-1500 AD – illustrations. The sixteenth-century woodcuts from Gesner, Belon and their contemporaries seem particularly apposite, being naive in a way comparable to, say, fourth-century fish plates. When depicting exotic animals, they also often have that air of strangeness that comes from reconstructions based on imperfect, second-hand information. Besides, the animal iconographers of the Renaissance were all working from Aristotle's texts.

The anatomical diagrams on pages 61, 64, 110 and 168 are all based on diagrams that Aristotle mentions. They were reconstructed by David Koutsogiannopoulos with the advice of a papyrologist, Grace Ioannidou. To do this, David began with the texts themselves, and then sought ancient models. No ancient Greek anatomical diagrams – Aristotelian or otherwise – have survived, but contemporary and Hellenistic papyri depicting geometrical diagrams and animals were a guide to technique. Fish plates gave a sense of the observed detail. After much experimentation, the result is a style that conveys the work not of an artist but of a thinker – one who thought, as any thinker does, with his pen, or rather his brush.*

* 'A scientist always carries a pen' – M. R. Rose to the author, *c.* 1986.

31   M. G. F. A. de Choiseul-Gouffier (1782–1822) *Voyage pittoresque en Grèce*, vol. 2, Paris.

31   Author photograph.

38   K. Gesner (1551–87) *Historia animalium*, Zurich.

45   K. Gesner (1551–87) *Historia animalium*, Zurich.

54   Modified from R. Pocock (1939) *The fauna of British India, including Ceylon and Burma: mammalia*, London.

58   J. Klein (1734) *Naturalis dispositio echinoderatum*, Danzig.

61   D. Koutsogiannopoulos.

64   D. Koutsogiannopoulos.

70   T. Gill (1906) Parental care in fishes. *Annual report of the Smithsonian Institution, Washington* for the year ending 30 June 1905:403–531.

71   H. Fischer (1894) Note sur le bras hectocotylis de l'*Octopus vulgaris* Lamarck. *Journal de Conchyliologie* 42:13–19, Paris.

71   K. Gesner (1551–87) *Historia animalium*, Zurich.

73   K. Gesner (1551–87) *Historia animalium*, Zurich.

76   K. Gesner (1551–87) *Historia animalium*, Zurich.

96   P. Belon (1551) *Histoire naturelle des estranges poissons*, Paris.

98   Author photograph.

110 D. Koutsogiannopoulos.

122 K. Gesner (1551–87) *Historia animalium*, Zurich.

126 M. A. Bell (1976) Evolution of phenotypic diversity in the *Gasterosteus aculeatus* superspecies on the Pacific coast of North America. *Systematic Zoology* 25:211–227. Modified; with permission.

134 K. Gesner (1551–87) *Historia animalium*, Zurich.

138 K. Gesner (1551–87) *Historia animalium*, Zurich.

141 R. Owen (1866) *Anatomy of vertebrates*, vol. 2, London.

142 T. Mortenson (1927) *Handbook of the echinoderms of the British Isles*.

152 K. Gesner (1551–87) *Historia animalium*, Zurich.

155 D. Koutsogiannopoulos.

168 L. H. Bojanus (1819–21) *Anatome testudinis Europaeae*, Vilnius.

173 Unidentified nineteenth-century lithograph.

180 J. Rueff (1554) *De Conceptu et generatione hominis*, Frankfurt. Wellcome Library, London.

186 D. Koutsogiannopoulos.

193 Unidentified nineteenth-century lithograph.

195 H. Fabricius *ab* Acquapendente (1604) *de Formatione ovo et pulli*, Padua. Wellcome Library, London.

196 M. K. Richardson *et al.* (1998) Haeckel, embryos and evolution. *Science* 280: 985–6. Modified; with permission.

204 K. Gesner (1551–87) *Historia animalium*, Zurich.

207 Author photograph.

218 Unkown nineteenth-century etchings.

224 M. Lister (1685) *Historiae sive synopsis methodicae conchyliorum*, London.

226 A. J. Dezallier d'Argenville (1772) *La conchyliologie, ou, Traité sur la nature des coquillages*, Paris.

236 Anon. (1792) *Natural history of insects compiled from Swammerdam, Brookes, Goldsmith & co.*, Perth.

238 P. Belon (1551) *Histoire naturelle des estranges poissons*, Paris.

240 D. Koutsogiannopoulos.

251 Anon. (1792) *Natural history of insects compiled from Swammerdam, Brookes, Goldsmith & co.*, Perth.

268 K. Gesner (1551–87) *Historia animalium*, Zurich.

282 G. Cuvier (1830) Considérations sur les mollusques et en particulier les Céphalopods. *Annales des Sciences Naturelles* 19:241-59.

283 R. Owen (1866) *Anatomy of vertebrates*, vol. 2, London.

294 A. Scilla (1670) *La vana speculazione disingannata dal senso*, Naples.

304 P. Belon (1551) *Histoire naturelle des estranges poissons*, Paris.

327 K. Gesner (1551–87) *Historia animalium*, Zurich.

331 Tunc Tezel. With permission.

344 T. Gaza (1552) Aristotelis et Theophrasti Historiae, Lyon

351 Author photograph.

377 D. Koutsogiannopoulos.

379 D. Koutsogiannopoulos.

ACKNOWLEDGEMENTS

I HAVE ACCUMULATED MANY debts while writing this book. My agent, Katinka Matson, and John Brockman at Brockman Inc. always saw what *The Lagoon* might become. I thank them as well as Rick Kot at Viking Penguin, Anna Simpson and, most of all, Michael Fishwick, my visionary editor, at Bloomsbury. Peter James, my wonderful copyeditor, saved me from many infelicities; doubtless even he has not saved me from them all.

Many people in Athens and Lesbos have answered specific queries: Makis Axiotis, Lara Barazai-Yeroulanos, Níkh Dimopoulou, George Filios (Scuba Lesvos), George Fotinos (Fotinos FishShells), Alkis Kalampokis, Dimitrios Karidis, Kostas Kostakis, Ignatis Manavis, Aleka Meliadou, Theodora and Eleni Panyotis, George Papadatos, Michaelis Stoupakis (sometime First Officer of F/B *Sappho*), Christos Samaras and Dimitra Vati.

Scientific colleagues, some from the University of the Aegean, Mytilene, answered zoological queries: Filios Akreotis, Ioannis Batjakis, Ioannis Bazos, Mike Bell, Tim Birkhead, Mick Crawley, Charles Godfray, Giorgos Kokkoris, Drosos Koutsoubas, Ioannis Leonardos, Sally Leys, Chris McDaniel, Ian Owens, Panyotis Panyotides, Vassilis Papasotiropoulos, Theodora Petanidou, Tommaso Pizzari, Michel Poulain, Mike Richardson, Sophia Spathari, Cleon Tsimabos, George Tsitiris and Nikolaos Zouros.

The classical philosophers and historians who truly know Aristotle have been generous and patient in helping me understand his thought; some, generously, commented on chapters: Keith Bemer, Istávan Bodnár, Nick Bunnin, Devin Henry, Wolfgang Kullmann, Jim Lennox, Mariska Leunissen, Geoffrey Lloyd, Diana Quarantotto, the late Bob Sharples, Alfred Stückelberger, Polly Winsor, Malcolm Wilson and Karen Zwier. A great Aristotelian, one of the kindest of all, died shortly before this book went to press. Allan Gotthelf would have argued with quite a bit of this book, but when you read of the functional analysis of the elephant, or how

Darwin compares to Aristotle, you are reading things that Allan made clear to me.

In 2009 I made a film for BBC4 about Aristotle and Lesbos called *Aristotle's Lagoon*. At the time of filming, I had already been working on this book for years. Many people contributed to that film, but my co-writer, Richard King, and director, Harry Killas, made it into the loveliest of all the films that I have worked on.

Emmanuelle Almira, Cassandra Coburn, Enrico Coen, Níkh Dimopoulou, Arnold Heumakers, Olivia Judson, David Koutso-giannopoulos, Marzena Pogorzaly, Jonathan Swire and, most of all, Clare Isacke and Rebecca Stott, read chapters or offered literary advice. David Angeli constructed the control diagram of the nutritive soul. David Koutsogiannopoulos advised me on Greek natural history, was my dive buddy and drew the Aristotelian figures (advised by Grace Ioannidou). Simon MacPherson, Classics Master at Harrow School, is credited with translating and transliterating the Greek, but did so much more than that. Giorgos Kokkoris introduced me to the island. He and Dimitra Filippopoulou have cared for me there ever since. This book began when Alkistis Kontou-Dimas told me that I *must* write it. I thank them all.

My greatest debts are to those close to me: friends – Austin Burt, Vasso Koufopanou, Daphne Burt, Olivia Judson, Jonathan Swire and Kaori Imoto, Michaelis Koutroumanidis and Katerina Ertsou; and family – Marie-France Leroi, Iracema Leroi, Harry Killas, Joseph Meagher and the Vancouver and Manchester branches of the NLS. There is no one to whom I, and this book, owe more than Clare Isacke.

London's currents have lately swept me to a Sargasso Sea, a sea-hoard composed of ambergris, rare inlays and strange spars of knowledge – the words are Pound's. But, were I to write my *Portrait d'une Femme*, I would insist that all these wonders are Jerry Hall's own. It is with love that I thank her for sharing them with me.

INDEX

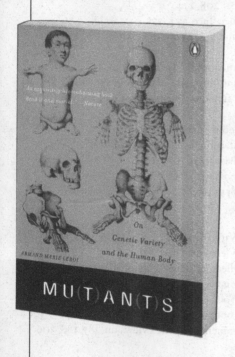